Lecture Notes in Mathematics 2145

More information about this series at
http://www.springer.com/series/304

Kunyu Guo • Hansong Huang

Multiplication Operators
on the Bergman Space

Kunyu Guo
School of Mathematical Sciences
Fudan University
Shanghai, China

Hansong Huang
Department of Mathematics
East China University of Science and
    Technology
Shanghai, China

ISSN 0075-8434   ISSN 1617-9692   (electronic)
Lecture Notes in Mathematics
ISBN 978-3-662-46844-9   ISBN 978-3-662-46845-6   (eBook)
DOI 10.1007/978-3-662-46845-6

Library of Congress Control Number: 2015943066

Mathematics Subject Classification (2010): 46E22, 47C15

Springer Heidelberg New York Dordrecht London

Springer-Verlag GmbH Berlin Heidelberg is part of Springer Science+Business Media
(www.springer.com)

Preface

The aim of this book is mainly to give an account of recent advancements in the study of multiplication operators on the Bergman space with a new point of view. The main focus will be on commutants and reducing subspaces of multiplication operators on function spaces along with relevant von Neumann algebras. Relevant techniques include complex analysis, complex geometry, operator theory, and group theory, which altogether yield a fascinating interplay and reveal some natural connections between them. The results and methods involved can be applied to most of the other analytic function spaces.

The last four decades have seen dramatic progress in the afore-mentioned topic on commutants and reducing subspaces of multiplication operators, including two main phases of development.

The first phase is mainly concerned with the topic of commutants and reducing subspaces of analytic multiplication operators on the Hardy space of the unit disk in the seventies of the last century. Several remarkable advances in this period were achieved mainly by Abrahamse and Douglas [AD], Cowen [Cow1, Cow2, Cow3], Baker et al. [BDU], Deddens and Wong [DW], Nordgren [Nor] and Thomson [T1, T2, T3, T4], etc.

A natural theme is to consider the case of the Bergman space. As well known, the approaches of the corresponding problems on the Bergman space depend heavily on metric structure of the space and relevant function-theoretic characters, and hence the case of the Bergman space diverges considerably from that of the Hardy space.

In the case of the Bergman space, the relevant topic began with Zhu's conjecture on numbers of minimal reducing subspaces of finite Blaschke product multiplication operators [Zhu1] in 2000. This research is presently experiencing a period of intense development. Most notably, during the past dozen years a lot of remarkable achievements had been made in this direction [DSZ, DPW, GH1, GH2, GH4, GSZZ, SZZ1, SZZ2], etc.

The topic of the Bergman space has been the focus of considerable attention from the authors in the past years. Briefly, the goal of this book is to give an account of the latest developments on commutants and reducing subspaces of multiplication operators on both the Hardy space and the Bergman space, and von Neumann algebras generated by multiplication operators on the Bergman space. It is shown that types of such von Neumann algebras turn out to be closely related to the geometric property of the symbols of the corresponding multiplication operators.

It is a pleasure to thank many people for their help and encouragements. Professor X. Chen deserves special gratitude for his suggestive advice and constant encouragements with the topics presented here. We would like to express our heartfelt thankfulness to Professor D. Zheng, who has put forward many thoughtful comments and suggestions for our study. We are deeply indebted to Professor S. Sun for numerous stimulating conversations. Professor R. Douglas is a man of great insight with whom communications enlarged our views on this topic. We thank Professor C. Cowen for his invaluable communications. Special thanks also go to Professors G. Yu, G. Zhang, K. Zhu, W. Qiu, R. Yang, K. Izuchi, X. Fang, K. Wang for their invaluable comments, which makes this book more readable. We would also like to express our debt to Dr. Ramon Peng, for his enthusiasm, indispensable editorial contribution in publishing this book in the Lecture Notes in Mathematics. We thank everyone who has made contributions in the publication of this book and corrections of the proof, including Manager Ramya Prakash, and other editors. Without their help, the publication of this book could not go well. The research for this work was partially supported by the NSFC in China.

Shanghai, China   Kunyu Guo
Shanghai, China   Hansong Huang
March 17, 2015

Contents

1 Introduction 1

2 Some Preliminaries 7
  2.1   Some Preliminaries in Complex Analysis 7
  2.2   The Notion of Capacity 26
  2.3   Local Inverse and Analytic Continuation 31
  2.4   Uniformly Separated Sequence 33
  2.5   Some Results in von Neumann Algebras 40
  2.6   Some Results in Operator Theory 46

3 Cowen-Thomson's Theorem 53
  3.1   Cowen-Thomson's Theorem on Commutants 53
  3.2   Facts from Real and Complex Analysis 58
  3.3   Proof of Cowen-Thomson's Theorem 64
  3.4   A Proposition on Singularities 76
  3.5   An Example not Satisfying Thomson's Condition 78
  3.6   Remarks on Chap. 3 84

4 Reducing Subspaces Associated with Finite Blaschke Products 87
  4.1   The Distinguished Reducing Subspace 87
  4.2   Abelian $\mathcal{V}^*(B)$ 95
  4.3   Representation for Operators in $\mathcal{V}^*(B)$ 100
  4.4   Further Consideration on Reducing Subspaces 105
  4.5   Proof of Proposition 4.4.6 111
  4.6   Abelian $\mathcal{V}^*(B)$ for Order $B = 5, 6$ 118
  4.7   Remarks on Chap. 4 122

5 Reducing Subspaces Associated with Thin Blaschke Products 125
  5.1   Properties of Thin Blaschke Products 125
  5.2   Representation for Operators in $\mathcal{V}^*(B)$ 129
  5.3   Geometric Characterization for $\mathcal{V}^*(B)$ 141
  5.4   Most M_B Are Irreducible 150
  5.5   The Construction of an Example 162

5.6   Another Proof for a Characterization on $\mathcal{V}^*(B)$ 170
5.7   Abelian $\mathcal{V}^*(B)$ for Thin Blaschke Products 177
5.8   Finite Blaschke Product Revisited 184
5.9   Remarks on Chap. 5 192

6   **Covering Maps and von Neumann Algebras** 193
6.1   Regular Branched Covering Maps and Orbifold Domains 193
6.2   Representations of Operators in $\mathcal{V}^*(\phi)$ 196
6.3   Abelian $\mathcal{V}^*(\phi)$ 207
6.4   Type II Factors Arising from Planar Domains 213
6.5 $\mathcal{V}^*(\phi)$ and Free Group Factors 218
6.6   Type II Factors and Orbifold Domains 226
6.7   Applications to Multi-variable Case 230
6.8   Representation of Operators in $\mathcal{V}_\alpha^*(\phi)$ 238
6.9   The Structure of $\mathcal{V}_\alpha^*(\phi)$ 242
6.10   Group-Like von Neumann Algebras 245
6.11   Weighted Bergman Spaces over the Upper Half Plane 250
6.12   Remarks on Chap. 6 251

7   **Similarity and Unitary Equivalence** 253
7.1   The Case of the Hardy Space 253
7.2   Unitary Equivalence on Analytic Multiplication Operators 258
7.3   Similarity of Analytic Toeplitz Operators 260
7.4   Remarks on Chap. 7 268

8   **Algebraic Structure and Reducing Subspaces** 269
8.1   Algebraic Structure of Essentially Normal Operators 269
8.2   Algebraic Structure and Reducing Subspaces 281
8.3   Monomial Case 283
8.4   More Examples in Multi-variable Case 291
8.5   Remarks on Chap. 8 299

A   **Berezin Transform** 301

B   **Nordgren's Results on Reducing Subspaces** 305

C   **List of Problems** 309

Bibliography 313

Index 321

Chapter 1
Introduction

These notes arose out of a series of research papers completed by the authors and others. This volume is devoted to recent developments in commutants, reducing subspaces and von Neumann algebras related to multiplication operators.

Let Ω be a bounded open subset of $\mathbb{C}^n$, and $\mathcal{H}$ be a Hilbert space of some holomorphic functions over Ω. If for every $\lambda \in \Omega$, the evaluation functional

$$E_\lambda : f \to f(\lambda), f \in \mathcal{H}$$

is bounded, then $\mathcal{H}$ is called *a reproducing kernel Hilbert space* of holomorphic functions on Ω. By Riesz's representation theorem, for each $\lambda \in \Omega$ there is a unique vector K_λ in $\mathcal{H}$ such that

$$f(\lambda) = \langle f, K_\lambda \rangle, \ \forall f \in \mathcal{H}.$$

This function K_λ is called *the reproducing kernel at* λ. Both the Hardy space $H^2(\mathbb{D})$ and the Bergman space $L_a^2(\mathbb{D})$ over the unit disk are known as classical reproducing kernel Hilbert spaces. Throughout this book, a *subspace* of a Hilbert space is a norm-closed linear space. Let T be a bounded linear operator on a Hilbert space $\mathcal{H}$. If M is a subspace of $\mathcal{H}$ satisfying $TM \subseteq M$, then M is called an *invariant subspace* of T. If in addition M is invariant under T^*, then M is called a *reducing subspace* of T. For a reproducing kernel Hilbert space $\mathcal{H}$ of holomorphic functions on a domain Ω of the complex plane, by an invariant subspace M we mean that M is invariant under the coordinate operator M_z.

Now consider a reproducing kernel Hilbert space $\mathcal{H}$ of holomorphic functions on Ω. Given a bounded holomorphic function ϕ on Ω, if the map $f \to \phi f, f \in \mathcal{H}$ defines a bounded operator, then define $M_\phi f = \phi f, f \in \mathcal{H}$, and M_ϕ is called a *multiplication operator* on $\mathcal{H}$. This function ϕ is called the *symbol* for M_ϕ. Recall that a von Neumann algebra is a unital C^*-algebra on a Hilbert space, which is closed in the weak operator topology. Throughout this book, let $\mathcal{W}^*(\phi)$ be the von

© Springer-Verlag Berlin Heidelberg 2015
K. Guo, H. Huang, *Multiplication Operators on the Bergman Space*,
Lecture Notes in Mathematics 2145, DOI 10.1007/978-3-662-46845-6_1

Neumann algebra generated by M_ϕ and denote $V^*(\phi) = W^*(\phi)'$, *the commutant algebra of* $W^*(\phi)$. We emphasize that *both* $W^*(\phi)$ *and* $V^*(\phi)$ *are von Neumann algebras*.

Let us turn to reducing subspaces for M_ϕ from the view of von Neumann algebras. For each reducing subspace M, denote by P_M the orthogonal projection onto M. Then it is easy to see that P_M commutes with both M_ϕ and M_ϕ^*. On the other hand, if P is a (self-adjoint) projection in $V^*(\phi)$, then the range of P is necessarily a reducing subspace for M_ϕ. Thus there is a one-to-one and onto correspondence: $M \mapsto P_M$, which maps all reducing subspaces of M_ϕ to projections in $V^*(\phi)$. Since a von Neumann algebra is generated by its projections, the study of the lattice of reducing subspaces for M_ϕ is to find projections in $V^*(\phi)$. Also note that by the von Neumann Bicommutant Theorem $V^*(\phi)' = W^*(\phi)$. This can help us to understand types of both $V^*(\phi)$ and $W^*(\phi)$ by reducing subspaces of M_ϕ.

There are many motivations to study the reducing subspaces of multiplication operators defined on reproducing kernel Hilbert spaces of holomorphic functions [GH3]. Firstly, it helps to understand connections of complex analysis and von Neumann algebras; secondly, given any bounded planar domain Ω, there always exists a holomorphic covering map ϕ from $\mathbb{D}$ onto Ω, and we will see the links between the von Neumann algebras $W^*(\phi)$, $V^*(\phi)$ and geometric properties of domains, as well as their fascinating connections to one of the long-standing problems in free group factors

$$\mathcal{L}(F_n) \overset{*}{\cong} \mathcal{L}(F_m)?$$

for $n \neq m$ and $n, m \geq 2$; here F_n denotes the free group on n generators, and $\mathcal{L}(F_n)$ is the von Neumann algebra generated by left regular representation of F_n on $l^2(F_n)$. Thirdly, the famous Invariant Subspace Problem is known to be equivalent to the problem of the invariant subspace lattice of the Bergman space. Precisely, the Invariant Subspace Problem asks that, if T is a bounded linear operator on a separable Hilbert space H and $\dim H = \infty$, then must T have a proper invariant subspace? It is well known that this problem can be reduced as follows: if M and N are invariant subspaces of $L_a^2(\mathbb{D})$ such that $N \subseteq M$ and $\dim M/N = \infty$, then is there another invariant subspace L satisfying $N \subsetneqq L \subsetneqq M$? More generally, it is shown that the Invariant Subspace Problem is equivalent to the problem whether there exists a saturated operator in the $A_{\aleph_0}$-class [BFP]. For a special class of multiplication operators M_ϕ on the Bergman space, we found that reducing subspace lattice of M_ϕ is saturated; and M_ϕ has rich invariant subspaces, see Chap. 6.

In general, it is difficult to determine the reducing subspaces of concrete operators, even of multiplication operators on the Hardy space of the unit disk. Nordgren gave a sufficient condition for M_ϕ having no nontrivial reducing subspaces [Nor]. Also, he showed that if $\phi = h \circ \psi$ where $h \in H^\infty$ and ψ is an inner function different from the Möbius transform, then M_ϕ admits a nontrivial reducing subspace, see Appendix B. By giving an example, Abrahamse showed that Nordgren's sufficient condition is not necessary [A1]. Since reducing subspaces of M_ϕ are exactly those projections in $V^*(\phi)$, the problem of finding reducing subspaces falls into that of determining the commutant of M_ϕ, $\{M_\phi\}'$. As commented in [Cow1], in studying an operator T on a Hilbert space, it is of interest to characterize its commutant, which helps in understanding the structure of T, and that of $V^*(T)$. Little is known about this question in general case unless the operator is normal. Even in the case of subnormal operators, such a question seems also difficult. Shields and Wallen [SWa] considered the commutants of a special class of operators which can be viewed as the coordinate operators on some reproducing kernel Hilbert spaces of holomorphic functions. Later, Deddens and Wong noticed that methods in [SWa] can be applied to show that if ϕ is a univalent function, then on the Hardy space $\{M_\phi\}' = \{M_z\}' = \{T_f : f \in H^\infty(\mathbb{D})\}$, and they raised six questions related to commutants of analytic multiplications in [DW] (see Remark 3.6), which had stimulated a lot of further work, see [AB, Nor, Cow1, Cow2, Cow3, T1, T2, T3, T4], and [AC, ACR, AD, Cl, CDG, CGW, Cow1, Cu, GW1, JL, Ro, SZ1, SZ2, Zhu1, Zhu2]. It is worthwhile to mention that Abrahamse and Ball [AB] constructed a counter-example arising from a covering map, and thus gave a negative answer to one of six questions. Based on Deddens and Wong's work, Baker, Deddens and Ullman [BDU] proved that on the Hardy space $H^2(\mathbb{D})$, for any entire function f there is always a positive integer k such that $\{T_f\}' = \{T_{z^k}\}'$. Also, they gave a geometric characterization for this integer k. Applying function theoretic methods, especially the techniques of local inverse and analytic continuation, Thomson gave more general conditions to ensure that the intersection of commutants or a single commutant equals $\{T_B\}'$ for some finite Blaschke product B, see [T1, T2, T3, T4]. An easier version of Thomson's results reads as follows: if ϕ is holomorphic on a neighborhood of $\overline{\mathbb{D}}$, then there is always a finite Blaschke product B and a function $g \in H^\infty(\mathbb{D})$ such that $\phi = g \circ B$ and $\{M_\phi\}' = \{M_B\}'$ [T1]. Cowen's work in [Cow1] was intended to shed light on the commutant problem for a special class of subnormal operators: the multiplication operators defined on function spaces, especially on the Hardy space. In spirit, he developed Thomson's techniques and extended some of Thomson's results in more general cases. Also, he gave a complete characterization for the commutant of T_ϕ when ϕ is a covering map. Furthermore, he established sufficient conditions on f such that $\{T_f\}'$ contains no zero compact operator [Cow1]. Along this line, some related work appeared in [Cow2, Cow3].

When we turn to the Bergman space, an analogue of Thomson's theorem on the commutant problem holds on the Bergman space. That is, if ϕ is holomorphic on a neighborhood of $\overline{\mathbb{D}}$, then there is always a finite Blaschke product B and a function $g \in H^\infty(\mathbb{D})$ such that $\phi = g \circ B$ and $\{M_\phi\}' = \{M_B\}'$ [T1]. In particular, $V^*(\phi) = V^*(B)$. Thus, our attention is drawn to studying M_ϕ where ϕ is an inner function. By Frostman's theorem [Ga], for each inner function ϕ, there is always a Möbius transform h such that $h \circ \phi$ is a Blaschke product. Then consideration for a general H^∞-symbol reduces to the case of a Blaschke product.

The first consideration naturally falls into those multiplication operators defined by finite Blaschke products. It is notable that the situation of the Bergman space is quite distinct from the Hardy space. By a simple analysis, the multiplication operator $M_{z^n}(n \geq 2)$ acting on the Hardy space of the unit disk, has infinitely many minimal reducing subspaces. However, M_{z^n} acting on the Bergman space has exactly n distinct minimal reducing subspaces. In general, if η is an inner function, not a Möbius transform, then the multiplication operator M_η on the Hardy space $H^2(\mathbb{D})$ has always infinitely many minimal reducing subspaces.

In the case of the Bergman space, and order $B = 2$, it was shown that M_B exactly has two distinct minimal reducing subspaces in [SW] and [Zhu1] independently. Motivated by this fact Zhu conjectured that for a finite Blaschke product B of order n, there are exactly n distinct minimal reducing subspaces [Zhu1]. In fact, by applying [SZZ2, Theorem 3.1] Zhu's conjecture holds only if $B(z) = \phi^n$ for some Möbius transform ϕ. Therefore, the conjecture is modified as follows: M_B has at most n distinct minimal reducing subspaces, and the number of nontrivial minimal reducing subspaces of M_B equals the number of connected components of the Riemann surface S_B of $B^{-1} \circ B$ on the unit disk [DSZ], here by a *Riemann surface*, we mean a complex manifold of complex dimension 1, not necessarily assumed to be connected. A simple reasoning shows that the modified conjecture is equivalent to assertion that $\mathcal{V}^*(B)$ is abelian. As proved in [HSXY, GSZZ], M_B has a distinguished reducing subspace when order $B \geq 2$, and hence $\mathcal{V}^*(B)$ is always nontrivial, and furthermore dim $\mathcal{V}^*(B) \leq$ order B. In the case of order $B = 3, 4, 5, 6$, the modified conjecture is demonstrated in [GSZZ, SZZ1, GH1]. By using the techniques of local inverse and group-theoretic methods, it was proved in [DSZ] that $\mathcal{V}^*(B)$ is abelian if order $B = 7, 8$. The latest progress is an affirmative answer to the modified conjecture due to Douglas, Putinar and Wang [DPW].

However, little is known about the structure of $\mathcal{V}^*(B)$ for an infinite Blaschke product B. Regarding the problem of reducing subspaces on the Bergman space, we first focus on two classes of special infinite Blaschke products. One of them is thin Blaschke products, and the other is covering maps. Both classes exist abundantly and share nice properties, though their behaviors are so far apart. Before continuing, the following question naturally arises:

For each infinite Blaschke product B, does M_B always have a nontrivial reducing subspace?

In Chap. 5 a negative answer to this question is presented by constructing a special thin Blaschke product. By applying the techniques of local inverse and analytic continuation, Chap. 5 investigates the geometry of the related Riemann surface S_B. It is shown that for any thin Blaschke product B, $\mathcal{V}^*(B)$ is abelian, and under a mild condition, $\mathcal{V}^*(B) = \mathbb{C}I$. Thus, in most cases of thin Blaschke products, $\mathcal{V}^*(B)$ is very simple. These results come from [GH4]. However, for another infinite Blaschke products, the structure of $\mathcal{V}^*(B)$ can still be very complicated. In fact, if ϕ is a holomorphic covering map from $\mathbb{D}$ onto a bounded planar domain Ω, then ϕ will be a Blaschke product where Ω is the unit disk minus a discrete subset E of $\mathbb{D}$ [Cl, Ga, GH2]. In this case, we will see that $\mathcal{V}^*(\phi)$ is $*$-isomorphic to the group von Neumann algebra $\mathcal{L}(\pi_1(\Omega))$, where $\pi_1(\Omega)$ denotes the fundamental group of Ω. In the case of $\Omega = \mathbb{D} - E$, with $n = \sharp E$, the cardinality of E(allowed to be infinity), $\pi_1(\Omega)$ is isomorphic to F_n, and thus $\mathcal{L}(\pi_1(\Omega)) = \mathcal{L}(F_n)$. Therefore, in this case, the reducing subspace problem has a close connection with the aforementioned open problem in free group factors, also see Sects. 6.5 and 6.6 of Chap. 6.

In this book, we focus primarily on the structures of $\mathcal{V}^*(\phi)$ and $\mathcal{W}^*(\phi)$ where ϕ run throughout three classes: finite Blaschke products, thin Blaschke products and bounded holomorphic covering maps. When ϕ varies in these different classes, the presentations for those operators in $\mathcal{V}^*(\phi)$ have similar forms. However, the structure of $\mathcal{V}^*(\phi)$ is closely related to the class to which ϕ belongs. We thus need different methods to study $\mathcal{V}^*(\phi)$ for these different classes.

The book is organized as follows.

In Chap. 2, some preliminaries are presented, along with some classical results from complex analysis, such as Rouche's theorem, Riemann mapping theorem, Koebe Uniformization Theorem. Two important notions are introduced: local inverse and analytic continuation, which will play an important role in Chaps. 3–5. Also, Chap. 2 furnishes some basic and useful results in the theory of von Neumann algebras and operator theory.

Chapter 3 presents the proof of Thomson's theorem on the commutant of multiplication operators arising from function spaces. Thomson's theorem says that: if ϕ is in $H^\infty(\mathbb{D})$ such that for uncountably many λ in $\mathbb{D}$ the inner part of $\phi - \phi(\lambda)$ is a finite Blaschke product, then there exists a finite Blaschke product B and an $H^\infty(\mathbb{D})$-function ψ such that $\phi = \psi(B)$ and $\{M_\phi\}' = \{M_B\}'$. Thomson's result was first done on the Hardy space, but it proves to be valid on the Bergman space. Also, Cowen gave an essential generalization of Thomson's result on the Hardy space. Precisely, if ϕ is in $H^\infty(\mathbb{D})$ such that for a single λ in $\mathbb{D}$ the inner part of $\phi - \phi(\lambda)$ is a finite Blaschke product, then the same conclusion holds. Cowen [Cow1] asked whether there is a function which satisfies Cowen's condition and fails Thomson's. Chapter 3 constructs such a function.

Chapter 4 is mainly devoted to the study of reducing subspaces of M_B for finite Blaschke products B. First, it is shown that if B is a finite Blaschke product with order $B \geq 2$, then M_B always has a nontrivial reducing subspace, called the distinguished reducing subspace [HSXY, GSZZ]. Douglas, Putinar and Wang [DPW] showed that for each finite Blaschke product B, $\mathcal{V}^*(B)$ is abelian, thus proving a modified conjecture mentioned before. There we present a different approach, which appears more geometric than the original one given by Douglas, Putinar and Wang. Some applications are given by using the techniques and methods developed in this chapter.

Chapter 5 investigates reducing subspaces of M_B for thin Blaschke products B. A representation for those operators in $\mathcal{V}^*(B)$ is furnished, and a geometric sufficient condition is obtained for when $\mathcal{V}^*(B)$ is trivial. Then it is shown that $\mathcal{V}^*(B)$ is always abelian for any thin Blaschke product B; furthermore, $\mathcal{V}^*(B)$ is trivial under a mild condition of B. We also provide the first example of an infinite Blaschke product B for which $\mathcal{V}^*(B)$ is trivial. As a byproduct, it is shown that for a finite Blaschke product B, $\mathcal{V}^*(B)$ is usually two-dimensional. This means that most M_B have exactly two minimal reducing subspaces for finite Blaschke products B. The results of this chapter comes mainly from [GH4].

Chapter 6 elaborates on those multiplication operators defined by covering maps and the related von Neumann algebras. A class of type II factors was constructed, which arise essentially from holomorphic coverings of bounded planar domains.

The types of such von Neumann algebras turn to have a close link with topological properties of planar domains. As a result, this chapter establishes a fascinating connections to one of the long-standing problems in free group factors. These results comes mainly from [GH2]. Such results are nontrivially generalized to the weighted Bergman space [Huang2]. We mention work by Abrahamse and Douglas [AD]. They constructed a class of subnormal operators related to multiply-connected domains, and considered the von Neumann algebras generated by such operators which act on vector-valued Hardy space, see Sect. 6.5 in Chap. 6. However, the techniques and ideas developed in Chap. 6 are quite different from that in [AD].

By using the methods developed in Chaps. 3–7 is to focus on the problems of similarity and unitary equivalence of multiplication operators, defined on both the Hardy space and the Bergman space. The relative materials come mainly from [Cow4, Cla1, Cla2, Cla3, GH1, GZ, JL, SY, SZZ2].

In most interesting cases, multiplication operators on function spaces are essentially normal. Chapter 8 is firstly devoted to discussion of the structure of essentially normal operators. Then we apply these results to study the structure of multiplication operators in the cases of both single variable and multi-variable. Some examples are provided.

In Appendices, some related results and proofs are presented. It is worthy to mention that some open problems and conjectures from this book are collected in Appendix C.

Chapter 2
Some Preliminaries

This chapter will present some basic facts from complex analysis, operator theory and von Neumann algebras. These results will be needed in the sequel.

2.1 Some Preliminaries in Complex Analysis

In this section, we will present some standard results in complex analysis. First, some basic notations will be provided. Let $\mathbb{C}$ denote the complex plane, and $\mathbb{D}$ denotes the unit disk in $\mathbb{C}$. By *the punctured disk* we refer to $\mathbb{D} - \{0\}$. For each positive integer d, $\mathbb{C}^d$ denotes the d-th Cartesian product of $\mathbb{C}$. We use $\mathbb{B}_d$ to denote *the unit ball* in $\mathbb{C}^d$, i.e.

$$\mathbb{B}_d = \{(z_1, \cdots, z_d) \in \mathbb{C}^d : \sum_{k=1}^{d} |z_k|^2 < 1\}.$$

Write

$$\mathbb{D}^d = \{(z_1, \cdots, z_d) \in \mathbb{C}^d : |z_k| < 1, k = 1, \cdots, d\},$$

called *the unit polydisk*.

A function f on some domain $\Omega(\Omega \subseteq \mathbb{C}^d)$ is called *holomorphic* if for each $z_0 \in \Omega$, there is some neighborhood V of z_0 on which f can be expressed as a uniformly convergent power series

$$f(z) = \sum_{\alpha} c_\alpha (z - z_0)^\alpha,$$

© Springer-Verlag Berlin Heidelberg 2015
K. Guo, H. Huang, *Multiplication Operators on the Bergman Space*,
Lecture Notes in Mathematics 2145, DOI 10.1007/978-3-662-46845-6_2

where α are multi-indices and each term $(z - z_0)^\alpha$ represents a monomial in $z_1 - z_{01}, \cdots, z_d - z_{0d}$. It is known [Hor, Kran] that if Ω is a domain in $\mathbb{C}^d (d \geq 2)$, then a function f on Ω is holomorphic if and only if f is holomorphic in each variable. Denote by $Hol(\Omega)$ the set of all holomorphic functions on Ω. All bounded functions in $Hol(\Omega)$ consist of a class, denoted by $H^\infty(\Omega)$, which is a Banach space with the sup-norm defined by

$$\|f\|_\infty = \sup_{z \in \Omega} |f(z)|, f \in H^\infty(\Omega).$$

In particular, $H^\infty(\mathbb{D})$ denotes the set of bounded holomorphic functions over $\mathbb{D}$. The uniform closure of polynomials on Ω is written by $A(\Omega)$. For example, one can refer to [Ru1, Ru2, Ru3] and [Ga, Hof1] for the study of $A(\mathbb{D})$, $A(\mathbb{D}^d)$, and $A(\mathbb{B}_d)$.

Let Ω be a domain on the complex plane $\mathbb{C}$. A holomorphic function f on Ω is called *univalent* if it is injective. If in addition f is onto, then f is called *biholomorphic*.

It is well-known that the terms "holomorphic" and "analytic" are synonym. In this book, we always take the term "holomorphic" except for the following cases: analytic continuation, analytic curve, and analytic Toeplitz operator. Sometimes we also mention the terms "conformal", "conformal map" and "conformal isomorphism", by which we mean the corresponding map is biholomorphic.

Rouche's theorem is a well-known result in complex analysis.

Theorem 2.1.1 (Rouche) *Suppose both f and g are holomorphic functions over a domain G on $\mathbb{C}$, and Ω is a sub-domain of G whose boundary $\partial\Omega$ consists of finitely many Jordan curves and $\partial\Omega \subseteq G$. If*

$$|f(z) - g(z)| < |f(z)|, z \in \partial\Omega,$$

then f and g have the same number of zeros in Ω, counting multiplicity.

Let Ω_0 and Ω be two domains in $\mathbb{C}^d$. A holomorphic map $\Phi : \Omega_0 \to \Omega$ is called *proper* if $\Phi^{-1}(E)$ is compact for every compact subset E of Ω. Equivalently, for any sequence $\{z_n\}$ $(z_n \in \Omega_0)$ without limit point in Ω_0, $\{\Phi(z_n)\}$ has no limit point in Ω. In particular, when $d = 1$, it is always an *n-folds map* [Mi, Appendix E]; that is, for every $z \in \Omega_0$, $\Phi - \Phi(z)$ has exactly n zeros in Ω_0, counting multiplicity. For example, a finite Blaschke product B with order n is a proper map from $\mathbb{D}$ onto $\mathbb{D}$. For each $w \in \mathbb{D}$, $B - w$ has n zeros in $\mathbb{D}$, counting multiplicity. In general, by applying Rouche's theorem one can give the following.

Proposition 2.1.2 *Suppose $\phi : \Omega_0 \to \Omega$ is a proper map with $\Omega_0, \Omega \subseteq \mathbb{C}$, then there is a constant integer n such that for any $w \in \Omega$, $\phi - w$ has exactly n zeros in Ω_0, counting multiplicity.*

Proof For two distinct points w_0 and w_1 in Ω, let γ be a curve in Ω connecting w_0 with w_1. Since the curve γ is compact and ϕ is proper, $\phi^{-1}(\gamma)$ is also compact. Then it is not difficult to construct a sub-domain Ω_1 of Ω_0 such that $\phi^{-1}(\gamma) \subseteq \Omega_1$

and $\partial \Omega_1$ is a subset of Ω_0 consisting of several smooth Jordan curves. By the compactness of $\phi^{-1}(\gamma)$, one can apply Rouche's theorem to show that for all $w \in \gamma$, $\phi - w$ has the same number of zeros in Ω_1. Since $\Omega_1 \supseteq \phi^{-1}(\gamma)$, $\phi - w$ has no zero on $\Omega_0 - \Omega_1$ for any $w \in \gamma$. Thus, $\phi - w$ has the same number of zeros in Ω_0. In particular, $\phi - w_0$ and $\phi - w_1$ has the same number of zeros in Ω_0, counting multiplicity. □

The following theorem characterizes all proper maps on $\mathbb{D}^n$ and $\mathbb{B}_n$, see [Ru1, Ru2].

Theorem 2.1.3

(I) *By omitting a permutation, every proper holomorphic map Φ of $\mathbb{D}^n$ into $\mathbb{D}^n$ must have the following form:*

$$\Phi(z) = (\varphi_1(z_1), \cdots, \varphi_n(z_n)), z \in \mathbb{D}^n$$

where all $\varphi_i(1 \leq i \leq n)$ are finite Blaschke products, see [Ru1]. In particular, every proper holomorphic map of $\mathbb{D}$ into $\mathbb{D}$ is a finite Blaschke product.

(II) *For $n > 1$, every proper holomorphic map Φ of $\mathbb{B}_n$ into $\mathbb{B}_n$ is a holomorphic automorphism of the unit ball $\mathbb{B}_n$, see [Ru2].*

For more about proper maps, one can refer to [Ru2, Chap. 15]. The remaining part of this section concerns with Blaschke products. In this book, we always write

$$\varphi_\lambda(z) = \frac{\lambda - z}{1 - \bar{\lambda} z},$$

with $\lambda, z \in \mathbb{D}$. By definition, a *Blaschke product* is a unimodular constant tuple of the following:

$$B(z) \overset{\Delta}{=} z^m \prod_n \frac{a_n}{|a_n|} \frac{a_n - z}{1 - \bar{a}_n z} \equiv z^m \prod_n \frac{a_n}{|a_n|} \varphi_{a_n}(z), z \in \mathbb{D}, m \in \mathbb{Z}_+$$

where $\{a_n\}$ is a nonzero sequence in $\mathbb{D}$ satisfying $\sum_n (1 - |a_n|^2) < \infty$. If the sequence $\{a_n\}$ contains $N - m$ elements, i.e. B has N zeros counting multiplicity, then B is called a *finite Blaschke product* with order N. A Blaschke product of order one is called a Blaschke factor. It is well-known that when $\{a_n\}$ is an infinite sequence, the product $B(z)$ converges and is holomorphic in $\mathbb{D}$ [Hof1]; in this case, B is called an *infinite Blaschke product*.

The class of inner functions enlarges the class of Blaschke products. Recall that if ϕ is a function in $H^\infty(\mathbb{D})$ admitting unimodular radial limits almost everywhere on $\mathbb{T}$ according to the arc length measure, then ϕ is called an *inner function*. Frostman's theorem states a relationship between inner functions and Blaschke products, for example, see [Ga, p. 75, Theorem 6.4].

Theorem 2.1.4 (Frostman) *Let f be a nonconstant inner function on $\mathbb{D}$. Then for all λ in $\mathbb{D}$ except possibly for a set of capacity zero, the function*

$$\varphi_\lambda(f) = \frac{\lambda - f}{1 - \bar{\lambda}f}$$

is a Blaschke product.

The definition of capacity will be illustrated in the next section.

The following gives a generalization of Frostman's theorem due to Rudin [Ru4].

Theorem 2.1.5 (Rudin) *For each nonconstant function $f \in H^\infty(\mathbb{D})$, the inner part of $f - f(\lambda)(\lambda \in \mathbb{D})$ is a Blaschke product with distinct zeros, except for λ in a set of capacity zero.*

By Frostman's theorem, the study of reducing subspaces for M_f acting on function spaces where f is an arbitrary inner function is equivalent to studying those for the multiplication operator defined by a Blaschke product $\varphi_\lambda(f)$; and this context will be considered in the next chapters.

It is well-known [Hof1] that a sequence $\{z_n\}$ in $\mathbb{D}$ is the zero sequence of a function $f \in H^\infty(\mathbb{D})$ if and only if

$$\sum_n (1 - |z_n|^2) < \infty.$$

This fact can be imported from the unit disk to the half plane.

Lemma 2.1.6 *Let f be a bounded holomorphic function defined on the half plane $\{z| \, Re\, z > \lambda\}$ with $\lambda \in \mathbb{R}$, and let $\lambda_1, \lambda_2, \cdots$ be all zeros of f on the half real axis $(\lambda, +\infty)$ (repeated according to multiplicity). Then for any $\delta > 0$,*

$$\sum_{0 \neq \lambda_i > \lambda + \delta} |\lambda_i|^{-1} < +\infty.$$

In particular, if $k = 1, 2, \cdots$ are the zeros of a bounded holomorphic function f, then f is necessarily zero [Hof1, p. 74, Exercise 3].

Proof of Lemma 2.1.6 The proof is from [Huang1]. Given $\delta > 0$, the mapping

$$w = \frac{z - \lambda - \delta}{z - \lambda + \delta}$$

is a bi-holomorphic map from $\{z| \, Re\, z > \lambda\}$ onto the unit disk. Let $z = z(w)$ be its inverse, and then $f(z(w))$ is in $H^\infty(\mathbb{D})$. Since $w(\lambda_1), w(\lambda_2), \cdots$ are the zeros of $f(z(w))$, they must satisfy the Blaschke condition:

$$\sum_i (1 - |z(\lambda_i)|^2) < \infty.$$

Therefore, $\sum_{\lambda_i > \lambda + \delta}(1 - |z(\lambda_i)|) < \infty$. That is,

$$28 \sum_{\lambda_i > \lambda + \delta} (\lambda_i - \lambda + \delta)^{-1} < \infty,$$

which easily leads to the desired conclusion. The proof of Lemma 2.1.6 is complete.   □

We shall adopt the notations from [Ga]. For a holomorphic function f on $\mathbb{D}$ and $\xi \in \mathbb{T}$, *the cluster set* of f at ξ is

$$\mathrm{Cl}(f, \xi) = \bigcap_{r>0} \overline{f(\mathbb{D} \cap O(\xi, r))},$$

where $O(\xi, r)) = \{z : |z - \xi| < r\}$. Clearly, $w \in \mathrm{Cl}(f, \xi)$ if and only if there is a sequence $\{z_n\}$ in $\mathbb{D}$ such that $z_n \to \xi$ and $f(z_n) \to w$ as n tends to infinity. The range set of f at ξ is defined to be

$$\mathcal{R}(f, \xi) = \bigcap_{r>0} f(\mathbb{D} \cap O(\xi, r)).$$

The following theorem gives a characterization for the singularities of inner functions, see [Ga, p. 80, Theorem 6.6].

Theorem 2.1.7 *Let f be an inner function on $\mathbb{D}$, and ξ is a singular point of f (i.e. a point at which f does not extend across the unit circle analytically). Then $\mathrm{Cl}(f, \xi) = \overline{\mathbb{D}}$ and $\mathcal{R}(f, \xi) = \mathbb{D} - L$, where L is a set of capacity zero.*

Proof The proof is from [Ga].

Since $\mathcal{R}(f, \xi) \subseteq \mathrm{Cl}(f, \xi) \subseteq \overline{\mathbb{D}}$, it suffices to show that $\mathcal{R}(f, \xi)$ equals $\mathbb{D}$ minus some set of capacity zero. By Theorem 2.1.4, there is a set L' of capacity zero such that $\varphi_\lambda(f)$ is always a Blaschke product whenever $\lambda \in \mathbb{D} - L'$. Now fix a point $\lambda \in \mathbb{D} - L'$. Since f is singular at ξ, so is the Blaschke product $\varphi_\lambda(f)$. Recall that a Blaschke product B with zero sequence $\{z_n\}$ can be extended analytically to $\mathbb{T} - E$, where E is the set of accumulation points of $\{z_n\}$ [Hof1]. Then it follows that there is a zero subsequence $\{z'_n\}$ of $\varphi_\lambda(f)$ tending to ξ. Thus, $\varphi_\lambda(f)(z'_n) = 0$, i.e. $f(z'_n) = \lambda$ for each n. This shows that $\lambda \in \mathcal{R}(f, \xi)$, and hence $\mathbb{D} - L' \subseteq \mathcal{R}(f, \xi) \subseteq \mathbb{D}$, completing the proof.   □

Concerning Blaschke products, two important classes will be introduced: interpolating and thin Blaschke products. On the unit disk, *the pseudohyperbolic distance* between two points z and λ is defined by

$$d(z, \lambda) = |\varphi_\lambda(z)|, \ z, \lambda \in \mathbb{D}.$$

An infinite Blaschke product B is called *an interpolating Blaschke product* if for any $(a_n) \in l^\infty$, there exists a function f in $H^\infty(\mathbb{D})$ such that

$$f(z_n) = a_n, \; n = 1, 2, \cdots,$$

where $\{z_n\}$ denotes its zero sequence. This is equivalent to the following:

$$\inf_k \prod_{j \geq 1, j \neq k} d(z_j, z_k) > 0,$$

see [Ga].

A Blaschke product B is called *a thin Blaschke product* [GM2] if the zero sequence $\{z_n\}$ of B satisfies

$$\lim_{k \to \infty} \prod_{j \geq 1, j \neq k} d(z_j, z_k) = 1.$$

Since

$$\prod_{j \geq 1, j \neq k} d(z_j, z_k) = (1 - |z_k|^2)|B'(z_k)|,$$

the above condition is equivalent to

$$\lim_{k \to \infty} (1 - |z_k|^2)|B'(z_k)| = 1.$$

A thin Blaschke product is never a covering map. This follows easily from the fact that if a Blaschke product B is a covering map whose zero sequence is $\{z_n\}$, then $(1 - |z_k|^2)|B'(z_k)|$ is a constant not depending on k [Cow1, Theorem 6]. Also, a thin Blaschke product has many good properties, see Sect. 5.2 in Chap. 5. Here, we just mention one of them as below, for example see [GM2, Lemma 3.2(3)], [Hof2, pp. 86, 106], or [Ga, pp. 404, 310].

Proposition 2.1.8 *Let B be a thin Blaschke product. Then the following hold:*

(1) *Each value in $\mathbb{D}$ can be achieved for infinitely many times by B.*
(2) *For every point w in $\mathbb{D}$, $\varphi_w \circ B$ is a thin Blaschke product.*

Note that in Proposition 2.1.8(1) is a direct consequence of (2). Besides, Proposition 2.1.8 also implies that a thin Blaschke product is never a covering map because its image is exactly the unit disk.

Another result on thin Blaschke products will also be used in the sequel, see [La, Ga] and [GM3, Lemma 4.2].

Proposition 2.1.9 *Suppose B is a thin Blaschke product with the zero sequence* $\{z_k\}$. *Then for each r with* $0 < r < 1$, *there exists an* $\varepsilon \in (0, 1)$ *such that* $|B(z)| \geq r$ *whenever* $d(z, z_n) \geq \varepsilon$ *for all n.*

Proof The proof is from [Ga, pp. 395–397].

Suppose B is a thin Blaschke product with the zero sequence $\{z_k\}$. For each $z \in \mathbb{D}$ and $\varepsilon \in (0, 1)$, put

$$\Delta(z, \varepsilon) \triangleq \{w \in \mathbb{D} : d(z, w) < \varepsilon\}.$$

To prove Proposition 2.1.9, it suffices to show for enough large $r(0 < r < 1)$, there is a constant $\varepsilon \in (0, 1)$ satisfying

$$B^{-1}(\Delta(0, r)) \subseteq \bigcup_{n \geq 0} \Delta(z_n, \varepsilon). \tag{2.1}$$

First, assume that there is a constant δ such that

$$\prod_{j; j \neq k} d(z_j, z_k) \geq \delta.$$

In this case, we will prove that for $r = r(\delta) \in (0, 1)$, $|B(z)| \geq r$ whenever $d(z, z_n) \geq \varepsilon$ for all n and $r(\delta) \to 1$ as $\delta \to 1^-$.

For each $a \in \mathbb{D}$, put

$$L_a(z) \triangleq -\varphi_a(-z) = \frac{z + a}{1 + \bar{a}z},$$

and set

$$h_n(\zeta) = B\left(\frac{\zeta + z_n}{1 + \bar{z}_n\zeta}\right) = B \circ L_{z_n}(\zeta).$$

By using the maximum value theorem, one can show that for each holomorphic function $f : \mathbb{D} \to \mathbb{D}$,

$$d(f(z), f(a)) \equiv \frac{|f(z) - f(a)|}{|1 - \overline{f(a)}f(z)|} \leq \frac{|z - a|}{|1 - \bar{a}z|}, \quad z \neq a \quad \text{and} \quad z, a \in \mathbb{D}. \tag{2.2}$$

Since $\|h_n\|_\infty = 1$ and $h_n(0) = 0$, then $\frac{h_n(\zeta)}{\zeta}$ is a holomorphic function from $\mathbb{D}$ into $\mathbb{D}$. Taking $f(\zeta) = \frac{h_n(\zeta)}{\zeta}$ and $a = 0$ in (2.2), we get

$$d\left(\frac{h_n(\zeta)}{\zeta}, h_n'(0)\right) \leq |\zeta|. \tag{2.3}$$

Write $\lambda = 1 - \sqrt{1-\delta} < \delta$. By (2.3), when $|\zeta| = \lambda$,

$$\left|\frac{h_n(\zeta)}{\zeta}\right| \geq \frac{|h'_n(0)| - |\zeta|}{1 - |\zeta||h'_n(0)|} \geq \frac{\delta - \lambda}{1 - \lambda\delta}.$$

For $|\zeta| = \lambda$,

$$|h_n(\zeta)| \geq \frac{\delta - \lambda}{1 - \lambda\delta}\lambda \triangleq r.$$

By the argument principle, for each w with $|w| < r$, $h_n - w$ and h_n has the same number of zeros in $\{\zeta : |\zeta| < \lambda\}$, counting multiplicity. Thus, $h_n - w$ has exactly one zero in $\{\zeta : |\zeta| < \lambda\}$, counting multiplicity. Put

$$V_n \triangleq L_{z_n}\left(h_n^{-1}(\Delta(0,r)) \cap \lambda\mathbb{D}\right),$$

and then B maps V_n biholomorphically onto $\Delta(0,r)$. Since

$$V_n \subseteq \Delta(z_n, \lambda),$$

(2.1) reduces to the following

$$B^{-1}(\Delta(0,r)) \subseteq \cup_n V_n. \tag{2.4}$$

Below, we will prove (2.4). For each $w(|w| < r)$ define

$$B_w(z) = \frac{B(z) - w}{1 - \overline{w}B(z)},$$

which has exactly one zero $z_n(w)$ in $V_n(n \geq 0)$ and $z_n(w)$ is a holomorphic function of $w(|w| < r)$ satisfying

$$z_n(0) = z_n.$$

To prove (2.4), we must show that B_w has no zero outside $\cup_n V_n$. Now let $H_w(z)$ denote the Blaschke product with the zero sequence $\{z_n(w)\}$, and then

$$B_w = H_w g_w,$$

where g_w is a bounded holomorphic function satisfying $\|g_w\|_\infty \leq 1$. Note that

$$|B_w(0)| = |g_w(0)| \prod_{n=1}^{\infty} |z_n(w)|.$$

Write

$$G(w) = \prod_{n=1}^{\infty} \frac{\overline{z_n}}{|z_n|} z_n(w),$$

This product converges on $\{w : |w| < r\}$ because its partial products are bounded by 1 and it converges at $w = 0$. Since $\|g_w\|_\infty \leq 1$,

$$|G(w)| = \prod_{n=1}^{\infty} |z_n(w)| \geq |B_w(0)| = \left| \frac{B(0) - w}{1 - \overline{B(0)}w} \right|.$$

Write

$$F(w) = \frac{B(0) - w}{1 - \overline{B(0)}w} G(w)^{-1}, \quad |w| < r,$$

and clearly

$$|F(w)| \leq 1 \quad \text{and} \quad |F(0)| = 1,$$

forcing $|F(w)| = 1$, $|w| < r$. This immediately shows that $|g_w(0)| = 1$, which shows that g_w is a constant because $\|g_w\|_\infty \leq 1$. Thus, B_w has the same zeros as H_w. Since all zeros $\{z_n(w)\}$ of H_w lie in $\cup_n V_n$ for each $w(|w| < r)$, then

$$B^{-1}(\{w : |w| < r\}) = \cup_n V_n \subseteq \cup_n \Delta(z_n, \delta),$$

as desired. The proof of (2.4) is complete. That is, with $r = r(\delta)$ we get (2.1).

In general, since B is a thin Blaschke product,

$$\prod_{j; j \neq k} d(z_j, z_k) \to 1, \quad (k \to \infty).$$

For a fixed number $\delta \in (0, 1)$, there is natural number N_0 such that

$$\prod_{j; j \neq N_0} d(z_j, z_k) > \delta.$$

Let B_1 denote the finite Blaschke product with zero sequence $\{z_k : 1 \leq k < N_0\}$, and B_2 denotes the Blaschke product with zero sequence $\{z_k : k \geq N_0\}$. Clearly, $B = B_1 B_2$. By the observations that

$$\{|B| < r^2(\delta)\} \subseteq \{|B_1| < r(\delta)\} \cup \{|B_2| < r(\delta)\},$$

and that $\lim_{\delta \to 1^-} r(\delta) = 1$, the remaining is an easy exercise.   □

To establish another result on thin Blaschke products, we first present the following lemma, see [Mas, Theorem 3.6] for example.

Lemma 2.1.10 *If ψ is a finite Blaschke product and $\lambda \in \mathbb{T}$, then*

$$|\psi'(\lambda)| = \lim_{z \to \lambda} \frac{1 - |\psi(z)|^2}{1 - |z|^2}.$$

Proof Assume ψ is a finite Blaschke product. We must show that for each $\lambda \in \mathbb{T}$,

$$|\psi'(\lambda)| = \lim_{z \to \lambda} \frac{1 - |\psi(z)|^2}{1 - |z|^2}.$$

The proof is divided into two part.

Step I. First, it will be shown that $\lim_{z \to \lambda} \frac{1-|\psi(z)|^2}{1-|z|^2}$ always exists for $\lambda \in \mathbb{T}$.

This will be handled by induction. In more detail, suppose ψ is a finite Blaschke product. If order $\psi = 1$, then one may write $\psi(z) = \varphi_a(z)$ with $a \in \mathbb{D}$. By computations, we have

$$1 - |\psi(z)|^2 = \frac{(1 - |a|^2)(1 - |z|^2)}{|1 - \bar{a}z|^2}.$$

Then

$$\frac{1 - |\psi(z)|^2}{1 - |z|^2} = \frac{1 - |a|^2}{|1 - \bar{a}z|^2} = \left| -\frac{1 - |a|^2}{(1 - \bar{a}z)^2} \right| = |\psi'(z)|.$$

By induction, assume that $\lim_{w \to \lambda} \frac{1-|\psi(z)|^2}{1-|z|^2}$ exists if order $\psi = k$. Now write that $\psi(z) = B(z)\varphi_a(z)$, where $a \in \mathbb{D}$ and B is a finite Blashcke product of order k. Note that

$$\frac{1 - |\psi(z)|^2}{1 - |z|^2} = \frac{1 - |B(z)|^2}{1 - |z|^2} + |B|^2 \frac{1 - |\varphi_a(z)|^2}{1 - |z|^2},$$

from which it follows that $\lim_{z \to \lambda} \frac{1-|\psi(z)|^2}{1-|z|^2}$ exists.

Step II. To complete the proof, it suffices to show that $|\psi'(\lambda)| = \lim_{k \to \infty} \frac{1-|\psi(w_k)|^2}{1-|w_k|^2}$ holds for some sequence $\{w_k\}$ tending to λ.

Put $z^* = \psi(\lambda) \in \mathbb{T}$, define

$$\gamma(t) = tz^*, \ 0 \le t \le 1,$$

and write $z_k = \gamma(1 - \frac{1}{k+1})$. It is known that the derivative of a finite Blaschke product never vanishes on $\mathbb{T}$. Then there is an $r_0 \in [0, 1)$ such that $\psi^{-1}(\gamma[r_0, 1])$ consists of

d (d = order ψ) disjoint arcs,

$$\gamma_1, \cdots, \gamma_d.$$

Without loss of generality, put $r_0 = 0$. For each z_k, $\psi^{-1}(z_k)$ consists of d different complex numbers, denoted by

$$w_k^1, \cdots, w_k^d,$$

where $w_k^j \in \gamma_j (1 \leq j \leq d)$ for each k. Without loss of generality, assume that $\lambda \in \gamma_1$. Rewrite w_k for w_k^1 and $w^* \equiv \lambda$. Note that z_k tends to z^* as k tends to infinity, and then w_k tends to w^* as k tends to infinity. Since $\psi'(w^*) \neq 0$, ψ is conformal at w^*. Then from

$$z_k = z^* - \frac{1}{k+1}z^*, k \to \infty,$$

it follows that

$$w_k = w^* - \delta_k w^* + o(1)\delta_k, k \to \infty,$$

where δ_k is a positive infinitesimal, and $o(1)$ denotes an infinitesimal. After some computations, we have

$$\frac{1 - |z_k|^2}{1 - |w_k|^2} \Big/ \frac{\frac{1}{k+1}}{\delta_k} \to 1, k \to \infty. \tag{2.5}$$

Since

$$\left| \frac{\psi(w_k) - \psi(w^*)}{w_k - w^*} \right| = \left| \frac{z_k - z^*}{w_k - w^*} \right| = \left| \frac{1 - z_k \overline{z^*}}{1 - w_k \overline{w^*}} \right|,$$

it is easy to verify that

$$\left| \frac{\psi(w_k) - \psi(w^*)}{w_k - w^*} \right| \Big/ \frac{\frac{1}{k+1}}{\delta_k} \to 1, k \to \infty.$$

Therefore, $|\psi'(w^*)| / \frac{\frac{1}{k+1}}{\delta_k} \to 1, k \to \infty$. Then by (2.5),

$$\lim_{k \to \infty} |\psi'(w^*)| \frac{1 - |w_k|^2}{1 - |z_k|^2} = 1,$$

and thus $|\psi'(\lambda)| = \lim_{k \to \infty} \frac{1 - |\psi(w_k)|^2}{1 - |w_k|^2}$. The proof is complete. $\qquad \square$

Now we are ready to state the following.

Proposition 2.1.11 *Given an infinite Blashcke product B and a finite Blaschke product ψ, B is a thin Blaschke product if and only if $B \circ \psi$ is a thin Blaschke product.*

Proof First, suppose B is an infinite Blaschke product with zero sequence $\{a_k\}_{k=1}^{\infty}$, ψ is a finite Blaschke product with order n, and $B \circ \psi$ is a thin Blaschke product. Put $\phi = B \circ \psi$, and write $\psi^{-1}(a_k)$ as a finite sequence

$$b_k^1, \cdots, b_k^n.$$

Since $B \circ \psi$ is a thin Blaschke product, we have

$$|\phi'(b_k^j)|(1 - |b_k^j|^2) \to 1, \ k \to \infty (j = 1, \cdots, n).$$

That is,

$$|B'(a_k)|(1 - |b_k^j|^2)|\psi'(b_k^j)| \to 1, \ k \to \infty (j = 1, \cdots, n).$$

Also, by Lemma 2.1.10 we have

$$\frac{(1 - |b_k^j|^2)|\psi'(b_k^j)|}{1 - |a_k|^2} \to 1, \ k \to \infty (j = 1, \cdots, n),$$

forcing

$$|B'(a_k)(1 - |a_k|^2)| \to 1, \ k \to \infty,$$

which shows that B is a thin Blaschke product.

The inverse direction follows from similar discussion as above. □

Remark 2.1.12 With a bit more effort, one can show that if B is a thin Blashcke product and $h \in H^{\infty}(\mathbb{D})$ satisfying $B = h \circ \psi$ with ψ a finite Blaschke product, then h is a thin Blashcke product.

In Chap. 6, geometric properties of thin Blaschke products are investigated in considerable detail. For that aim, we present Böttcher's theorem which studies the behavior of a holomorphic function over a neighborhood of 0. It is of independent interest, see [Mi, Theorem 9.1] or [CaG, p. 33, Theorem 4.1].

Theorem 2.1.13 (Böttcher) *Suppose*

$$f(z) = a_n z^n + a_{n+1} z^{n+1} + \cdots,$$

where $a_n \neq 0$ for $n \geq 2$. Then there exists a local holomorphic change of coordinate $w = \varphi(z)$ such that $\varphi(0) = 0$ and $\varphi \circ f \circ \varphi^{-1}(w) = w^n$ holds on a neighborhood of 0. Furthermore, φ is unique up to multiplication by an $(n - 1)$-th root of unity.

Proof The proof is from [CaG].

First, we prove the existence of φ. By the Fourier expansion of f at 0, there are two constants δ and C satisfying $0 < \delta < 1 < C$, $C\delta < 1$, and

$$|f(z)| \le C|z|^n, \quad |z| < \delta.$$

By induction, write $f^{[1]} = f$ and $f^{[k+1]} = f^{[k]} \circ f (k \ge 2)$. Also, we have

$$|f^{[k]}(z)| \le (C|z|)^{n^k} \le 1, \quad |z| < \delta, \; k = 1, 2, \cdots . \tag{2.6}$$

If we change variables by setting $\zeta = cz$ where $c^{n-1} = \frac{1}{a_n}$, then we have conjugated f to the form $f(\zeta) = \zeta^n + \cdots$, where suspension points "$\cdots$" denote terms of higher degrees here and below. Thus, we may assume that $a_n = 1$. In this case, we will prove that there is a map $\varphi(z) = z + \cdots$ satisfying

$$\varphi(f(z)) = \varphi(z)^n.$$

Put

$$\varphi_k(z) \triangleq f^{[k]}(z)^{n^{-k}} = \sqrt[n^k]{(z^{n^k} + \cdots)} = z\sqrt[n^k]{1 + \cdots},$$

which is well-defined in a neighborhood of 0. These φ_k's satisfy

$$\varphi_{k-1} \circ f = (f^{[k-1]} \circ f)^{n^{-k+1}} = \varphi_k^n.$$

If one can show that on some small disk Δ centered at 0, $\{\varphi_k\}$ converges uniformly to a function φ, then φ is necessarily holomorphic and satisfies that

$$\varphi(f(z)) = \varphi(z)^n, \; z \in \Delta.$$

Also,

$$\varphi'(0) = \lim_{k\to\infty} \varphi_k'(0) = 1,$$

which will finish the proof. Now it remains to show that $\{\varphi_k\}$ converges. To see this, recall that $f^{[k+1]} = f^{[k]} \circ f$. Note that by (2.6)

$$\frac{\varphi_{k+1}(z)}{\varphi_k(z)} = \left(\frac{\varphi_1 \circ f^{[k]}(z)}{f^{[k]}(z)}\right)^{n^{-k}}$$

$$= (1 + O(|f^{[k]}(z)|))^{n^{-k}}$$

$$= 1 + O(n^{-k})O(|f^{[k]}(z)|)$$

$$= 1 + O(n^{-k})O(1)$$

$$= 1 + O(n^{-k}), \quad |z| < \delta.$$

Thus, there is an enough large number N_0, such that the product $\prod_{k=N_0}^{\infty} \frac{\varphi_{k+1}}{\varphi_k}$ converges uniformly on the disk $\{z : |z| < \delta\}$, which implies that $\{\varphi_k\}$ converges uniformly on the disk $\{z : |z| < \delta\}$, to a function φ as desired. The proof for the existence of φ is complete.

Next, it remains to deal with the uniqueness of φ. If ψ is a biholomorphic map on a neighborhood of 0 such that $\psi(0) = 0$ and $\psi \circ f \circ \psi^{-1}(w) = w^n$ holds on a neighborhood of 0. Since

$$\varphi \circ f \circ \varphi^{-1}(w) = w^n,$$

then

$$\varphi^{-1} \circ T \circ \varphi(w) = f(w),$$

where $T(w) = w^n$. Thus,

$$\psi \circ \varphi^{-1} \circ T \circ \varphi \circ \psi^{-1}(w) = T(w)$$

holds on a neighborhood V of 0. Write $h = \varphi \circ \psi^{-1}$, and then

$$T(h(w)) = h(T(w)), \ w \in V. \tag{2.7}$$

Write $h(z) = c_1 z + c_2 z^2 + \cdots$, and by (2.7) one can show $c_1^{n-1} = 1$ and

$$c_k = 0, \ k = 2, 3, \cdots.$$

This shows that φ is unique up to multiplication by an $(n-1)$-th root of unity. $\square$

Roughly speaking, Theorem 2.1.13 tells us that if f is a non-constant function that is holomorphic at z_0 and $f'(z_0) = 0$, then on an enough small neighborhood of z_0, it behaves no more complicated than $z \mapsto z^n$ at $z = 0$.

Below, we present the definitions of covering maps and branched covering maps.

Let Ω be a bounded planar domain. A mapping $\phi : \mathbb{D} \to \Omega$ is called *a holomorphic covering map* if for each point of Ω there exists a connected open neighborhood U in Ω such that ϕ maps each component of $\phi^{-1}(U)$ conformally onto U. That is, the holomorphic function ϕ is topologically a covering map. It is well-known that such a map ϕ always exists, and ϕ is unique up to a conformal automorphism of the unit disk [Gol, Mi, V], see as follows.

Theorem 2.1.14 (The Koebe Uniformization Theorem) *Given a point z_0 in $\mathbb{D}$ and w_0 in Ω with the cardinality $\sharp \partial \Omega \geq 2$, there is a unique holomorphic covering map ϕ of $\mathbb{D}$ onto Ω with $\phi(z_0) = w_0$ and $\phi'(z_0) > 0$.*

Thus, covering maps over $\mathbb{D}$ exist abundantly. For example, let E be a discrete subset of $\mathbb{D}$, and ϕ is a holomorphic covering map from $\mathbb{D}$ onto $\mathbb{D} - E$. Then one can show that ϕ is an interpolating Blaschke product if $0 \notin E$, see Example 6.3.9 for details.

It is notable that Riemann mapping theorem can be regarded as a special case of Theorem 2.1.14. It appears as a classical result in standard textbooks of complex analysis, see [Ne, p. 175] and [BG, p. 174, Theorem 2.6.21] for example.

Theorem 2.1.15 (Riemann Mapping Theorem) *Let Ω be a simple connected domain whose boundary consists of more than one point. Then for each fixed point $w_0 \in \Omega$ and $z_0 \in \mathbb{D}$, there is a unique conformal map ϕ from $\mathbb{D}$ onto Ω satisfying $\phi(z_0) = w_0$ and $\phi'(z_0) > 0$.*

As follows, we introduce the notion of branched covering map [Mi, Appendix E], which is a natural generalization of covering map. Given a holomorphic map ϕ : $\mathbb{D} \to \Omega$, write

$$G(\phi) = \{\gamma; \gamma \text{ is a conformal automorphism of } \mathbb{D} \text{ satisfying } \phi \circ \gamma = \phi\},$$

which is called *the deck transformation group* of ϕ. For this map $\phi : \mathbb{D} \to \Omega$, we pose the following conditions:

(1) every point of Ω has a connected open neighborhood U such that each connected component V of $\phi^{-1}(U)$ maps onto U by a proper map $\phi|_V$;
(2) for any $z_1, z_2 \in \mathbb{D}$, $\phi(z_1) = \phi(z_2)$ implies that there is a member $\gamma \in G(\phi)$ such that $\gamma(z_1) = z_2$.

A holomorphic map $\phi : \mathbb{D} \to \Omega$ is called *a branched covering map* if it satisfies (1); and it is called *regular* if (2) is satisfied. For each $w \in \Omega$, $\phi^{-1}(w)$ is called a *fiber* over w. Note that condition (2) tells us that the deck transformation group $G(\phi)$ acts transitively on each fiber.

Regular branched covering maps share several good properties. Let Ω be a bounded planar domain, and let $\phi : \mathbb{D} \to \Omega$ be such a map. Then it is not difficult to check that for each $w \in \Omega$, the multiplicities at $z(z \in \phi^{-1}(w))$ for zeros of $\phi - w$ only depend on w [Mi, Appendix E]. Also, ϕ has the following property [Mi, BG].

Proposition 2.1.16 *For a planar domain Ω, let $\phi : \mathbb{D} \to \Omega$ be a holomorphic regular branched covering map, and put*

$$\mathcal{E}_\phi = \{\phi(z) : z \in \mathbb{D} \text{ and } \phi'(z) = 0\}.$$

Then both $\mathcal{E}_\phi$ and $\phi^{-1}(\mathcal{E}_\phi)$ are discrete in Ω and $\mathbb{D}$, respectively.

Proof Assume that $\phi : \mathbb{D} \to \Omega$ is a holomorphic regular branched covering map. We first show that $\mathcal{E}_\phi$ is discrete in Ω. For this, it suffices to show that for each $w \in \Omega$, there is a neighborhood V of w such that $V \subseteq \Omega$ and $\mathcal{E}_\phi \cap V$ is finite. Now for a given $w \in \Omega$, there exists an enough small neighborhood V $(V \subseteq \Omega)$ of w such that

$$\phi^{-1}(V) = \bigsqcup_{i \geq 0} V_i,$$

where V_i are components of $\phi^{-1}(V)$; and for each i, $\phi|_{V_i} : V_i \to V$ is a proper map. In particular, $\phi(V_i) = V$ for each i. Also, we may assume that at least one V_i satisfies that $\overline{V_i} \subseteq \mathbb{D}$, say,

$$\overline{V_0} \subseteq \mathbb{D}.$$

Put

$$F_0 \triangleq \{\phi(z) : z \in V_0 \text{ and } \phi'(z) = 0\}.$$

Since $Z(\phi') \cap V_0$ is a finite set, so is F_0.

In addition, by regularity of ϕ, for any $z_1, z_2 \in \mathbb{D}$, $\phi(z_1) = \phi(z_2)$ implies that there is a $\gamma \in G(\phi)$ such that $\gamma(z_1) = z_2$. This implies that if $\phi'(z_0) = 0$ and $\phi(z_0) \in V$ for some $z_0 \in \mathbb{D}$, then there is a member ρ in $G(\phi)$ such that $\rho(z_0) \in V_0$, and thus $\phi'(\rho(z_0)) = 0$ since $\phi \circ \rho = \phi$. By arbitrariness of z_0,

$$\mathcal{E}_\phi \cap V = F_0,$$

which is finite, as desired. Thus, $\mathcal{E}_\phi$ is discrete in Ω.

It remains to show that $\phi^{-1}(\mathcal{E}_\phi)$ is discrete in $\mathbb{D}$; equivalently, $\phi^{-1}(\mathcal{E}_\phi) \cap r\mathbb{D}$ is finite for any $r \in (0, 1)$. With r fixed, the discrete property of $\mathcal{E}_\phi$ shows that $\phi(r\mathbb{D}) \cap \mathcal{E}_\phi$ is finite, and hence $\phi^{-1}(\mathcal{E}_\phi) \cap r\mathbb{D}$ is finite. By arbitrariness of r, $\phi^{-1}(\mathcal{E}_\phi)$ is discrete in $\mathbb{D}$, as desired. The proof is complete. $\qquad\square$

For example, a finite Blaschke product B is a branched covering map over $\mathbb{D}$, but it is rarely regular. If B is also regular, then there is a $\lambda \in \mathbb{D}$ such that $B = c\varphi_\lambda^n$ for some unimodular constant c. In general, a finite Blaschke product B with order n is always an n-folds map. When restricted on $\mathbb{D} - B^{-1}(\mathcal{E}_B)$, B becomes a covering map.

The following result comes from the proof of [Cow1, Theorem 6].

Proposition 2.1.17 *Suppose $\phi : \mathbb{D} \to \Omega$ is a bounded holomorphic regular branched covering map. For each $a_0 \in \mathbb{D} - \phi^{-1}(\mathcal{E}_\phi)$, let B denote the Blaschke product whose zero set is $\phi^{-1}(\phi(a_0))$. Then both $\dfrac{\phi - \phi(a_0)}{B}$ and $\dfrac{B}{\phi - \phi(a_0)}$ are in $H^\infty(\mathbb{D})$.*

Proof The proof is from [Cow1].

It is enough to deal with the case of $\phi : \mathbb{D} \to \Omega$ being a bounded holomorphic covering map. As will be seen later, the proof for general case is similar. Now fix $a_0 \in \mathbb{D} - \phi^{-1}(\mathcal{E}_\phi)$. Without loss of generality, assume that $\phi(a_0) = 0$, and the deck transformation group $G(\phi)$ of ϕ is infinite. Then there is an enough small $r > 0$ such that $r\mathbb{D} \subseteq \Omega$ and

$$\phi^{-1}(r\mathbb{D}) = \bigsqcup_{n=0}^{\infty} V_n,$$

where $\phi|_{V_n} : V_n \to r\mathbb{D}$ are biholomorphic for all n. Write $\{a_n\}_{n=0}^{\infty} = \phi^{-1}(\phi(a_0))$, where $a_n \in V_n$ for each n, and let B denote the Blaschke product whose zero sequence is $\{a_n\}_{n=0}^{\infty}$. Clearly, $\dfrac{\phi}{B} \in H^{\infty}(\mathbb{D})$, see [Hof1].

To see $\dfrac{B}{\phi} \in H^{\infty}(\mathbb{D})$, it suffices to show that $\left|\dfrac{\phi}{B}\right|$ is bounded below. For this, note that if $z \in \mathbb{D} - \bigsqcup_{n=0}^{\infty} V_n$, then

$$\left|\frac{\phi(z)}{B(z)}\right| \geq |\phi(z)| \geq r.$$

Now consider $\dfrac{\phi}{B}$ on $\overline{V_0}$. Since a_0 is the simple zero of B and ϕ, then by theory of complex analysis there is a constant $c > 0$ satisfying

$$\left|\frac{\phi(z)}{B(z)}\right| \geq c, \, z \in \overline{V_0}.$$

In general, let ρ be a member of $G(\phi)$ satisfying $\rho(a_k) = a_0$, and then ρ maps V_k onto V_0. Since $B \circ \rho$ is also a Blaschke product, whose zero sequence equals $\{a_k\}_{k=0}^{\infty}$, we have

$$|B(\rho(z))| = |B(z)|, \, z \in \mathbb{D}.$$

This, combined with $\phi \circ \rho = \phi$, yields that

$$\left|\frac{\phi(z)}{B(z)}\right| = \left|\frac{\phi(\rho(z))}{B(\rho(z))}\right|, \, z \in V_k,$$

forcing $\left|\dfrac{\phi}{B}\right| \geq c$ on V_k. Therefore,

$$\left|\frac{\phi(z)}{B(z)}\right| \geq c, \, z \in \bigsqcup_{n=0}^{\infty} V_n.$$

This leads to the conclusion that $\left|\dfrac{\phi}{B}\right|$ is bounded below, forcing $\dfrac{B}{\phi} \in H^{\infty}(\mathbb{D})$, as desired. The proof is complete.   □

Next, we would like to mention Runge's theorem, see [Hor, Theorem 1.3.1].

Theorem 2.1.18 (Runge) *Let Ω be an open set in $\mathbb{C}$ and K a compact subset of Ω. The following are equivalent.*

(1) Every function which is holomorphic on a neighborhood of K can be approximated uniformly on K by functions in $A(\Omega)$;

(2) *The open set $\Omega - K$ has no component which is relatively compact in Ω;*

(3) *For every $z \in \Omega - K$, there is a function $f \in A(\Omega)$ such that*

$$|f(z)| > \sup_{w \in K} |f(w)|.$$

The following is obtained by taking $\Omega = \mathbb{C}$, also see [Hor].

Corollary 2.1.19 *Every function which is holomorphic on a neighborhood of the compact set K can be approximated by polynomials uniformly on K if and only if $\mathbb{C} - K$ is connected, if and only if for every $z \notin K$ there is a polynomial f such that*

$$|f(z)| > \sup_{w \in K} |f(w)|.$$

The multi-variable version of Runge's theorem is Oka-Weil theorem [Oka, Weil]. Before stating it, we recall the notion of convex hull, or in short, hull. Let K be a compact set in some domain G in $\mathbb{C}^d$, and $\mathcal{F}$ be a family of holomorphic functions over G. The *hull* $\hat{K}$ of K with respect to $\mathcal{F}$ is defined to be

$$\{z \in G : |f(z)| \leq \|f\|_{K,\infty}, f \in \mathcal{F}\},$$

where $\|f\|_{K,\infty} = \sup\{|f(z)| : z \in K\}$.

Theorem 2.1.20 (Oka-Weil) *Let G be a domain in $\mathbb{C}^d$, and suppose the compact set $K(K \subseteq G)$ coincides with its hull with respect to the algebra $\mathrm{Hol}(G)$ of all holomorphic functions on G. Then for any function f holomorphic in a neighborhood of K, and for any $\varepsilon > 0$, there is a function $F \in \mathrm{Hol}(G)$ such that*

$$\max_{z \in K} |f(z) - F(z)| < \varepsilon.$$

Finally, some results are presented from real analysis, which will be used in the sequel.

For a topological space X, $C(X)$ always denotes the algebra of all continuous complex-valued functions over X. If X is a compact Hausdorff space, then $C(X)$ is equipped with the maximal-norm:

$$\|h\| \triangleq \max_{x \in X} |h(x)|, \ h \in C(X).$$

Theorem 2.1.21 (Stone-Weierstrass Theorem) *Let X be a compact Hausdorff space and let $\mathcal{S}$ be a subset of $C(X)$ which separates points in X. That is, for any $x, y \in X$ with $x \neq y$, there is a function $f \in \mathcal{S}$ such that $f(x) \neq f(y)$. Then the complex unital $*$-algebra generated by $\mathcal{S}$ is dense in $C(X)$.*

Given a function $h \in L^1(\mathbb{T})$, expand its Fourier series as follows:

$$h \sim \sum_{n=-\infty}^{\infty} a_n e^{i\theta},$$

where the right hand side only represents a formal series. Write $s_k = \sum_{n=-k}^{k} a_n e^{i\theta}$, and put

$$\sigma_k = \frac{s_0 + \cdots + s_{k-1}}{k},$$

called the *k-th Cesaro mean* of h, see [Hof1]. It is well known that Cesaro means have good property in approximation. For example, if in addition $h \in L^p(\mathbb{T})$ for some $p \in [1, +\infty)$, then $\{\sigma_k\}$ converges to h in $L^p(\mathbb{T})$-norm. Besides, if $h \in L^\infty(\mathbb{T})$, then

$$\|\sigma_k\|_\infty \le \|h\|_\infty, \quad k = 1, 2, \cdots.$$

In this case, $\{\sigma_k\}$ converges to h almost everywhere with respect to the arc-length measure, and thus $\{\sigma_k\}$ converges to h in the weak*-topology of $L^\infty(\mathbb{T})$.

Theorem 2.1.22 (Lebesgue's Dominated Convergence Theorem) *Let $\{f_n\}$ be a sequence of measurable functions on a complete measure space (X, Σ, μ). Suppose there is a non-negative $g \in L^1(X, \mu)$ such that $|f_n| \le g\,(n \ge 1)$ hold almost everywhere and one of the following holds:*

(1) $\{f_n\}$ converges to f almost everywhere;
(2) $\{f_n\}$ converges to f in measure.

Then f is integrable and

$$\lim_{n \to \infty} \int_X |f_n - f| d\mu = 0.$$

In particular, $\lim_{n \to \infty} \int_X f_n d\mu = \int_X f d\mu$.

Also presented is Lusin's theorem, which is well known in real analysis. One can refer to [Hal1, p. 242] or [Ru3, Theorem 2.24].

Theorem 2.1.23 (Lusin's Theorem) *Suppose X is a locally compact Hausdorff space, and $\mathcal{B}$ is the class of all Borel subsets of X. Let $(X, \mathcal{B}, \mu)$ be a Radon measure space, and let f be a Borel measurable function on $E(E \in \mathcal{B})$ with $\mu E < \infty$, then for each $\varepsilon > 0$, there is a compact subset F of E such that $f|_F$ is continuous, and $\mu(E - F) < \varepsilon$.*

For a Radon measure space $(X, \mathcal{B}, \mu)$, by definition μ is inner regular and for each $x \in X$, there exists a open neighborhood O_x of x such that $\mu(O_x) < \infty$. It is known that for a measure space $(X, \mathcal{B}, \mu)$, if X is a locally compact Hausdorff space such that every open set in X is a countable union of compact sets and $\mu(K)$ is finite for every compact set K, then μ is always regular and locally finite [Ru3, Theorem 2.18]. Recall that a measure μ is called *regular* if for any measurable set E, we have

$$\mu E = \inf\{\mu U : U \text{ is an open set containing } E\} \qquad \text{(outer regularity)}$$

and

$$\mu E = \sup\{\mu F : F \text{ is a compact subset of } E\} \qquad \text{(inner regularity)}.$$

For example, if μ is the Lebesgue measure m on $\mathbb{R}^n$, then μ is a regular measure. In this case, it is well-known that for any Lebesgue measurable function f, there is a Borel measurable function $\hat{f}$ satisfying $f = \hat{f}$ a.e. with respect to m. Thus, Lusin's theorem also holds for Lebesgue measurable functions.

In Lusin's theorem, since X is a locally compact Hausdorff space, then applying Tietze extension theorem shows that there is a continuous function g on X such that $g|_F = f$. Also, in Theorem 2.1.23 one can require $\|g\|_\infty \leq \|f\|_\infty$ provided that $f \in L^\infty(X, \mu)$.

Below, we present Tietze Extension Theorem, see [Arm, Ke]. Given a topological space X, if every pair (K_1, K_2) of disjoint closed sets in X can be separated by two disjoint open sets U_1 and U_2, satisfying $U_1 \supseteq K_1$ and $U_2 \supseteq K_2$, then X is called *a normal space*.

Theorem 2.1.24 *Let K be a closed subset of X, a normal topological space, and f is a continuous function from K into $[a, b]$ with $a, b \in \mathbb{R}$. Then there exists a continuous function g from X into $[a, b]$ such that*

$$g(x) = f(x), \quad x \in K.$$

2.2 The Notion of Capacity

In this section, we will give the notion of capacity. This section mainly consults [Ga, Ru1] and [CL].

Let K be a compact set in the complex plane, and μ a positive measure supported on K with $\mu \neq 0$. Define

$$U_\mu(z) = \int_K \ln \frac{1}{|\zeta - z|} d\mu(\zeta), \ z \in \mathbb{C},$$

called *the logarithmic potential* of μ. The following displays equivalent conditions for a compact set in $\mathbb{C}$ to have positive (logarithmic) capacity, refer to [Ga, pp. 78, 79] and [Ru1, pp. 56, 67].

Proposition 2.2.1 *Suppose K is a compact subset of $\mathbb{C}$. Then the following are equivalent:*

(1) there is a nonzero positive measure μ on K whose logarithmic potential is bounded on some neighborhood of K.
(2) K carries a nonzero positive measure μ whose logarithmic potential is continuous in $\mathbb{C}$.

If either (1) or (2) holds, then K is called to have positive capacity. In particular, if $K \subseteq \mathbb{D}$, then K has positive capacity if and only if K supports a nonzero positive measure ν for which the function

$$G_\nu(z) = \int_K \ln \left| \frac{1 - \bar{\zeta}z}{\zeta - z} \right| d\nu(\zeta)$$

is bounded on $\mathbb{D}$.

In general, a set $E(E \subset \mathbb{C})$ is called to *have positive capacity* if there is some compact subset K of E having positive capacity. If E does not have positive capacity, then we say *E has capacity zero*.

It is known that a capacity-zero subset K of $\mathbb{R}$ (regarded as a subset of $\mathbb{C}$) must have null Lebesgue measure [Ru1, p. 57]. In general, a set of capacity zero is of linear measure zero [CL]. Here, a subset E of $\mathbb{C}$ has linear measure zero if and only if for any $\varepsilon > 0$, there is a sequence of disks $O(z_n, r_n)$ whose union covers E and $\sum_n r_n < \varepsilon$; that is, the one-dimensional Hausdorff measure of E is zero. However, the converse does not hold. As mentioned in [CL, p. 10], the standard Cantor's ternary set P is of positive capacity [Nev, Fr]. However, P is a perfect set with null Lebesgue measure. Recall that a set is called *perfect* if it is a closed set with no isolated point.

Let Ω be a domain in $\mathbb{C}$ and E is a compact subset of Ω. Then E is called H^∞-*removable* if every bounded holomorphic function on $\Omega - E$ can be analytically extended to Ω. It should be pointed out that the property of H^∞-removable does not depend on the choice of Ω. One may refer to [Ma, Du] for an account on H^∞-removable property.

The following theorem is known in the theory of complex analysis, which was shown by Painlevé, and later by Besicovitch [Bes].

Theorem 2.2.2 *If E is a compact subset in $\mathbb{C}$ whose one-dimensional Hausdorff measure is zero, then E is H^∞-removable. In particular, if E is a compact set of capacity zero, then E is H^∞-removable.*

The above paragraphs have provided some descriptions for E having positive capacity. Below, to each subset E of $\mathbb{C}$ one can assign a precise value cap E, which is exactly the logarithmic capacity of E. The following content is from [CL]. Denote by $\mathcal{M}$ the set of all Borel probability measures μ supported on E, i.e. all Borel measures μ satisfying

$$\int_E d\mu(\zeta) = 1.$$

Let F be a compact subset of $\mathbb{C}$. If $\mathbb{C}-F$ were not connected, then the linear measure of F would never be zero, which means that in some sense F were "large". Now, assume that F is a compact subset of $\mathbb{C}$ such that $\mathbb{C} - F$ is connected. Let μ be a non-negative Borel measure over $\mathbb{C}$ and $\mu \in \mathcal{M}$, and set

$$u(z) = \int_F \ln \frac{1}{|z-\zeta|} d\mu(\zeta), z \notin F,$$

which is harmonic. Define

$$V_F = \inf_{\mu \in \mathcal{M}} \left(\sup_{z \notin F} u(z) \right),$$

which is called the *equilibrium potential* of F, and the *capacity of the set F* is defined to be

$$\operatorname{cap} F \overset{\triangle}{=} \exp(-V_F).$$

Here, $\exp(-\infty)$ is assigned to be 0. In general, the capacity of a Borel set E is defined to be the supremum of capacities of all compact subsets of E. From the definition, it follows that $\operatorname{cap} A \le \operatorname{cap} B$ whenever $A \subseteq B$. Also, it is clear that the union of two capacity-zero sets is of capacity zero. Furthermore, the union of a countable family of capacity-zero sets is of capacity zero [Nev, Fr]. In particular, any countable subset of the complex plane has capacity zero. Concerning with capacity, there are many interesting results arising from function theory, see [CL, Theorem 1.7] and [Ru2, 3.6.2].

The following property of closed capacity-zero sets proves useful.

Lemma 2.2.3 ([CL, Theorem 1.7]) *If E is a closed, bounded set of capacity zero, then there exists a probability measure μ on E such that the potential*

$$u(z) = \int_E \ln \frac{1}{|z-\zeta|} d\mu(\zeta), z \notin E$$

tends to $+\infty$ as z tends to an arbitrary point of E.

The following result will be needed in the sequel, which collects two known results from [Ga] and [CL]. The reader can also consult [GM2, Theorem 1.1].

Proposition 2.2.4 ([Ga, CL]) *If $f : \mathbb{D} \to \Omega$ is a holomorphic covering map, then f is an inner function if and only if $\Omega = \mathbb{D} - E$, where E is a relatively closed subset of $\mathbb{D}$ with capacity zero.*

Proof Since the image of a non-constant holomorphic map is open, the "only if" part follows directly from [Ga, p. 80, Theorem 6.6] or Theorem 2.1.7. Thus it remains to deal with the "if" part, whose proof comes from [CL, pp. 37, 38]. Here, we include the proof for completeness. Now assume that $f : \mathbb{D} \to \Omega$ is a holomorphic covering map, where $\Omega = \mathbb{D} - E$, with E being a relatively closed, capacity-zero subset of $\mathbb{D}$. We will show that f is an inner function. For this, we first make the following claim:

Claim If, on a set $F(F \subseteq \mathbb{T})$ of positive measure, the radial limit values of f lie in a set E of capacity zero, then f is identically constant.

The proof reduces to the case of E being a compact set of capacity zero. For this, assume that on a set $F(F \subseteq \mathbb{T})$ of positive measure, f admits radial limits in E, a set of capacity zero. By standard analysis, the function $f|_F$ (taking radial limits) is Lebesgue measurable over F. By Lusin's theorem, there is a compact subset F_0 of F with positive measure such that $f|_{F_0}$ is continuous, and thus $f(F_0)$ is a compact subset of E with capacity zero. Then one can replace E with $f(F_0)$.

Now we may assume that E is compact. By Lemma 2.2.3, there is a probability measure μ on E such that the potential

$$u(w) = \int_E \ln \frac{1}{|w - \zeta|} d\mu(\zeta) \quad (w \notin E)$$

tends to $+\infty$ as w tends to any given point of E. Set

$$u_1(w) = \int_E \ln \frac{2}{|w - \zeta|} d\mu(\zeta),$$

which is a non-negative harmonic function in $\mathbb{D} - E$. Also, $u_1(w)$ tends to $+\infty$ as w tends to any point of E. Define $U(z) = u_1 \circ f$, a non-negative harmonic function in $\mathbb{D}$. By our assumption on f, there is a subset J of $\mathbb{T}$ with positive measure such that

$$\{f(\zeta) : \zeta \in J\} \subseteq E,$$

where $f(\zeta)$ is defined to be the radial limit of f at ζ, i.e. $\lim_{r \to 1^-} f(r\zeta)$. Thus for each $\zeta \in J$, $\lim_{r \to 1^-} U(r\zeta) = +\infty$.

Let V denote a harmonic conjugate of U, and $U + iV$ is holomorphic in $\mathbb{D}$. Define

$$F = \exp(-U - iV),$$

a bounded holomorphic function on $\mathbb{D}$. Since for each $\zeta \in J$, $\lim_{r \to 1^-} U(r\zeta) = +\infty$, $\lim_{r \to 1^-} F(r\zeta) = 0$. By a uniqueness theorem of Riesz [Hof1], F is identically zero, which is a contradiction. Therefore, f is identically constant. The proof of the claim is complete.

To show that f is an inner function, it suffices to prove that for almost everywhere $\zeta \in \mathbb{T}$, the radial limit $f(\zeta)$ at ζ exists and lies in $\{z; |z| = 1\}$. Assume conversely that f possesses radial limits w with $|w| < 1$ on a set of positive measure on $\mathbb{T}$. Then we would see that *those radial limits w lie in E*. If so, then applying the above claim yields that f is a constant, which is a contradiction. Therefore, f is an inner function, as desired.

To complete the proof, it remains to show that if the radial limit w lies in the unit disk, then $w \in E$. Otherwise, there would be some $t \in \mathbb{R}$ such that

$$\lim_{r \to 1^-} f(re^{it}) = w_0 \in \Omega = \mathbb{D} - E.$$

Since f is a covering map from $\mathbb{D}$ onto Ω, there is a disk O containing w_0 such that $O \subseteq \Omega$ and

$$f^{-1}(O) = \bigsqcup_n U_n$$

where U_n are all connected components, and $f : U_n \to O$ is a biholomorphic map for each n. Since $\lim_{r \to 1^-} f(re^{it}) = w_0$, there is an enough small $\delta > 0$ satisfying

$$f(re^{it}) \in O, \; r \in (1 - \delta, 1).$$

Then there is some integer n_0 such that

$$re^{it} \in U_{n_0}, \; r \in (1 - \delta, 1).$$

Since $f : U_{n_0} \to O$ is a bijection, there is a point $a \in U_{n_0}$ such that $f(a) = w_0$. Then there is a $r_0 > 0$ and $\varepsilon > 0$ such that

$$\overline{O(a, r_0)} \subseteq U_{n_0} \quad \text{and} \quad O(w_0, \varepsilon) \subseteq f(O(a, r_0)) \subseteq O.$$

Noting $\lim_{r \to 1^-} f(re^{it}) = w_0$, we deduce that there is an $r_1 \in (1 - \delta, 1)$ such that

$$f(r_1 e^{it}) \in O(w_0, \varepsilon) \quad \text{and} \quad r_1 e^{it} \notin O(a, r_0).$$

Since $f(O(a, r_0)) \supseteq O(w_0, \varepsilon)$, it follows that there is a point $a' \in O(a, r_0)$ satisfying $f(a') = f(r_1 e^{it})$ and $a' \neq r_1 e^{it}$, which is a contradiction to the bijectivity of the map $f : U_{n_0} \to O$, completing the proof.   □

Remark 2.2.5 By similar discussion as above, one can prove a similar version of Proposition 2.2.4 for regular branched covering maps; that is, if $\phi : \mathbb{D} \to \Omega$ is a regular branched covering map, then ϕ is an inner function if and only if $\Omega = \mathbb{D} - E$, where E is a relatively closed subset of $\mathbb{D}$ with capacity zero.

For more results on singularities and sets of capacity zero, we call the reader's attention to [CL, pp. 10–13].

2.3 Local Inverse and Analytic Continuation

This section mainly introduces the notations of local inverse and analytic continuation, which proves useful in the analysis of geometric property of holomorphic functions, see Chaps. 3–5.

Let Ω_0 be a domain of the complex plane and f be a holomorphic function on Ω_0. If ρ is a map defined on some sub-domain V of Ω_0 such that $\rho(V) \subseteq \Omega_0$ and $f(\rho(z)) = f(z), z \in V$, then ρ is called *a local inverse* of f on V [T1]. In this book, we take $\Omega_0 = \mathbb{D}$ in most cases.

For example, put $f(z) = z^n$ ($z \in \mathbb{D}$) and $\rho(z) = \xi z$ ($z \in \mathbb{D}$), where ξ is one of the n-th root of unit. Then ρ is a local inverse of f on $\mathbb{D}$. To see more examples, assume that ϕ is a covering map over $\mathbb{D}$. Then each local inverse of ϕ can be analytically extended to an automorphism of $\mathbb{D}$, precisely, a member in the deck transformation group of ϕ.

We need some definitions from [Ru3, Chap. 16]. *A function element* is an ordered pair (f, D), where D is a simply-connected open set and f is a holomorphic function on D. Two function elements (f_0, D_0) and (f_1, D_1) are called *direct continuations* if $D_0 \cap D_1$ is not empty and $f_0 = f_1$ holds on $D_0 \cap D_1$. By a *curve* or a *path*, we mean a continuous map from $[0, 1]$ into $\mathbb{C}$. A *loop* is a path σ satisfying $\sigma(0) = \sigma(1)$. Given a function element (f_0, D_0) and a curve γ with $\gamma(0) \in D_0$, if there is a partition of $[0, 1]$:

$$0 = s_0 < s_1 < \cdots < s_n = 1$$

and function elements $(f_j, D_j)(0 \le j \le n)$ such that

1. (f_j, D_j) and (f_{j+1}, D_{j+1}) are direct continuation for all j with $0 \le j \le n - 1$;
2. $\gamma[s_j, s_{j+1}] \subseteq D_j (0 \le j \le n - 1)$ and $\gamma(1) \in D_n$,

then (f_n, D_n) is called *an analytic continuation of* (f_0, D_0) *along* γ; and (f_0, D_0) is called to *admit* an analytic continuation along γ. In this case, we write $f_0 \sim f_n$. Clearly, this is an equivalence and we write $[f]$ for *the equivalent class* of f.

In Chaps. 4 and 5, we consider the case $f = B$, a thin or finite Blaschke product. In such a situation, put

$$E = \mathbb{D} - B^{-1}(\mathcal{E}_B),$$

where $\mathcal{E}_B$ denotes the critical value set; that is

$$\mathcal{E}_B = \{B(z) : z \in \mathbb{D} \text{ and } B'(z) = 0\}.$$

All functions mentioned in the last paragraph (such as f_j, ρ and σ, γ) are well-defined on some subsets of E.

The following theorem is well-known, which states that the analytic continuation along a curve must be unique, see [Ru3, Theorem 16.11].

Theorem 2.3.1 *If (f, D) is a function element and γ is a curve which starts at the center of D, then (f, D) admits at most one analytic continuation along γ.*

As follows we present some examples, which come from standard textbooks of complex analysis.

Example 2.3.2 Let D_0 be the unit disk $\mathbb{D}$, and D_1 be the upper half plane $\{z : Imz > 0\}$. Write

$$f_0(z) = \sum_{n=0}^{\infty} z^n, \ z \in D_0,$$

and put

$$f_1(z) = \frac{1}{1-z}, \ z \in D_1.$$

Then by direct computation one sees that (f_0, D_0) and (f_1, D_1) are direct continuation.

Set $V_0 = \{z : |z - 1| < 1\}$, define

$$g_0(z) = -\sum_{n=1}^{\infty} \frac{(1-z)^n}{n}, \ z \in V_0.$$

That is, $g_0(z) = \ln z$, where $\ln 1 = 0$. With $V_1 = D_1$, the upper half plane, put

$$g_1(z) = \int_1^z \frac{1}{\zeta} d\zeta, \ z \in V_1,$$

where the integral is along any curve in V_1. Then it is not difficult to verify that for any $x \in (0, 1)$, $g_0(x) = g_1(x)$, and hence by the uniqueness theorem $g_0 = g_1$ on $V_0 \cap V_1$. Therefore, (g_0, V_0) and (g_1, V_1) are direct continuation.

Example 2.3.3 Let V_2 and V_3 denote the left and lower half plane, respectively. That is,

$$V_2 = \{z : Rez < 0\} \quad \text{and} \quad V_3 = \{z : Imz < 0\}.$$

Similarly, define

$$g_2(z) = \int_{-1}^{z} \frac{1}{\zeta} d\zeta + \pi i, \ z \in V_2,$$

where the integral is along any curve in $\overline{V_2} - \{0\}$. Define

$$g_3(z) = \int_{1}^{z} \frac{1}{\zeta} d\zeta + 2\pi i, \ z \in V_3,$$

where the integral is along any curve in $\overline{V_3} - \{0\}$. Then one can check that (g_i, V_i) and (g_{i+1}, V_{i+1}) are direct continuation for $i = 0, 1, 2$. However, (g_3, V_3) and (g_1, V_1) are not direct continuation since $g_1(1) \neq g_3(1)$.

Furthermore, let γ be a curve in $\mathbb{C} - \{0\}$ with $\gamma(0) \in V_1$. Then it is not difficult to show that there is a unique analytic continuation of g_0. Inspired by this fact, one can present more examples. For instance, define

$$z^\alpha = \exp(\alpha \ln z),$$

where α is an irrational real number. Along γ there always exists an analytic continuation of $(\exp(\alpha g_1), V_1)$.

To enclose this section, an important result on analytic continuations will be presented, known as the monodromy theorem, see [Ru3] for example.

Theorem 2.3.4 (The Monodromy Theorem) *Suppose Ω is a simply connected domain, (f, D) is a function element with $D \subseteq \Omega$, and (f, D) can be analytically continued along every curve in Ω that starts at the center of D. Then there exists $g \in Hol(\Omega)$ such that $g(z) = f(z)$, $z \in D$.*

For example, let L be a simple curve connecting 0 and ∞, say $L = [0, +\infty)$. Then set $\Omega = \mathbb{C} - L$, a simply connected domain. It is easy to see that the function element (g_2, V_2) in Example 2.3.3 satisfies the assumptions in Theorem 2.3.4. Therefore, there is a holomorphic function g in Ω satisfying $g|_{V_2} = g_2$. Precisely,

$$g(z) = \int_{-1}^{z} \frac{1}{\zeta} d\zeta + \pi i, \ z \in \Omega,$$

where the integral is along any smooth curve in Ω.

2.4 Uniformly Separated Sequence

As usual, let $\mathbb{D}$ denote the open unit disk in the complex plane $\mathbb{C}$, and let dA be the normalized area measure on $\mathbb{D}$. Denote by $L_a^2(\mathbb{D})$ the *Bergman* space over $\mathbb{D}$ consisting of all holomorphic functions over $\mathbb{D}$, which are square integrable with respect to dA. In general, for a domain Ω in $\mathbb{C}^d$, denote by $L_a^2(\Omega)$ the *Bergman space over* Ω, which consists of all holomorphic functions over Ω that are square integrable with respect to the volume measure dV on Ω. We also introduce the weighted Bergman space. Precisely, for each $\alpha > -1$, denote by $L_{a,\alpha}^2(\mathbb{D})$ the *weighted Bergman space*, which consists of all holomorphic functions over $\mathbb{D}$ that are square integrable with respect to the normalized measure $(\alpha+1)(1-|z|^2)^\alpha dA(z)$. When $\alpha = 0$, $L_{a,0}^2(\mathbb{D})$ is exactly the usual Bergman space over $\mathbb{D}$. Denote by $L_{a,\alpha}^p(\mathbb{D})(0 < p < \infty)$ the space of all holomorphic functions f over $\mathbb{D}$ satisfying

$$\|f\|_p = \left(\int_\mathbb{D} |f(z)|^p (1-|z|^2)^\alpha dA(z) \right)^{\frac{1}{p}} < \infty.$$

When $p \geq 1$, it is well-known that $L_{a,\alpha}^p(\mathbb{D})$ is a Banach space.

Another classical model of reproducing kernel Hilbert space is the Hardy space $H^2(\mathbb{D})$. In general, for $0 < p < +\infty$, $H^p(\mathbb{D})$ consists of all holomorphic functions f on $\mathbb{D}$ which satisfies

$$\|f\|_p = \sup_{0<r<1} \left(\int_0^{2\pi} |f_r(\theta)|^p d\theta \right)^{\frac{1}{p}} < \infty.$$

For $1 \leq p < \infty$, $H^p(\mathbb{D})$ is a Banach space. The Hardy space $H^2(\mathbb{D}^n)$ over the polydisk $\mathbb{D}^n$ is defined to be the class of all holomorphic functions f on $\mathbb{D}^n$ which satisfies

$$\sup_{0<r<1} \left(\int_{\mathbb{T}^n} |f(r\zeta)|^2 dm_n(\zeta) \right)^{\frac{1}{2}} < \infty,$$

where dm_n denotes the normalized Lebesgue measure on $\mathbb{T}^n$. Note that each holomorphic function f on $\mathbb{D}^n$ has the following expansion:

$$f(z) = \sum_{\alpha \in \mathbb{Z}_+^n} c_\alpha z^\alpha,$$

and it is well-known that $f \in H^2(\mathbb{D}^n)$ if and only if the coefficients c_α are square-summable.

A sequence $\{z_k\}$ in $\mathbb{D}$ is called *uniformly separated* if there is a numerical constant $\delta > 0$ such that

$$\delta_k = \prod_{j \neq k} d(z_j, z_k) \equiv \prod_{j \neq k} \left| \frac{z_j - z_k}{1 - \bar{z}_j z_k} \right| \geq \delta, \quad k = 1, 2, \cdots.$$

A sequence $\{z_k\}$ is uniformly separated if and only if it is an interpolating sequence for $H^\infty(\mathbb{D})$; namely, for each bounded complex sequence $\{w_k\}$ there is an f in $H^\infty(\mathbb{D})$ satisfying $f(z_k) = w_k$ [Ga, Hof1].

A sequence $\{z_k\}$ of distinct points in $\mathbb{D}$ is called *an interpolating sequence for* $L_a^p(\mathbb{D})(0 < p < \infty)$ if the equations $f(z_k) = w_k$ for $k = 1, 2, \cdots$ have a common solution f in $L_a^p(\mathbb{D})$ whenever

$$\sum_{k=1}^{\infty}(1 - |z_k|^2)^2|w_k|^p < \infty.$$

A sequence $\{z_k\}$ is called *an interpolating sequence for* $H^p(\mathbb{D})(0 < p < \infty)$ if the equations $f(z_k) = w_k$ for $k = 1, 2, \cdots$ has a solution f in $H^p(\mathbb{D})$ whenever

$$\sum_{k=1}^{\infty}(1 - |z_k|^2)|w_k|^p < \infty.$$

The following result is known[DS, pp. 157, 175].

Theorem 2.4.1 *A uniformly separated sequence is an interpolating sequence for both $L_a^p(\mathbb{D})$ and $H^p(\mathbb{D})$, where $0 < p < \infty$.*

In addition, if a sequence $\{z_k\}$ is interpolating for some $H^p(\mathbb{D})$, then it is uniformly separated, and hence interpolating for any $H^p(\mathbb{D})$. For details, see [DS, Sect. 6.2]. However, a similar version for $L_a^p(\mathbb{D})$ fails.

A sequence $\{z_k\}$ in $\mathbb{D}$ is called a $L_{a,\alpha}^2(\mathbb{D})$-*interpolating sequence* if for any sequence $\{w_k\}$ of complex numbers satisfying

$$\sum_{k=0}^{\infty}(1 - |z_k|^2)^{2+\alpha}|w_k|^2 < +\infty,$$

there exists an $h \in L_{a,\alpha}^2(\mathbb{D})$ satisfying

$$h(z_k) = w_k, k = 1, 2 \cdots .$$

In the case of $p = 2$, Theorem 2.4.1 has a generalization, which essentially comes from Seip [Se1].

Proposition 2.4.2 *A uniformly separated sequence $\{z_k\}$ is interpolating for all* $L_{a,\alpha}^2(\mathbb{D})(\alpha > -1)$.

Proof Assume that $\{z_k\}$ is a uniformly separated sequence. By the method in [Se1], a sequence $\{\lambda_k\}$ is interpolating for $L_{a,\alpha}^2(\mathbb{D})$ if and only if $\{\lambda_k\}$ is separated and $D^+(\{\lambda_k\}) < \frac{\alpha+1}{2}$, where $D^+(\{\lambda_k\})$ denotes the upper density of $\{\lambda_k\}$ (for details, refer to [DS] and [HKZ, Theorem 5.22]). When $\{\lambda_k\}$ is uniformly separated,

$$D^+(\{\lambda_k\}) = 0,$$

see [DS, pp. 174, 175]. Thus, a uniformly separated sequence $\{z_k\}$ is interpolating for all $L^2_{a,\alpha}(\mathbb{D})(\alpha > -1)$, as desired. $\qquad\square$

By a careful verification, one can show that all results on [DS, pp. 103–109] hold for $L^p_{a,\alpha}(\mathbb{D})$. In particular, one gets the following, refer to [Ho, McS, Sh] and [DS].

Theorem 2.4.3 *Let B denote the Blaschke product whose zero sequence is $\{z_k\}$. Then the following statements are equivalent:*

(1) *The Blaschke sequence $\{z_k\}$ is a finite union of uniformly separated sequences;*
(2) $\sum_{k=1}^{\infty}(1 - |z_k|^2)|f(z_k)|^p < \infty$ *for all f in $H^p(\mathbb{D})$ for some $p \in (0, \infty)$;*
(3) $\sup_{\lambda \in \mathbb{D}} \sum_{k=1}^{\infty}(1 - |\varphi_\lambda(z_k)|) < \infty$;
(4) *The Blaschke product B is a universal divisor of $L^p_a(\mathbb{D})$ for some $p \in (0, \infty)$;*

   that is, $\dfrac{f}{B} \in L^p_a(\mathbb{D})$ for every function f in $L^p_a(\mathbb{D})$ which vanishes on $\{z_k\}$;

(5) *The Blaschke product B is a universal divisor of $L^p_{a,\alpha}(\mathbb{D})(0 < p < \infty)$;*
(6) *The multiplication operator M_B is bounded below on some $L^p_a(\mathbb{D})$ with $p > 0$;*
   that is, there is a positive constant c such that $\|Bf\|_p \geq c\|f\|_p$ for all $f \in L^p_a(\mathbb{D})$;
(7) *The multiplication operator M_B is bounded below on some $L^p_{a,\alpha}(\mathbb{D})$ $(0 < p < \infty)$.*

Let B be such a Blaschke product as in Theorem 2.4.3, and denote by $\mathcal{N}$ the subspace of all functions in $L^p_{a,\alpha}(\mathbb{D})(0 < p < \infty, \alpha > -1)$ that vanish on the zero set of B, counting multiplicity. Then by Theorem 2.4.3(7),

$$\mathcal{N} = BL^p_{a,\alpha}(\mathbb{D}).$$

A sequence $\{z_k\}$ of distinct points in $\mathbb{D}$ is called *uniformly discrete* [DS] if

$$\inf_{j \neq k} d(z_j, z_k) > 0.$$

It is worthwhile to mention that in [HKZ] an equivalent definition of uniformly discrete is formulated.

The following result comes essentially from [Has, Lu], also see [Zhu3, Theorem 2.25].

Theorem 2.4.4 *Given $p > 0$ and $\alpha > -1$, if $\{z_k\}$ is uniformly discrete sequence in $\mathbb{D}$, then there is a constant $C > 0$ such that*

$$\sum_{k=1}^{\infty}(1 - |z_k|^2)^{2+\alpha}|f(z_k)|^p \leq C\|f\|^p_p, \ f \in L^p_{a,\alpha}(\mathbb{D}).$$

In particular, if $\{z_k\}$ is uniformly discrete, then

$$\sum_{k=1}^{\infty}(1 - |z_k|^2)^{2+\alpha}|f(z_k)|^2 \leq C\|f\|^2_2, \ f \in L^2_{a,\alpha}(\mathbb{D}).$$

In the case of $\alpha = 0$, Theorem 2.4.4 is a partial result of [DS, p. 70, Theorem 15].

To establish Theorem 2.4.4, we first present an estimate from [Zhu3, Lemma 2.20] as follows.

Fix $r \in (0, 1)$. Given any $z \in \mathbb{D}$, $\Delta(z, r)$ denotes the pseudohyperbolic disk centered at z with radius r. For $w \in \Delta(z, r)$, write $\lambda = \varphi_z(w)$, and then $w = \varphi_z(\lambda)$. Since the pseudohyperbolic metric d is invariant under Möbius map,

$$d(0, \lambda) = d(z, w) < r, \ i.e. \ |\lambda| < r.$$

By computations,

$$1 - |w|^2 = \frac{(1 - |z|^2)(1 - |\lambda|^2)}{|1 - \overline{\lambda}z|^2};$$

that is,

$$\frac{1 - |w|^2}{1 - |z|^2} = \frac{1 - |\lambda|^2}{|1 - \overline{\lambda}z|^2}.$$

Since $|z| \leq 1$ and $|\lambda| \leq r$, then there is a constant $C_0 > 0$ such that

$$\frac{1}{C_0} \leq \frac{1 - |\lambda|^2}{|1 - \overline{\lambda}z|^2} \leq C_0,$$

and thus

$$\frac{1}{C_0} \leq \frac{1 - |w|^2}{1 - |z|^2} \leq C_0, \quad w \in \Delta(z, r). \tag{2.8}$$

Since

$$\begin{aligned}
(1 - |w|^2)(1 - |z|^2) &= 1 - (|z|^2 + |w|^2) + |z|^2|w|^2 \\
&\leq 1 - 2|z||w| + |z|^2|w|^2 \\
&= (1 - |z||w|)^2 \\
&\leq |1 - \overline{w}z|^2,
\end{aligned}$$

then by (2.8), there is a constant C_1 such that

$$|1 - \overline{w}z| \geq C_1(1 - |z|^2), \quad w \in \Delta(z, r). \tag{2.9}$$

Proof of Theorem 2.4.4 The proof of Theorem 2.4.4 comes from [Zhu3, pp. 57–61].

Let f be a holomorphic function in $\mathbb{D}$. Since $|f|^p$ is subharmonic, then for all $a \in \mathbb{D}$ and $r \in (0, 1 - |a|)$, we have

$$|f|^p(a) \le \int_0^{2\pi} |f|^p(a + r\rho e^{i\theta}) \frac{d\theta}{2\pi}, \quad 0 \le \rho \le 1.$$

By integration in polar coordinates, one gets

$$|f|^p(a) \le \int_{\mathbb{D}} |f|^p(a + rz) dv_\alpha(z),$$

where

$$dv_\alpha(z) = (\alpha + 1)(1 - |z|^2)^\alpha dA(z).$$

In particular, for any $f \in L_{a,\alpha}^p(\mathbb{D})$,

$$|f|^p(0) \le \frac{1}{v_\alpha(\Delta(0, r))} \int_{\Delta(0,r)} |f(w)|^p dv_\alpha(w).$$

Replacing f with $f \circ \varphi_z$ gives

$$|f|^p(z) \le \frac{1}{v_\alpha(\Delta(0, r))} \int_{\Delta(0,r)} |(f \circ \varphi_z)(w)|^p dv_\alpha(w).$$

With the change of variable $w = \varphi_z$, one gets

$$|f|^p(z) \le \frac{1}{v_\alpha(\Delta(0, r))} \int_{\Delta(z,r)} |f(w)|^p \frac{(1 - |z|^2)^{2+\alpha}}{|1 - \overline{w}z|^{2(2+\alpha)}} dv_\alpha(w).$$

From (2.9) it follows that when $r(0 < r < 1)$ is fixed, there is a constant C_2 depending on both p and α, satisfying

$$|f|^p(z) \le \frac{C_2}{(1 - |z|^2)^{2+\alpha}} \int_{\Delta(z,r)} |f(w)|^p dv_\alpha(w).$$

That is,

$$(1 - |z|^2)^{2+\alpha} |f|^p(z) \le C_2 \int_{\Delta(z,r)} |f(w)|^p dv_\alpha(w). \qquad (2.10)$$

Since $\{z_k\}$ is uniformly discrete,

$$\delta \triangleq \inf_{j \ne k} d(z_j, z_k) > 0.$$

Write $r = \frac{\delta}{2}$. We have $\Delta(z_j, r) \cap \Delta(z_k, r) = \emptyset$ for $j \neq k$. By (2.10),

$$\sum_{k=1}^{\infty}(1 - |z_k|^2)^{2+\alpha}|f(z_k)|^p \leq C_2 \int_{\bigsqcup_k \Delta(z_k, r)} |f(w)|^p dv_\alpha(w) \leq C_2 \|f\|_p^p, \; f \in L_{a,\alpha}^p(\mathbb{D}),$$

with $C = C_2$, the proof of Theorem 2.4.4 is complete. □

Combining Proposition 2.4.2 with Theorem 2.4.4 yields the following.

Proposition 2.4.5 *For a uniformly separated sequence $\{z_j\}$, let B be the Blaschke product for $\{z_j\}$. Then for $\alpha > -1$,*

$$h \mapsto \{(1 - |z_j|^2)^{\frac{2+\alpha}{2}} h(z_j)\} \tag{2.11}$$

is a bounded invertible linear map from $L_{a,\alpha}^2(\mathbb{D}) \ominus BL_{a,\alpha}^2(\mathbb{D})$ onto l^2.

Proof The proof comes from [Huang2].

To prove Proposition 2.4.5, it suffices to show that (2.11) defines a bounded linear bijection. Assume that $\{z_j\}$ is a uniformly separated sequence. Let B denote the Blaschke product for $\{z_j\}$, and $\mathcal{N}$ denotes the closed subspace of those functions in $L_{a,\alpha}^2(\mathbb{D})$ which vanish on $\{z_j : j \in \mathbb{Z}_+\}$. By the comments below Theorem 2.4.3, $\mathcal{N} = BL_{a,\alpha}^2(\mathbb{D})$, and hence

$$L_{a,\alpha}^2(\mathbb{D}) = BL_{a,\alpha}^2(\mathbb{D}) \oplus \left(L_{a,\alpha}^2(\mathbb{D}) \ominus BL_{a,\alpha}^2(\mathbb{D})\right). \tag{2.12}$$

By Theorem 2.4.4, (2.11) is a map into l^2. By Proposition 2.4.2, $\{z_j\}$ is interpolating for $L_{a,\alpha}^2(\mathbb{D})(\alpha > -1)$, which shows that (2.11) is a surjective linear map from $L_{a,\alpha}^2(\mathbb{D})$ onto l^2, and by (2.12) it is also a surjective map from $L_{a,\alpha}^2(\mathbb{D}) \ominus BL_{a,\alpha}^2(\mathbb{D})$ onto l^2. Its injectivity follows directly from the identity $\mathcal{N} = BL_{a,\alpha}^2(\mathbb{D})$. Thus (2.11) is a bijection.

We claim that the linear map (2.11)

$$h \mapsto \{(1 - |z_j|^2)^{\frac{2+\alpha}{2}} h(z_j)\}, \, h \in L_{a,\alpha}^2(\mathbb{D}) \ominus BL_{a,\alpha}^2(\mathbb{D})$$

is bounded. For this, write A for this map. Suppose that (h_n, Ah_n) is a sequence tending to $(h, \mathbf{d})$, where all h_n and h are in $L_{a,\alpha}^2(\mathbb{D}) \ominus BL_{a,\alpha}^2(\mathbb{D})$, and $\mathbf{d}$ is in l^2. Soon one will see that $Ah = \mathbf{d}$. Since $\{h_n\}$ converges to h in norm, it follows that $\{h_n\}$ converges to h at each point in $\mathbb{D}$, and hence $(Ah_n)_j$ tends to $(Ah)_j$ for each j. Since Ah_n converges to $\mathbf{d}$, then for each j we have $(Ah)_j = \mathbf{d}_j$. Then $Ah = \mathbf{d}$, as desired. By applying the closed graph theorem, the map A is bounded. Therefore

$$h \mapsto \{(1 - |z_j|^2)^{\frac{2+\alpha}{2}} h(z_j)\}$$

is a bounded linear bijection from $L_{a,\alpha}^2(\mathbb{D}) \ominus BL_{a,\alpha}^2(\mathbb{D})$ to l^2, as desired. □

2.5 Some Results in von Neumann Algebras

This section provides some preliminaries from the theory of von Neumann algebras.

We first review some common terminology. As mentioned in the introduction, *a von Neumann algebra* $\mathcal{A}$ is a unital C^*-algebra on a Hilbert space $\mathcal{H}$, which is closed in the weak operator topology [Con1]. By a *projection* P in $\mathcal{A}$, we mean a self-adjoint operator P satisfying $P^2 = P$. If in addition, P commutes with each member in $\mathcal{A}$, then P is called *a central projection*. A projection P is called *minimal* if the only nonzero projection majorized by P is itself. By *the rank of a projection*, we mean the dimension of its range. For each closed subspace M, P_M always denote the orthogonal projection onto M.

The following results are basic and useful, see [Con1, Proposition 13.3].

Proposition 2.5.1 *Let $\mathcal{A}$ be a von Neumann algebra in $B(H)$ and let $A \in \mathcal{A}$.*

(a) *If A is normal and ϕ is a bounded Borel function on the spectrum of A, then $\phi(A) \in \mathcal{A}$.*

(b) *The operator A is the linear combination of four unitary operators that belong to $\mathcal{A}$.*

(c) *If E and F are the projections onto the closure of $Range\,A$ and $\ker A$, respectively, then $E, F \in \mathcal{A}$. Here, $\ker A \triangleq \{x \in H : Ax = 0\}$.*

(d) *If $A = W|A|$ is the polar decomposition of A, then both W and $|A|$ belong to $\mathcal{A}$.*

(e) *A von Neumann algebra is the norm closed linear span of its projections.*

Two projections $P, Q \in \mathcal{A}$ are called *equivalent*, if there is an operator V in $\mathcal{A}$ such that $V^*V = P$ and $VV^* = Q$. This operator V must be a partial isometry, and in this case we write $P \sim Q$. A projection $P \in \mathcal{A}$ is called finite if there exists no projection $Q \in \mathcal{A}$ such that $Q < P$ and $Q \sim P$. Otherwise P is called *infinite*. A von Neumann algebra $\mathcal{A}$ is called *finite* if its identity is finite; otherwise, $\mathcal{A}$ is called infinite. By a simply reasoning one can prove that a von Neumann algebra $\mathcal{A}$ is finite if and only if $\mathcal{A}$ contains no non-unitary isometry. A projection P in $\mathcal{A}$ is called *abelian* if $P\mathcal{A}P$ is an abelian algebra.

As we mention the dimension of a von Neumann algebra (or a C*-algebra) $\mathcal{A}$, we refer to the algebraic dimension of $\mathcal{A}$.

Recall that the center $Z(\mathcal{A})$ of a von Neumann algebra $\mathcal{A}$ is the set consisting of all members of $\mathcal{A}$ that commute with each operator in $\mathcal{A}$; that is,

$$Z(\mathcal{A}) = \{A \in \mathcal{A} : AB = BA, \ \forall B \in \mathcal{A}\}.$$

A projection in $Z(\mathcal{A})$ is called a *cental projection* in $\mathcal{A}$.

A von Neumann algebra is called *homogeneous* if there is a family of orthogonal abelian projections that are mutually equivalent and whose sum is the identity, see [Con1, p. 285]. The following characterizes the structure of homogenous von Neumann algebras, see [Con1, Proposition 50.15] and its corollary.

Theorem 2.5.2 *If A is a homogeneous von Neumann algebra in $B(\mathcal{H})$, and let $\{E_i\}$ be a collection of pairwise orthogonal, mutually equivalent projections in A with $\sum_i E_i = I$. If $\{E_i\}$ has cardinality n, then A is unitarily isomorphic to $M_n(\mathcal{B})$, where $\mathcal{B} = A|_{E_1 \mathcal{H}}$ is $*$-isomorphic to $Z(A)$.*

A linear map $\tau : A \to Z(A)$ is called a *faithful, center-valued trace* on A if τ satisfies the following:

(1) $\tau(I) = I$;
(2) for any $A \in A$ with $A \geq 0$, we have $\tau(A) \geq 0$; and $\tau(A) = 0$ if and only if $A = 0$;
(3) $\tau(AB) = \tau(BA)$ for all A and B in A.

The following result is well-known. The reader may refer to Corollaries 50.13 and 55.9 in [Con1] for example.

Theorem 2.5.3 *A von Neumann algebra is finite if and only if it has a faithful, center-valued trace.*

If $Z(A) = \mathbb{C} I$, then the von Neumann algebra A is called *a factor*.

(1) type I factor– if there is a minimal projection $E \neq 0$, i.e. a projection E such that there is no nonzero projection F satisfying $F < E$;
(2) type II factor–if there is no minimal projection but there is a non-zero finite projection. By a II_1 factor we mean that it is a type II factor and its identity is finite, otherwise, II_∞.
(3) type III factor–if it does not contain any nonzero finite projection at all.

The following result tell us that for any factor A, both A and A' must share the same type, see [Con1, Corollary 48.17] and [Bla].

Proposition 2.5.4 *A factor A is of type I, II or III if and only if A' is of type I, II or III, respectively. Moreover, any factor is exactly one of the types I_n, I_∞, II_1, II_∞ or III.*

For the definition of type I_n factor, see the paragraph before Theorem 2.5.10.

It is well-known that to a great extent the study of von Neumann algebras reduces to the study of factors [Di, Con1, Jon]. A type I factor is always unitarily equivalent to the tensor product of the operator algebra $B(\mathcal{H})$ and I, see Theorem 2.5.10. By Theorem 2.5.3, one can give an equivalent characterization for type II_1 factors: a factor is a type II_1 factor if and only if A is an infinite dimensional factor and A has a faithful, finite, complex-valued trace. This definition can be reformulated as follows [Jon, Definition 6.1.10]: A is called a *type II_1 factor* if A is an infinite dimensional factor and there is a nonzero linear map $tr : A \to \mathbb{C}$ such that for $A, B \in A$,

(i) $tr(AB) = tr(BA)$;
(ii) $tr(A^*A) \geq 0$;
(iii) tr is ultraweakly continuous; that is, tr is continuous under the weak* topology, where the weak* topology is induced by semi-norm family

$\{\tau_A \; : \; A \text{ is in the trace class } \}$, and $\tau_A(B) = |Tr(BA)|$, $B \in B(\mathcal{H})$, and Tr is the classical trace on trace class.

The above map tr turns out to be faithful, and it is unique if we require that $tr(I) = 1$.

Familiar type II_1 factors arise from group von Neumann algebras, $\mathcal{L}(G)$ and $\mathcal{R}(G)$, where G is a countable, discrete group. For this context, see [Con1] or Sect. 6.5 in Chap. 6 of this book. In this book, we would encounter concrete factors of type I and II arising from multiplication operators defined on classical reproducing kernel Hilbert spaces, the Bergman space and the Hardy space; and our attention is mainly focused on the Bergman space.

Now we turn back to some technical results in the theory of von Neumann algebra. The following seems likely to be known.

Lemma 2.5.5 *Suppose P is a minimal projection in a von Neumann algebra $\mathcal{A}$. Then for each projection Q, either $PQ = 0$ or there is some projection Q_0 such that $Q_0 \leq Q$ and $Q_0 \sim P$.*

Proof Suppose P is a minimal projection in $\mathcal{A}$, whose range is denoted by M. Now assume that Q is a projection onto N satisfying $PQ \neq 0$, and set $W = PQ$. Let $W = U|W|$ be the polar decomposition. Since P is minimal, the partial isometry U in $\mathcal{A}$ satisfies

$$UU^* = P \quad \text{and} \quad U^*U = Q_0,$$

where Q_0 denotes the orthogonal projection onto the range of U^*(and hence, $Q_0 \leq Q$). That is, $Q_0 \sim P$ and $Q_0 \leq Q$. The proof is complete.   □

The following result is a direct consequence of Lemma 2.5.5.

Corollary 2.5.6 *For two minimal projections E and F in a von Neumann algebra $\mathcal{A}$, either $E \perp F$ or $E \sim F$.*

Also, we require the following. For a collection $\mathcal{E}$ of projections, let $\bigvee_{P \in \mathcal{E}} P$ denote the supremum of $\mathcal{E}$ [Con1, pp. 242, 243], which proves to be the orthogonal projection onto the closed space spanned by the ranges of P, where $P \in \mathcal{E}$.

Corollary 2.5.7 *Suppose P is a minimal projection in a von Neumann algebra $\mathcal{A}$. Then the projection $\bigvee_{Q \sim P} Q$ (where Q run over all projections in $\mathcal{A}$) is in the center of $\mathcal{A}$.*

Proof Suppose P is a minimal projection in a von Neumann algebra $\mathcal{A}$. By [Con1, Proposition 43.3], $\bigvee_{Q \sim P} Q$ is in $\mathcal{A}$. By Proposition 2.5.1, a von Neumann algebra is the norm-closed span of its projections. Then it suffices to show that $\bigvee_{Q \sim P} Q$ commutes with each projection P_0 in $\mathcal{A}$.

To see this, rewrite $\hat{P}$ for $\bigvee_{Q \sim P} Q$. First, we assume that there is no projection Q_0 satisfying $Q_0 \leq P_0$ and $Q_0 \sim P$. In this case, by Lemma 2.5.5 we get $P_0 \perp P$; similarly, by the minimality of P we have $P_0 \perp Q$ whenever $Q \sim P$. This implies that $P_0 \perp \hat{P}$, and we are done.

Otherwise, there is a projection Q_0 in $\mathcal{A}$ satisfying $Q_0 \leq P_0$ and $Q_0 \sim P$. In this case, consider $P_0 - Q_0$. By the same reasoning as above, either $P_0 - Q_0 \perp \hat{P}$ or there is a projection Q_1 in $\mathcal{A}$ satisfying $Q_1 \leq P_0 - Q_0$ and $Q_1 \sim P$. Applying Zorn's Lemma shows that there is a maximal family of mutually orthogonal projections Q_i such that $\bigvee_i Q_i \leq P_0$ with $Q_i \sim P$ for each i. Then there is no projection Q satisfying $Q \leq P_0 - \bigvee_i Q_i$ and $Q \sim P$, and hence by Lemma 2.5.5 $(P_0 - \bigvee_i Q_i) \perp P$. By the same reasoning, if P is replaced with any $\tilde{P}$ satisfying $\tilde{P} \sim P$, then $(P_0 - \bigvee_i Q_i) \perp \tilde{P}$, and thus

$$\left(P_0 - \bigvee_i Q_i\right) \perp \bigvee_{\tilde{P} \sim P} \tilde{P}.$$

That is,

$$\left(P_0 - \bigvee_i Q_i\right) \perp \hat{P}.$$

Also, we have $\bigvee_i Q_i \leq \hat{P}$, which implies that $\hat{P}$ commutes with P_0 since

$$P_0 = \left(P_0 - \bigvee_i Q_i\right) + \bigvee_i Q_i.$$

The proof is complete.   □

By the proof of Corollary 2.5.7, we have the following consequence.

Corollary 2.5.8 *Suppose P is a minimal projection in a von Neumann algebra $\mathcal{A}$ and put $\hat{P} = \bigvee_{Q \sim P} Q$ (where Q run over all projections in $\mathcal{A}$). Then there is a family of mutually orthogonal projections $\{Q_i\}$ such that $Q_i \sim P$ and*

$$\hat{P} = \bigvee_i Q_i.$$

The following will be concerned with the structure of type I factors. It is clear that the full matrix algebra $M_n(\mathbb{C})$ is a factor of type I. Here, we allow n to be ∞, and $M_\infty(\mathbb{C})$ represents the algebra of all bounded operators on an ∞-dimensional separable Hilbert space, say l^2.

To investigate the structure of type I factors, we need the following result [Jon], which shows that all minimal projections in a type I factor has the same rank.

Proposition 2.5.9 *Let P and Q be nonzero projections in a factor $\mathcal{A}$. Then there exists a unitary operator $U \in \mathcal{A}$ such that $PUQ \neq 0$. Furthermore, if P, Q are minimal, then $P \sim Q$.*

Proof Suppose conversely that $PUQ = 0$ for any unitary operator U in $\mathcal{A}$. Then $U^*PUQ = 0$, and hence

$$\left(\bigvee_{U \in \mathcal{A}} U^*PU \right)Q = 0, \tag{2.13}$$

where U run over all unitary operators in $\mathcal{A}$. Since $\bigvee_{U \in \mathcal{A}} U^*PU$ commutes with any unitary operator in $\mathcal{A}$, $\bigvee_{U \in \mathcal{A}} U^*PU$ lies in $\mathcal{A}' \cap \mathcal{A}$. Since $\mathcal{A}$ is a factor, $\bigvee_{U \in \mathcal{A}} U^*PU = I$, which is a contradiction to (2.13). Thus, there exists a unitary operator $U \in \mathcal{A}$ such that $PUQ \neq 0$.

Assume that P and Q are minimal projections. Pick a unitary operator U such that $PUQ \neq 0$, and let V be the partial isometry in the polar decomposition of PUQ. Then

$$VV^* \leq P \quad \text{and} \quad V^*V \leq Q.$$

By the minimality of P and Q,

$$VV^* = P \quad \text{and} \quad V^*V = Q.$$

That is, $P \sim Q$.   □

Now assume $\mathcal{A}$ is a type I factor, and so is $\mathcal{A}'$. By Proposition 2.5.9, the rank of a minimal projection in $\mathcal{A}$ is a constant integer, which does not depend on the choice of the minimal projection. Then set $n_1 = $ rank of a minimal projection in $\mathcal{A}$, and $n_2 = $ rank of a minimal projection in $\mathcal{A}'$. It is easy to verify that n_1 is equal to the cardinality of a maximal family of mutually orthogonal minimal projections in $\mathcal{A}$, and similar is true for n_2. A *type I_n factor* is by definition one for which $n = n_2$, and the integer n_1 is called *the multiplicity of the factor*. The following theorem can be found in most books on operator algebra, see [Jon] for instance. It describes how a type I factor looks like.

Theorem 2.5.10 *Assume $\mathcal{A}$ is a type I factor on a Hilbert space H. Then there exist Hilbert spaces H_1, H_2 with $\dim H_1 = n_2$, $\dim H_2 = n_1$ and a unitary operator $U : H \to H_1 \otimes H_2$ such that $U\mathcal{A}U^* = B(H_1) \otimes I_{H_2}$.*

For two operators A and B defined on Hilbert spaces H and K, respectively, A is called *unitarily equivalent* to B if there is a unitary operator $U : H \to K$ such that $B = UAU^*$. Write $Ad_U : B(H) \to B(K), A \to UAU^*$. Given two von Neumann algebras $\mathcal{A}$ and $\mathcal{B}$, if there is a unitary operator U such that $\mathcal{B} = Ad_U(\mathcal{A})$, then $\mathcal{A}$ is called *unitarily isomorphic* to $\mathcal{B}$, or *spatially isomorphic* to $\mathcal{B}$.

The following is an immediate consequence of Theorem 2.5.10, which determines the structure of all finite dimensional von Neumann algebras, see [Da, Theorem III.1.2] or [Jon].

Theorem 2.5.11 *Assume that $\mathcal{A}$ is a finite dimensional von Neuamm algebra on a Hilbert space H. Then $\mathcal{A}$ is $*$-isomorphic to $\bigoplus_{k=1}^{r} M_{n_k}(\mathbb{C})$, where r equals the dimension of the center $Z(\mathcal{A})$. Precisely, $\mathcal{A}$ is unitarily isomorphic to the direct sum*

$$\bigoplus_{k=1}^{r}\left(M_{n_k}(\mathbb{C}) \otimes I_{H_k}\right),$$

where H_k are subspaces of H.

Proof Note that $Z(\mathcal{A})$ is a finite dimensional abelian von Neumann algebra and that if P is a minimal projection in $Z(\mathcal{A})$, then $P\mathcal{A}P$ is a type I factor on PH. Let $\{P_1, \cdots, P_r\}$ be a maximal family of mutually orthogonal minimal projections in $Z(\mathcal{A})$. Then we have the decomposition:

$$\mathcal{A} = \bigoplus_{k=1}^{r} P_k \mathcal{A} P_k,$$

and hence the conclusion follows directly from Theorem 2.5.10. □

In von Neumann algebras, an important result is *the von Neumann Bicommutant Theorem*. This theorem relates the closure of a set of bounded operators on a Hilbert space in certain topologies to the bicommutant of that set. In essence, it is a connection between the algebraic and topological sides of operator theory.

Let $\mathcal{A}$ be a subset of $B(H)$, and the commutant of $\mathcal{A}$ be defined as

$$\mathcal{A}' = \{S \in B(H) : AS = SA, \forall A \in \mathcal{A}\}.$$

Then $\mathcal{A}'$ is a WOT-closed subalgebra of $B(H)$. The set $\mathcal{A}$ is called self-adjoint if $A^* \in \mathcal{A}$ for all A in $\mathcal{A}$. Let $\mathcal{A}_s$ and $\mathcal{A}_w$ be the SOT-closure and the WOT-closure of $\mathcal{A}$, respectively.

The formal statement of the bicommutant theorem is as follows:

Theorem 2.5.12 (von Neumann Bicommutant Theorem) *Let $\mathcal{A}$ be a self-adjoint subalgebra of $B(H)$ and $I \in \mathcal{A}$, then*

$$\mathcal{A}'' = \mathcal{A}_w = \mathcal{A}_s.$$

Proof The proof comes from [Ar2]. By the fact that $\mathcal{A}_s \subseteq \mathcal{A}_w \subseteq \mathcal{A}''$, we only need to prove that for each $B \in \mathcal{A}''$, and any $\varepsilon > 0$, and $h_1, \cdots, h_n \in H$, there exists $A \in \mathcal{A}$, such that

$$\sum_{k=1}^{n} \|(B - A)h_k\|^2 < \varepsilon^2.$$

Given any $h \in H$, it is easy to see that the closure $\overline{Ah}$ of Ah is a reducing subspace of A, and hence the orthogonal projection P onto $\overline{Ah}$ belongs to A'. Therefore, $PB = BP$. Since $1 \in A$, and $Ph = h$, this implies that $Bh \in \overline{Ah}$. It follows that for any $\varepsilon > 0$, there exists $A \in A$ such that $\|(B - A)h\| < \varepsilon$.

Next we will use the above reasoning to complete the proof. Write $H_n = H \oplus \cdots \oplus H$, and let $A_n = \{S \oplus \cdots \oplus S : S \in A\}$, then A_n is a unital self-adjoint subalgebra on H_n. It is easy to show that

$$A'_n = \{[T_{ij}]_{n \times n} : T_{ij} \in A'\},$$

and $[S_{ij}]$ commutes with each element in A'_n if and only if there exists $S \in A''$ such that

$$[S_{ij}] = S \oplus \cdots \oplus S.$$

This implies that

$$A''_n = \{S \oplus \cdots \oplus S : S \in A''\}.$$

It follows that $B_n = B \oplus \cdots \oplus B \in A''_n$ if $B \in A''$. Set $\mathbf{h} = h_1 \oplus \cdots \oplus h_n$, the above reasoning shows that $B_n \mathbf{h} \in \overline{A_n \mathbf{h}}$, and hence there exists $A \in A$ such that

$$\|B_n \mathbf{h} - A_n \mathbf{h}\| < \varepsilon,$$

where $A_n = A \oplus \cdots \oplus A$, that is, $\sum_{k=1}^{n} \|(B - A)h_k\|^2 < \varepsilon^2$. $\square$

Corollary 2.5.13 *Let A be a self-adjoint subalgebra of $B(H)$ and $I \in A$, then A is a von Neumann algebra if and only if $A = A''$.*

2.6   Some Results in Operator Theory

In this section, we present some operator-theoretic results.

For any bounded holomorphic function ϕ over $\mathbb{D}$, let M_ϕ be the multiplication operator defined on the Bergman space $L_a^2(\mathbb{D})$ with the symbol ϕ. As done before, let $W^*(\phi)$ denote the von Neumann algebra generated by M_ϕ and put $V^*(\phi) \triangleq W^*(\phi)'$, the commutant algebra of $W^*(\phi)$. It is well-known that $V^*(\phi)$ equals the von Neumann algebra generated by the orthogonal projections onto M, where M run over all reducing subspaces of M_ϕ. If M_ϕ has no nontrivial reducing subspace, then M_ϕ is called *irreducible*; in this case, $V^*(\phi) = \mathbb{C}I$, and by von Neumann bi-commutant theorem $W^*(\phi) = V^*(\phi)' = B(L_a^2(\mathbb{D}))$, all bounded linear operators on $L_a^2(\mathbb{D})$.

Now, let us have a look at the reducing subspaces from the view of von Neumann algebra. Given two reducing subspaces M and N of M_ϕ, if there exists a unitary operator U from M onto N and U commutes with M_ϕ, then M is called to be *unitarily equivalent* to N. In this case we can extend U to $\tilde{U}$ such that $\tilde{U}|_M = U$ and $\tilde{U}|_{M^\perp} = 0$. It follows that $\tilde{U}$ commutes with both M_ϕ and M_ϕ^*. Write P and Q for the orthogonal projections from $L_a^2(\mathbb{D})$ onto M and N, respectively. Observe that $P = \tilde{U}^*\tilde{U}$ and $Q = \tilde{U}\tilde{U}^*$. That is, two projections P and Q are equivalent in $V^*(\phi)$. In this way, *the unitary equivalence between reducing subspaces can be identified with the equivalence between projections in $V^*(\phi)$.*

A more general setting is presented as follows. Let $\mathbf{T} = (T_1, \cdots, T_d)$ be a commuting operator tuple acting on a separable Hilbert space H. Write $W^*(\mathbf{T})$ for von Neumann algebra generated by $T_1, \cdots, T_d$, and $V^*(\mathbf{T})$ for the commutant algebra of $W^*(\mathbf{T})$, i.e. $V^*(\mathbf{T}) = (W^*(\mathbf{T}))'$, which is also a von Neumann algebra. A closed subspace M of H is called *a reducing subspace* for the tuple $\mathbf{T}$ if M is invariant for both T_i and T_i^*, $i = 1, \cdots, d$; equivalently, both M and $M^\perp$ are invariant for all T_i. A reducing subspace M is called *minimal* if there is no nonzero reducing subspace N satisfying $N \subsetneqq M$. To put it in another way, P_M is a minimal projection in $V^*(\mathbf{T})$. Two reducing subspaces M_1 and M_2 are called *unitarily equivalent* if there exists a unitary operator U from M onto N and U commutes with $T_i(1 \le i \le d)$. In this case, we write

$$M_1 \overset{U}{\cong} M_2.$$

One can show that $M_1 \overset{U}{\cong} M_2$ if and only if P_{M_1} and P_{M_2} are equivalent in $V^*(\mathbf{T})$. Later in Chap. 8, we take $\mathbf{T}$ to be a tuple of multiplication operators acting on a function space, such as the Hardy space, the Bergman space, and etc.

Now set $H_0 = H \ominus (T_1 H + \cdots + T_d H)$, and put $Q = P_{H_0}$, the orthogonal projection onto H_0. We claim that $Q \in W^*(\mathbf{T})$. To see this, note that each $T_i \in W^*(\mathbf{T})$, and thus the range projection of $T_i T_i^*$ is in $W^*(\mathbf{T})$, i.e.

$$P_{\overline{T_i H}} \in W^*(\mathbf{T}).$$

Then

$$\bigvee_{1 \le i \le d} P_{\overline{T_i H}} \in W^*(\mathbf{T}).$$

That is, $I - P_{H_0} \in W^*(\mathbf{T})$, forcing $Q \in W^*(\mathbf{T})$.

Set $\mathcal{R}(\mathbf{T}) = Q W^*(\mathbf{T}) Q$. By the theory of von Neumann algebras, $\mathcal{R}(\mathbf{T})$ is a von Neumann algebra on H_0, and

$$\mathcal{R}'(\mathbf{T}) = Q(W^*(\mathbf{T}))' Q = Q V^*(\mathbf{T}) Q.$$

For the tuple $\mathbf{T} = (T_1, \cdots, T_d)$ acting on a Hilbert space H, write $\mathcal{H}_n$ for all homogeneous polynomials in $\mathbf{T}$ with degree n, $n = 0, 1, \cdots$. Now define a map $\tau : \mathcal{V}^*(\mathbf{T}) \to Q\mathcal{V}^*(\mathbf{T})Q$ by setting

$$\tau(A) = QA \equiv QAQ, \quad A \in \mathcal{V}^*(\mathbf{T}).$$

Since $Q \in \mathcal{W}^*(\mathbf{T})$, it is easy to see that τ is a $*$-homomorphism. Furthermore, we have the following result, due to Guo [Guo5].

Theorem 2.6.1 *If $\cap_n \overline{\mathcal{H}_n H} = 0$, then the map τ is a $*$-isomorphism.*

Proof It suffices to show that τ is injective. Assume that there exists an operator $A \in \mathcal{V}^*(\mathbf{T})$ such that $QAQ = 0$. This implies that $A|_{H_0} = 0$. Since

$$H = H_0 \oplus \overline{(T_1 H + \cdots + T_d H)}$$

and $AT_i = T_i A$ for $i = 1, \cdots, d$, then we have

$$AH \subseteq \overline{A(T_1 H + \cdots + T_d H)}$$
$$= \overline{T_1 AH + \cdots + T_d AH}$$
$$\subseteq \overline{\sum_{i,j} T_i T_j AH}$$
$$\subseteq \overline{\sum_{i,j,k} T_i T_j T_k AH}$$
$$\subseteq \cdots .$$

This immediately gives that $AH \subseteq \cap_n \overline{\mathcal{H}_n H}$, and hence $AH = 0$, i.e. $A = 0$. □

The following result, due to Guo [Guo5], characterizes when $\mathcal{V}^*(\mathbf{T})$ is abelian.

Proposition 2.6.2 *Suppose $\cap_n \overline{\mathcal{H}_n H} = 0$, and*

$$k \triangleq \dim H \ominus (T_1 H + \cdots + T_d H) < \infty.$$

Then the following are equivalent:

(1) $\mathcal{V}^*(\mathbf{T})$ is abelian.
(2) any two distinct minimal projections in $\mathcal{V}^*(\mathbf{T})$ are orthogonal;
(3) there exist at most k minimal reducing subspaces for the tuple $\mathbf{T}$;
(4) there exist at most finitely many minimal reducing subspaces for $\mathbf{T}$.

Proof By Theorem 2.6.1,

$$\dim \mathcal{V}^*(\mathbf{T}) = \dim Q\mathcal{V}^*(\mathbf{T})Q \le (\dim QH)^2 = k^2 < \infty.$$

That is, $V^*(\mathbf{T})$ is a finite dimensional von Neumann algebra. Then the equivalence between (1)–(4) in Proposition 2.6.2 follows directly from Theorem 2.5.11, which states that any finite dimensional von Neumann algebra is unitarily isomorphic to the direct sum

$$\bigoplus_{k=1}^{r} \left(M_{n_k}(\mathbb{C}) \otimes I_{H_k} \right),$$

where H_k are subspaces of H. The proof is complete. □

The following is an immediate consequence [Guo5].

Corollary 2.6.3 *If* $\bigcap_n \overline{\mathcal{H}_n H} = 0$*, and* $k = \dim H \ominus (T_1 H + \cdots + T_d H) < \infty$*, then* $V^*(\mathbf{T})$ *is finite dimensional, and* $\dim V^*(\mathbf{T}) \le k^2$*.*

By Corollary 2.5.6, we have the following.

Proposition 2.6.4 *Let* M_1 *and* M_2 *be two minimal reducing subspaces for* T*. If* M_1 *is not orthogonal to* M_2*, then* $M_1 \stackrel{U}{\cong} M_2$*. Equivalently, if two minimal projections* P, Q *in* $V^*(\mathbf{T})$ *satisfy* $PQ \ne 0$*, then* $P \sim Q$*.*

Proof Here, we provide a different proof.

By assumption, $P_{M_1} P_{M_2} \ne 0$, and hence by spectral decomposition, there are positive constants λ_1 and λ_2 such that

$$(P_{M_1} P_{M_2})(P_{M_1} P_{M_2})^* = P_{M_1} P_{M_2} P_{M_1} = \lambda_1 P_{M_1},$$

and

$$(P_{M_1} P_{M_2})^* (P_{M_1} P_{M_2}) = P_{M_2} P_{M_1} P_{M_2} = \lambda_2 P_{M_2}.$$

This leads to the identity

$$\lambda_1^2 P_{M_1} = P_{M_1} (P_{M_2} P_{M_1} P_{M_2}) P_{M_1} = \lambda_2 P_{M_1} P_{M_2} P_{M_1} = \lambda_1 \lambda_2 P_{M_1},$$

forcing $\lambda_1 = \lambda_2$. Write $V = \frac{1}{\sqrt{\lambda_1}} P_{M_1} P_{M_2}$, and then

$$P_{M_1} = VV^*, \quad P_{M_2} = V^*V.$$

This is $P_{M_1} \sim P_{M_2}$. Equivalently, $M_1 \stackrel{U}{\cong} M_2$. □

By applying Theorem 2.5.11, one can show that if $\dim V^*(\mathbf{T}) < \infty$, then $V^*(\mathbf{T})$ is abelian if and only if for any distinct projections P and Q, P is never equivalent to Q. The following example is from [Guo5, Example 3], which shows that $V^*(\mathbf{T})$ is not necessarily abelian in general.

Example 2.6.5 Given $p_1, \cdots, p_n \in C[z_1, z_2]$, let $M_{p_1}, \cdots, M_{p_n}$ be multiplication operators on the Hardy space $H^2(\mathbb{D}^2)$. Assume that the common zeros $\bigcap_{k=1}^n Z(p_k) \bigcap \mathbb{D}^2$ is finite and nonempty, then by [CG, Theorem 2.2.15, Coroallry 2.26],

$$l = \dim H^2(D^2) \ominus (p_1 H^2(D^2) + \cdots + p_n H^2(D^2)) < \infty,$$

and l is equal to the cardinality of the common zeros (counting multiplicities). By Corollary 2.6.3,

$$\dim V^*(M_{p_1}, \cdots, M_{p_n}) \leq l^2.$$

In general, the von Neumann algebra $V^*(M_{p_1}, \cdots, M_{p_n})$ is not abelian. For example, put $p_1 = z^2$, $p_2 = w^3$, and then $V^*(M_{p_1}, M_{p_2})$ is *-isomorphic to

$$V^*(M_{p_1}) \otimes V^*(M_{p_2}),$$

where $V^*(M_{p_i})$ are defined over $H^2(\mathbb{D})$ for $i = 1, 2$. Note that none of $V^*(M_{p_i})$ is abelian.

However, if $V^*(M_{z^2}, M_{w^3})$ is defined over $L_a^2(\mathbb{D}^2)$, then by the same reasoning shows that $V^*(M_{z^2}, M_{w^3})$ is abelian. This is because on the Bergman space $L_a^2(\mathbb{D})$, $V^*(M_{z^n})$ is abelian for all positive integer n.

The following describes the structure of $V^*(S)$ in the case of S being a pure isometry.

Example 2.6.6 Let S be a pure isometry, i.e., $\bigcap_n S^n H = 0$. Set $H_0 = H \ominus SH$, and hence $Q = I - SS^*$. We have $QV^*(S)Q = B(H_0)$, and hence by Theorem 2.6.1, $V^*(S)$ is *-isomorphic to $B(H_0)$.

To see this, since

$$S^*Q = 0, \quad QS = 0,$$

and

$$S^{*m}S^n = S^{n-m} \text{ if } n \geq m; \quad S^{*m}S^n = S^{(n-m)*} \text{ if } n < m,$$

we see

$$QW^*(S)Q = \mathbb{C}Q,$$

and hence

$$QV^*(S)Q = (QW^*(S)Q)' = B(H_0).$$

This example says that there exists a one-to-one and onto correspondence between reducing subspaces of S and closed subspaces of H_0.

The next example is presented by Guo.

Example 2.6.7 Let $f \in H^\infty(\mathbb{D})$, and let M_f be multiplication operator on $L_a^2(\mathbb{D})$ defined by f. If f has zero points in $\mathbb{D}$, then $\bigcap_n \overline{f^n L_a^2(\mathbb{D})} = 0$. We assume $k = \dim L_a^2(\mathbb{D}) \ominus f L_a^2(\mathbb{D}) < \infty$. Corollary 2.6.3 implies that $V^*(M_f)$ is finite dimensional, and $\dim V^*(M_f) \leq k^2$. In particular, if $k = 1$, then M_f has no nontrivial reducing subspace.

Now let us turn back to Proposition 2.6.2. Consider $\mathbf{T} = M_B$, the multiplication operator defined by a finite Blaschke product B on the Bergman space $L_a^2(\mathbb{D})$. Then one has the following conclusion, also see [GH1].

Proposition 2.6.8 *Let B be a finite Blaschke product of order n. Then the following are equivalent.*

(1) $V^*(B)$ has at most n distinct minimal projections;
(2) $V^*(B)$ is abelian;
(3) All minimal projections in $V^*(B)$ are mutually orthogonal.

Zhu conjectured that for a finite Blaschke product B of order n, there are exactly n distinct minimal reducing subspaces [Zhu1]. This is equivalent to the fact that $V^*(B)$ has exactly n minimal projections. In fact, by applying [SZZ2, Theorem 3.1] Zhu's conjecture holds only if $B(z) = \phi^n$ for some Möbius transform ϕ. Therefore, the conjecture is modified as follows: M_B has at most n distinct minimal reducing subspaces [DSZ]. Therefore the modified conjecture is equivalent to assertion that $V^*(B)$ is abelian. In the case of order $B = 3, 4, 5, 6$, the modified conjecture is demonstrated in [GSZZ, SZZ1, GH1]. By using the techniques of local inverse and group-theoretic methods, it was proved that $V^*(B)$ is abelian if order $B = 7, 8$ in [DSZ]. The latest progress is an affirmative answer to the modified conjecture due to Douglas et al. [DPW], see Chap. 4.

Chapter 3
Cowen-Thomson's Theorem

The root of study of reducing subspaces for multiplication operators on function spaces, as will be illustrated by subsequent chapters, lies in work on the commutants of analytic Toeplitz operators on the Hardy space $H^2(\mathbb{D})$, essentially initiated by Thomson and Cowen [T1, T2, Cow1, Cow2]. In considerable detail, this chapter gives an account of Cowen-Thomson's theorem on commutants of those operators. Also presented is Thomson's original proof of this theorem, with some modifications. In the end of this chapter, we provide a brief review on some topics closely associated with commutants on the Hardy space, which stimulated much further work. The material of this chapter mainly comes from [T1, T2] and [Cow1].

3.1 Cowen-Thomson's Theorem on Commutants

This section mainly introduces Cowen-Thomson's theorem on commutant of analytic Toeplitz operators.

In this chapter, we shall adopt the notations in [DW] and [BDU]. All multiplication operators M_ϕ and M_B will be written as T_ϕ and T_B respectively, which stand for analytic Toeplitz operators. Let $\mathcal{H}$ denote the Hardy space $H^2(\mathbb{D})$ or the Bergman space $L_a^2(\mathbb{D})$.

In [DW] Deddens and Wong raised several questions about the commutants for analytic Toeplitz operators defined on the Hardy space $H^2(\mathbb{D})$, see Sect. 3.6 in Chap. 3. One of them asks for a function $\phi \in H^\infty(\mathbb{D})$, whether there is an inner function ψ such that $\{T_\phi\}' = \{T_\psi\}'$ and that $\phi = h \circ \psi$ for some h in $H^\infty(\mathbb{D})$. Baker, Deddens and Ullman [BDU] proved that for an entire function f, there is a positive integer k such that $\{T_f\}' = \{T_{z^k}\}'$. By using function-theoretic techniques, Thomson [T1, T2] gave more general sufficient conditions that a commutant equals some $\{T_B\}'$ for a finite Blaschke product B. Later, Cowen made an essential generalization for Thomson's result, see [Cow1]. Let us state this as Cowen-Thomson's theorem.

© Springer-Verlag Berlin Heidelberg 2015
K. Guo, H. Huang, *Multiplication Operators on the Bergman Space*,
Lecture Notes in Mathematics 2145, DOI 10.1007/978-3-662-46845-6_3

Theorem 3.1.1 (Cowen-Thomson) *Suppose $\phi \in H^\infty(\mathbb{D})$, and there exists a point λ in $\mathbb{D}$ such that the inner part of $\phi - \phi(\lambda)$ is a finite Blaschke product. Then there exists a finite Blaschke product B and an H^∞-function ψ such that $\phi = \psi(B)$ and $\{T_\phi\}' = \{T_B\}'$ holds on $\mathcal{H}$.*

The following is an immediate consequence, see [T1].

Corollary 3.1.2 *Let ϕ be a nonconstant function holomorphic on the closed unit disk $\overline{\mathbb{D}}$. Then there exists a finite Blaschke product B and a $\psi \in H^\infty(\overline{\mathbb{D}})$ such that $\phi = \psi(B)$ and $\{T_\phi\}' = \{T_B\}'$ holds on $\mathcal{H}$. In particular, if ϕ is entire, then ψ is entire and $B(z) = z^n$ for some positive integer n.*

Let us have a look at Theorem 3.1.1. The Blaschke product B is unique in the following sense: if there is another Blaschke product B_1 satisfying $\phi = h \circ B_1$ and $\{T_{B_1}\}' = \{T_B\}' = \{T_\phi\}'$, then by the proof of Theorem 3.1.1 there is a Möbius map m satisfying $B_1 = m(B)$. Thus, the order n of the Blaschke product B depends only on ϕ. A question was raised in [BDU]: if $\phi \in H^\infty(\overline{\mathbb{D}})$ and B is as in Corollary 3.1.2 with order $B = n$, then does it hold that

$$n = \min\{|k|; k = \text{the winder number of } \phi(e^{it}) \text{ around } a \in \phi(\mathbb{D}) - \phi(\mathbb{T}), k \neq 0\}?$$

For an entire function ϕ, it is known that the above identity is true [BDU]. Note that in Corollary 3.1.2, if ϕ is an entire function, then there is an entire function ψ such that $\phi(z) = \psi(z^n)$ and $\{T_\phi\}' = \{T_{z^n}\}'$. In this case, this integer n is maximal; that is, if there is another entire function h such that $\phi(z) = h(z^{n'})$, then $n \geq n'$. This observation shows that if the entire function ϕ expands as $\phi(z) = \sum_{j=0}^\infty c_j z^j$, then the integer n is exactly

$$\gcd\{m; c_m \neq 0\}.$$

Before continuing, we present two conditions on functions in $H^\infty(\mathbb{D})$. It is convenient to call the assumption in Theorem 3.1.1 *Cowen's condition*. That is, if h is a function in $H^\infty(\mathbb{D})$ such that for some λ in $\mathbb{D}$ the inner part of $h - h(\lambda)$ is a finite Blaschke product, then h is said to satisfy *Cowen's condition*. Similarly, for a function ϕ in $H^\infty(\mathbb{D})$, if there are uncountably many λ in $\mathbb{D}$ such that the inner part of $\phi - \phi(\lambda)$ is a finite Blaschke product, then ϕ is said to satisfy *Thomson's condition*. All bounded holomorphic functions satisfying Thomson's condition consist of a set, called *Thomson's class*. Similarly, the set of all bounded holomorphic functions over $\mathbb{D}$ satisfying Cowen's condition, is then called *Cowen's class*.

Theorem 3.1.1 was first proved by Thomson [T2] for Thomson's class, and later it was essentially generalized to Cowen's class by Cowen [Cow1]. Also, on the Hardy space Cowen [Cow1] gave an extension of Theorem 3.1.1, see as follows.

Theorem 3.1.3 (Cowen) *Let $\mathcal{F}$ be a family of $H^\infty(\mathbb{D})$-functions. If for some point $a \in \mathbb{D}$, the greatest common divisor of the inner parts of $\{h - h(a) : h \in \mathcal{F}\}$ is a finite*

Blaschke product, then there is a finite Blaschke product B such that on the Hardy space $H^2(\mathbb{D})$, $\bigcap_{h \in \mathcal{F}} \{T_h\}' = \{T_B\}'$, and for each $h \in \mathcal{F}$, there is an $H^\infty(\mathbb{D})$-function ψ such that $h = \psi(B)$.

In particular, if $\mathcal{F}$ contains exactly one function, then Theorem 3.1.3 is exactly the Hardy space version of Theorem 3.1.1. It is remarkable that a Bergman-space version of Theorem 3.1.3 holds, see Remark 3.3.8.

It is worthy to mention that Theorem 3.1.1 comes from [Cow1, Theorem 5, Theorem 1, Theorem 2], whose proof is in spirit a considerable development of Thomson's proof in [T2].

Remark 3.1.4 In [Cow1], the proof of Theorem 3.1.3 depends on the structure of the Hardy space $H^2(\mathbb{D})$.

On the other hand, as pointed out in [DSZ] Thomson's proof for Theorem 3.1.1 is also valid on the Bergman space at least for Thomson's class. By a close look at the ideas in [T2] and [Cow1], it turns out that a combination of Thomson and Cowen's proofs gives the Bergman-space version of Theorem 3.1.1; and this is the main content of Sects. 3.2 and 3.3 in Chap. 3, from which one will see that Corollary 3.1.2 also holds on the Dirichlet space.

It was aforementioned that Cowen made an essential generalization of Thomson's result on the commutants of analytic Toeplitz operators. In more detail, in Theorem 3.1.1 he replaced Thomson's condition with Cowen's condition. One natural question is whether these two conditions are equal. In [Cow1], Cowen raised it as a question precisely as follows:

Is there a function in Cowen's class which does not satisfy Thomson's condition?

This is equivalent to ask whether Thomson's class is properly contained in Cowen's class. The following example (see [GH5], to appear elsewhere) provides an affirmative answer, and the details will be delayed in Sect. 3.5 of this chapter according to its length.

Example 3.1.5 Pick $a \in \mathbb{D} - (-1, 1)$, and denote by B the thin Blaschke product with only simple zeros: a and $1 - \frac{1}{n!} (n \geq 2)$. By Riemann mapping theorem, there is a conformal map h from the unit disk onto $\mathbb{D} - [0, 1)$. For such a function h, define $\phi = B \circ h$ and put $a' = h^{-1}(a)$. Later in Sect. 3.5 of this chapter, it will be demonstrated in detail that the inner part of $\phi - \phi(a')$ is a finite Blaschke product; and for any $\lambda \in \mathbb{D} - \{a'\}$, the inner part of $\phi - \phi(\lambda)$ is never a finite Blaschke product. Thus, ϕ fails to satisfy Thomson's condition, though it lies in Cowen's class.

Of course, Cowen's condition fails for any infinite Blaschke product B. One may ask whether the disk algebra $A(\mathbb{D})$ is contained in Cowen's class. To the best of our knowledge, this question is open. The following is due to Guo and Huang, which indicates that there may be some function f in $A(\mathbb{D})$ not lying in Cowen's class. Also refer to [T2, Example] which affords a function f in $A(\mathbb{D})$ such that $f(\mathbb{T}) = \overline{\mathbb{D}}$.

Example 3.1.6 Below we will construct a function f in $A(\mathbb{D})$, which attains each value w in $\overline{\mathbb{D}}$ for infinitely many times on $\mathbb{T}$ and satisfies $\|f\|_\infty = 1$.

The idea is from [T2, Example 1]. Let K be the Cantor set. Each x in K has a unique ternary expansion:

$$x = (0.a_1a_2 \cdots a_n \cdots)_3$$

if we require that each a_n takes value in $\{0, 2\}$. Define

$$h(x) = ((0.\hat{a}_2\hat{a}_5 \cdots \hat{a}_{3n+2} \cdots)_2, \ (0.\hat{a}_3\hat{a}_6 \cdots \hat{a}_{3n} \cdots)_2),$$

where $\hat{a}_n = \frac{a_n}{2}$ and $(y)_2$ means the binary expansion of y. Then one can verify that h is a continuous map from K onto the unit square $[0, 1] \times [0, 1]$. Let v be a continuous map from the unit square onto the closed unit disk. For example, set $v(x, y) = x + yi$ if $x^2 + y^2 < 1$(here i denotes the imaginary unit); otherwise, put

$$v(x, y) = \frac{x + yi}{\sqrt{x^2 + y^2}}.$$

Write $g = v \circ h$, a continuous map from K onto $\overline{\mathbb{D}}$. Clearly, each value w in $\overline{\mathbb{D}}$ is attained by g for infinitely many times. Now set

$$\tilde{g}(e^{i\pi t}) = g(t), \ t \in K,$$

and put $\tilde{K} = \{e^{i\pi t} : t \in K\}$. Then $\tilde{K}$ is a closed subset of Lebesgue measure zero on $\mathbb{T}$, and by Fatou's theorem [Hof1, p. 80] $\tilde{K}$ is a zero set for $A(\mathbb{D})$. That is, there is a function in $A(\mathbb{D})$ which vanished precisely on $\tilde{K}$. Since a zero set is also a peak interpolation set [Ru1, p. 132, Theorem 6.1.2], there is a function f in $A(\mathbb{D})$ such that $f|_{\tilde{K}} = \tilde{g}$ and $\|f\|_\infty = \|\tilde{g}\|_\infty = 1$.

The remaining of this section will present some consequences of Theorem 3.1.1. One is the following [T2].

Corollary 3.1.7 *With the same assumptions in Theorem 3.1.1, ϕ has inner-outer factorization $\phi = \chi F$, then the following hold:*

(1) $\{T_\phi\}' = \{T_\chi\}' \cap \{T_F\}'$;
(2) The only compact operator commuting with T_ϕ is the zero operator.

In particular, both (1) and (2) hold if ϕ is holomorphic on $\overline{\mathbb{D}}$.

Proof We first prove (1). For any $\phi \in H^\infty$ with $\phi = \chi F$, we have $\{T_\phi\}' \supseteq \{T_\chi\}' \cap \{T_F\}'$. On the other hand, Theorem 3.1.1 shows that there is a finite Blaschke product B satisfying $\phi = \psi \circ B$ and $\{T_B\}' = \{T_\phi\}'$. Soon we will see that both χ and F are functions of B. In fact, let $\psi = \tilde{\chi}\tilde{F}$ be the inner-out decomposition in the Hardy space $H^2(\mathbb{D})$. Since B is a finite Blaschke product, B' never vanishes on the unit circle, which gives that $\tilde{\chi} \circ B$ is a bounded holomorphic function with radial limit lying in $\mathbb{T}$ almost everywhere, and hence $\tilde{\chi} \circ B$ is an inner function. Since $\tilde{F}$ is an outer function, there is a sequence $\{p_n\}$ of polynomial such that $\lim\limits_{n\to\infty}\|\tilde{F}p_n-1\|=0$, and then by conformality of B, one can verify that

$$\lim_{n\to\infty}\|\tilde{F}\circ Bp_n\circ B-1\|=0,$$

forcing $\tilde{F}\circ B$ to be an outer function, as desired. Noting that $\phi=\psi\circ B$, and by omitting a unimodular constant, we have

$$\psi=\tilde{\psi}\circ B \quad\text{and}\quad F=\tilde{F}\circ B,$$

as desired. For an alternative approach, see [Ba, Theorem 1].

Since both χ and F are functions of B,

$$\{T_\chi\}'\cap\{T_F\}'\supseteq\{T_B\}'=\{T_\phi\}'.$$

Thus, $\{T_\chi\}'\cap\{T_F\}'=\{T_\phi\}'$, completing the proof of (1).

As for (2), if T_ϕ is defined on the Hardy space, then there is a finite Blaschke product B satisfying $\{T_B\}'=\{T_\phi\}'$. Note that T_B is unitarily equivalent to $T_z\otimes I_n$ on $H^2(\mathbb{D})\otimes\mathbb{C}^n$ ($n=\text{order}\,B$), and it is well-known that any compact operator commuting with T_z is the zero operator. Then any compact operator commuting with T_ϕ must be the zero operator. If T_ϕ is defined on the Bergman space, then by Theorem 3.1.1 there is a finite Blaschke product B satisfying $\{T_B\}'=\{T_\phi\}'$. By [JL, Main Theorem] or Corollary 7.3.3 in Chap. 7, T_B is similar to $T_z\otimes I_n$ on $L^2_a(\mathbb{D})\otimes\mathbb{C}^n$. The remaining discussion is similar. The proof is complete. □

A holomorphic function ϕ on Ω is called p-valent if for each $\lambda\in\Omega$, $\phi-\phi(\lambda)$ has no more than p zeros, counting multiplicity; and this integer p can be attained for some λ. In particular, a 1-valent function is just a univalent function.

Theorem 3.1.1 also has some direct consequences [T2].

Corollary 3.1.8 *Let ϕ be in H^∞. If one of the following holds:*

(1) ϕ is p-valent for some positive integer p;
(2) $\phi\in A(\mathbb{D})$, and there exists a point λ in $\mathbb{D}$ such that $\phi(\lambda)\notin\phi(\mathbb{T})$, (in particular, the latter condition holds if $\phi(\mathbb{T})$ has area measure zero)

then $\{T_\phi\}'=\{T_B\}'$, where B is a finite Blaschke product such that $\phi=\psi(B)$ for some $\psi\in H^\infty$.

Proof
(1) Let $q(z)$ be the number of zeros of $\phi-\phi(z)$ counting multiplicity, write $p=\max\{q(z):z\in\mathbb{D}\}$, and set

$$W=\{z\in\mathbb{D}:q(z)=p\}.$$

Soon we will see that W is an open set and furthermore, if $z_0\in W$, then $\phi-\phi(z_0)$ is bounded away from zero on a neighborhood of $\mathbb{T}$. The reasoning is as follows.

Since $\phi - \phi(z_0)$ has exactly p zeros in $\mathbb{D}$, counting multiplicity, then there is an $r \in (0, 1)$ such that $\phi - \phi(z_0)$ has exactly p zeros in $r\mathbb{D}$. Applying Rouche's theorem shows that there is an enough small $\varepsilon > 0$ such that whenever $|w| < \varepsilon$, $\phi - \phi(z_0) - w$ has exactly p zeros in $r\mathbb{D}$, counting multiplicity. This shows W is open. In addition, by the definition of p it follows that $\phi - \phi(z_0) - w$ has no zero in $\mathbb{D} - r\mathbb{D}$ when $|w| < \varepsilon$. Therefore,

$$|\phi(z) - \phi(z_0)| \geq \varepsilon, \ z \in \mathbb{D} - r\mathbb{D},$$

and thus the above inequality also holds on $\mathbb{T}$ as desired.

Since $\phi - \phi(z_0)$ is bounded away from zero on a neighborhood of $\mathbb{T}$, the inner part of $\phi - \phi(z_0)$ is a finite Blaschke product. Applying Theorem 3.1.1 leads to the desired conclusion.

(2) If $\phi(\lambda) \notin \phi(\mathbb{T})$, then there is some neighborhood V of λ such that for every $a \in V, \phi - \phi(a)$ is bounded away from zero on a neighborhood of $\mathbb{T}$, and hence the inner part of $\phi - \phi(a)$ is a finite Blaschke product. Applying Theorem 3.1.1 implies the desired conclusion. $\square$

Note that these two function classes in Corollary 3.1.8 lie in Thomson's class.

3.2   Facts from Real and Complex Analysis

In this section, we collect some basic results from real and complex analysis.

In this book, when we say a set is *countable*, it is either finite or countably infinite, and the term "countably many" has the same meaning. The following two statements are classical results in real analysis.

Proposition 3.2.1 *Suppose $\{O_\alpha : \alpha \in \Lambda\}$ is a family of open sets in $\mathbb{R}^n$ or $\mathbb{C}^n$. Then there is a sequence of open sets O_n in this family satisfying*

$$\bigcup_n O_n = \bigcup_{\alpha \in \Lambda} O_\alpha.$$

Proof We only deal with the case of $\mathbb{R}^n$. For each open set O_α, consider those open ball $O(\lambda, r_\lambda)$ where $\lambda \in \mathbb{R}^n$ is of rational coordinates and r_λ is an enough small positive rational number such that $O(\lambda, r_\lambda) \subseteq O_\alpha$. Then the union of all such $O(\lambda, r_\lambda)$ equals O_α, and hence the union of all these $O(\lambda, r_\lambda)$ equals $\cup_{\alpha \in \Lambda} O_\alpha$. Since all these $O(\lambda, r_\lambda)$ consist of countably many open sets, we rewrite them by V_n. For each V_n, there is at least one $\alpha = \alpha(n)$ such that $V_n \subseteq O_{\alpha(n)}$. Then the union of all $O_{\alpha(n)}$ equals $\cup_{\alpha \in \Lambda} O_\alpha$. The proof is complete. $\square$

Note that Proposition 3.2.1 can be represented as follows: if A is a subset of $\mathbb{R}^n$ or $\mathbb{C}^n$ and $\{O_\alpha : \alpha \in \Lambda\}$ is an open covering of A, then there is a sequence O_n in this family whose union contains A.

Proposition 3.2.2 *If F is an uncountable subset of* $\mathbb{C}$, *then F contains uncountably many accumulation points. Furthermore, the conclusion remains true if we require that each accumulation point z has the following property: each neighborhood of z contains uncountably many points of F.*

Proof Let F be an uncountable subset of $\mathbb{C}$. For each possible isolated point w, there is an open disk O_w containing only one point w in F. By Proposition 3.2.1, one can pick at most countably many open disks from all O_w such that they have the same union. This immediately shows that the set of isolated points of F is countable. Since F consists of only isolated points and accumulation points, and F is uncountable, then F must contains uncountably many accumulation points z.

Furthermore, if we classify the accumulation points z in F by two classes: the first class consists of those z such that there is a neighborhood of z containing countably infinite points of F; the second class is the remaining, i.e. it consists of points $z \in F$ whose neighborhood always contains uncountably many points of F. Consider the first class. By carefully choosing a disk O_z centered at z, such that $O_z \cap F$ is countable. Following the same discussion as above shows that the first class is countable. Thus, the remaining class is uncountable, as desired. $\square$

Also, we need a classical result in real analysis. Recall that a subset E of $\mathbb{R}^n$ is called *perfect* if E is a closed set with no isolated point.

Lemma 3.2.3 *If E is a nonempty perfect subset of* $\mathbb{R}^n$, *then the cardinality* card $E = \aleph$, *the continuum. That is,* card $E =$ card $\mathbb{R}$.

Proof For $n = 1$, the proof essentially comes from [Na, pp. 51, 52].

Assume that E is a nonempty perfect subset of $\mathbb{R}$. Since card $E \leq \aleph$, it suffices to show that card $E \geq \aleph$. Without loss of generality, assume that

E contains no sub-interval. $(\star)$

Furthermore, it can be required that E is bounded. Otherwise, one can pick $\alpha, \beta \in \mathbb{R} - E$ such that

$$[\alpha, \beta] \cap E \neq \emptyset.$$

Then it is easy to see that $[\alpha, \beta] \cap E$ is a bounded nonempty perfect set, and one can replace E with $[\alpha, \beta] \cap E$.

Now E is a nonempty bounded perfect subset of $\mathbb{R}$. Pick $a, b \in E$ such that $E \subseteq [a, b]$. Clearly,

$$[a, b] - E = (a, b) \cap (\mathbb{R} - E).$$

Since each open set in $\mathbb{R}$ can be written as a union of countably many disjoint open intervals, and so does $\mathbb{R} - E$, then by $(\star)$ there must be infinitely many disjoint open intervals U_k satisfying

$$[a, b] - E = \bigsqcup_{k=0}^{\infty} U_k, \tag{3.1}$$

where

$$\partial U_m \cap \partial U_n = \emptyset \quad \text{for} \quad m \neq n, \tag{3.2}$$

because E has no isolated point.

Write

$$[a,b] - U_0 = \Delta_0 \bigsqcup \Delta_1, \tag{3.3}$$

where Δ_0 and Δ_1 are closed intervals, with Δ_0 lying on the left side of U_0, and Δ_1 on the right side of U_0. Observe that none of Δ_0 and Δ_1 is degenerated to be a single point since E has no isolated point and

$$a, b \in E - U_0.$$

Let U_0' and U_1' be the first members in $\{U_k\}$ satisfying

$$U_i' \subseteq \Delta_i, \quad i = 0, 1.$$

By (3.2) and (3.3), one sees that

$$\partial U_i' \cap \partial \Delta_i = \emptyset, \quad i = 0, 1.$$

Then there are disjoint closed intervals $\Delta_{i,j}(i,j = 0,1)$ satisfying

$$\Delta_i - U_i' = \Delta_{i,0} \bigsqcup \Delta_{i,1}, \quad i = 0, 1,$$

where $\Delta_{i,0}$ lies on the left side of U_i', and $\Delta_{i,1}$ on the right side of U_i' for $i = 0, 1$. Similarly, for $i, j = 0, 1$ let $U_{i,j}'$ be the first member in $\{U_k\}$ satisfying

$$U_{i,j}' \subseteq \Delta_{i,j}.$$

Then by similar reasoning, there are disjoint closed intervals $\Delta_{i,j,0}$ and $\Delta_{i,j,1}$ satisfying

$$\Delta_{i,j} - U_{i,j}' = \Delta_{i,j,0} \bigsqcup \Delta_{i,j,1}, \quad i, j = 0, 1,$$

where $\Delta_{i,j,0}$ lies on the left side of $U_{i,j}'$, and $\Delta_{i,j,1}$ on the right side of $U_{i,j}'$ for $i = 0, 1$. After the n-th step, we have 2^n distinct open intervals in $\{U_k\}$

$$U_{i_1, \cdots, i_n}'; \, i_k = 0, 1, \, k = 1, \cdots, n,$$

and closed intervals $\Delta_{i_1,\cdots,i_n}$, $\Delta_{i_1,\cdots,i_n,0}$ and $\Delta_{i_1,\cdots,i_n,1}$ satisfying

$$\Delta_{i_1,\cdots,i_n} - U'_{i_1,\cdots,i_n} = \Delta_{i_1,\cdots,i_n,0} \bigsqcup \Delta_{i_1,\cdots,i_n,1},$$

where $\Delta_{i_1,\cdots,i_n,0}$ lies on the left side of $U'_{i_1,\cdots,i_n}$, and $\Delta_{i_1,\cdots,i_n,1}$ on the right side of $U'_{i_1,\cdots,i_n}$. By the above construction, we deduce that

$$U_0;\ U'_0, U'_1;\ U'_{0,0}, U'_{0,1}, U'_{1,0}, U'_{1,1}; \cdots$$

is a rearrangement of the sequence $\{U_k\}$. Also, observe that given an infinite sequence

$$i_1, i_2, \cdots, i_k, \cdots$$

with $i_k = 0, 1$ for all k, we have

$$\Delta_{i_1} \supseteq \Delta_{i_1,i_2} \supseteq \Delta_{i_1,i_2,i_3} \supseteq \cdots .$$

Thus the intersection

$$\bigcap_{k=1}^{\infty} \Delta_{i_1,\cdots,i_k}$$

is a nonempty set, which has no intersection with any interval $U'_{j_1,\cdots,j_k}$ in $\{U_n\}$. By arbitrariness of n,

$$\left(\bigcap_{k=1}^{\infty} \Delta_{i_1,\cdots,i_k}\right) \cap \left(\bigcup_{n=0}^{\infty} U_n\right) = \emptyset.$$

Then by (3.1) we have

$$\bigcap_{k=1}^{\infty} \Delta_{i_1,\cdots,i_k} \subseteq E. \tag{3.4}$$

From the above construction, it is not difficult to verify that for two different sequences $\{i_k\}$ and $\{j_k\}$ in $\{0,1\}$,

$$\left(\bigcap_{k=1}^{\infty} \Delta_{i_1,\cdots,i_k}\right) \cap \left(\bigcap_{k=1}^{\infty} \Delta_{j_1,\cdots,j_k}\right) = \emptyset. \tag{3.5}$$

Denote by $\mathcal{J}$ the set of all infinite sequences in $\{0, 1\}$. For each sequence $\{i_k\}$ in $\mathcal{J}$, there exists at least one point $z_{\{i_k\}}$ in $\bigcap_{k=1}^{\infty} \Delta_{i_1, \cdots, i_k}$; in particular by (3.4),

$$z_{\{i_k\}} \in E.$$

By Zorn's lemma, this induces a map h from $\mathcal{J}$ to E, mapping $\{i_k\}$ to $z_{\{i_k\}}$; and by (3.5), h is injective. Therefore,

$$\text{card } E \geq \text{card } \mathcal{J}.$$

On the other hand,

$$\{i_k\} \mapsto (0.i_1 i_2 \cdots)_2, \quad \left((0.i_1 i_2 \cdots)_2 \text{ denotes the binary expansion} \right)$$

defines a map from $\mathcal{J}$ onto $[0, 1]$, and hence

$$\text{card } \mathcal{J} \geq \text{card } [0, 1] = \aleph.$$

Thus, card $E \geq \aleph$. In the case of $n = 1$ the proof is finished.

In general, assume that Lemma 3.2.3 holds for $n = 1, \cdots, k$. Now let E be a nonempty perfect subset of $\mathbb{R}^{k+1}$, and we must prove that card $E = \aleph$, which reduces to show card $E \geq \aleph$.

Now consider the map

$$f : \mathbb{R}^{k+1} \longrightarrow \mathbb{R},$$
$$(x_1, \cdots, x_{k+1}) \mapsto \sum_{1 \leq i \leq k+1} x_i^2.$$

Note that $f(E)$ is a closed set in $\mathbb{R}$. If $f(E)$ has no isolated point, then $f(E)$ is a nonempty perfect subset of $\mathbb{R}$, and hence by the above discussion

$$\text{card } E \geq \text{card } f(E) \geq \aleph,$$

as desired. Otherwise, $f(E)$ has an isolated point, say a. Then put

$$F = \{x \in E : f(x) = a\},$$

which proves to be a nonempty perfect set. Consider the map

$$g : \mathbb{R}^{k+1} \longrightarrow \mathbb{R}^k$$
$$(x_1, \cdots, x_{k+1}) \mapsto (x_1, \cdots, x_k).$$

Since F is a perfect set and any point in F has the form

$$(x_1, \cdots, x_k, \left[a^2 - \sum_{1 \le i \le k} x_i^2\right]^{\frac{1}{2}}) \quad \text{or} \quad (x_1, \cdots, x_k, -\left[a^2 - \sum_{1 \le i \le k} x_i^2\right]^{\frac{1}{2}}),$$

it is easy to see that $g(F)$ has no isolated point, and hence $g(F)$ is a nonempty perfect subset of $\mathbb{R}^k$. By induction, we have

$$\text{card} E \ge \text{card } F \ge \text{card } g(F) \ge \aleph.$$

Therefore, in either case card $E = \aleph$, finishing the proof. $\qquad\qquad\square$

In the proof of Lemma 3.2.3, for each infinite sequence $\{i_k\}$

$$\bigcap_{k=1}^{\infty} \Delta_{i_1, \cdots, i_k}$$

consists of exactly one point. Otherwise, this intersection would be a closed interval, and by (3.4) E would contain a sub-interval, which is a contradiction to $(\star)$.

The following result is a special case of Theorem 2.2.2, and we would like to include a proof.

Proposition 3.2.4 *If F is a countable, relatively closed subset of $\mathbb{D}$, then any bounded holomorphic function f over $\mathbb{D} - F$ can be extend analytically to $\mathbb{D}$.*

Proof Before giving the proof, we make a claim:

For a nonempty relatively closed subset A of $\mathbb{D}$, if A contains no isolated point, then A must be uncountable.

In fact, since A is nonempty, there is an $r(0 < r < 1)$ such that $A \cap r\mathbb{D}$ is nonempty. Write A_r for $A \cap r\mathbb{D}$ and consider the closure $\overline{A_r}$. Since A is relatively closed, $\overline{A_r}$ is a subset of A. Since A_r contains no isolated point, $\overline{A_r}$ is a perfect set. By Lemma 3.2.3, $\overline{A_r}$ is uncountable, and so is A.

Let f be a bounded holomorphic function on $\mathbb{D} - F$. Consider those extensions $\tilde{f}$ of f whose definition domain $D(\tilde{f})$ is a sub-domain Ω of $\mathbb{D}$. By Zorn's lemma, we can pick one $\tilde{f}$ such that $D(\tilde{f})$ is maximal: there is no extension g of f such that

$$D(\tilde{f}) \subsetneqq D(g) \subseteq \mathbb{D}.$$

It suffices to show that $D(\tilde{f}) = \mathbb{D}$. Otherwise, assume that $E \triangleq \mathbb{D} - D(\tilde{f})$ is not empty. Clearly, E is relatively closed in $\mathbb{D}$ and has no isolated point (otherwise, applying Riemann's theorem shows that this isolated point must be a removable singular point of $\tilde{f}$, a contradiction with the maximality of $D(\tilde{f})$). Then by the above claim, E is uncountable, and so is F. This is a contradiction. Therefore, any bounded holomorphic function f over $\mathbb{D} - F$ can be extend analytically to $\mathbb{D}$, completing the proof. $\qquad\qquad\square$

Proposition 3.2.4 can be generalized to the case of f lying in $L_a^2(\mathbb{D} - F)$ with the same proof.

A countable, closed set F can be very complicated. For example, set $F_0 = \{0\} \cup \{\frac{1}{2^n} : n \geq 1\}$, and

$$F_1 = \{0\} \cup \bigcup_{n \geq 1} (\frac{1}{2^n} + \frac{1}{2^n} F_0).$$

In general, for each n define

$$F_{n+1} = \{0\} \cup \bigcup_{n \geq 1} (\frac{1}{2^n} + \frac{1}{2^n} F_n).$$

Clearly, each F_n is a closed subset of $\mathbb{D}$. Put

$$F = \{0\} \cup \bigcup_{n \geq 0} (\frac{1}{3^{n+1}} F_n + \frac{1}{3^{n+1}} i),$$

where i denotes the imaginary unit. One can show that F is a closed subset of $\mathbb{D}$.

3.3   Proof of Cowen-Thomson's Theorem

In the last two sections, an introduction of Cowen-Thomson's theorem is presented, along with its applications and some examples. Also, we made some preparations. In this section, the proof of Theorem 3.1.1 will be provided [T1, T2]. As we will see, some notable modifications are made of the original one. For convenience, Theorem 3.1.1 is restated as follows.

Theorem 3.3.1 *Suppose $\phi \in H^\infty(\mathbb{D})$, and there exists a point λ in $\mathbb{D}$ such that the inner part of $\phi - \phi(\lambda)$ is a finite Blaschke product. Then there exists a finite Blaschke product B and an H^∞-function ψ such that $\phi = \psi(B)$ and $\{T_\phi\}' = \{T_B\}'$ holds on $\mathcal{H}$.*

Because of the length of the proof, we divide it into several steps.

Step I. We will use local inverse to give a local representation for each member $T \in \{T_\phi\}'$. Precisely, the following will be established.

Lemma 3.3.2 *Suppose ϕ belongs to Cowen's class or Thomson's class. Then there is a disk Δ on which finitely many local inverses $\{\rho_j\}$ of ϕ are well-defined and for each $T \in \{T_\phi\}'$ satisfying*

$$Tf(z) = \sum_j s_j(z) f(\rho_j(z)), \quad z \in \Delta, f \in \mathcal{H}$$

Proof Here we only present the proof in the case of ϕ belonging to Thomson's class, due to Thomson [T2]. In general, in the case of Cowen's class, the proof given by Cowen [Cow1] uses more techniques in complex analysis and it is more difficult. Thus, Cowen's proof is deferred to the end of this section.

Let Z denote the critical points of ϕ in $\mathbb{D}$, i.e.

$$Z = \{z \in \mathbb{D} : \phi'(z) = 0\},$$

and write

$$E = \mathbb{D} - \phi^{-1}(\phi(Z)).$$

Note that $\phi^{-1}(\phi(Z))$ is countable. Let Y_j denote the set of $z \in E$ such that the inner part of $\phi - \phi(z)$ is a finite Blaschke product whose order equals j. By assumption the union of Y_j is uncountable. This implies that there is some integer p such that Y_p is uncountable. Then by Proposition 3.2.2, Y_p contains an accumulation point, say a. Since $a \in Y_p$, there are exactly p points in $\phi^{-1}(\phi(a))$. Since $a \notin \phi^{-1}(\phi(Z))$, it follows that there is an open disk Δ centered at a and p local inverses $\{\rho_j\}_{j=1}^p$ of ϕ defined on Δ such that $\rho_j(\Delta)$ are pairwise disjoint. For each $z \in Y_p \cap \Delta$, let $\phi - \phi(z) = B_0 F$ be the inner-outer decomposition, where $B_0 = \prod_{1 \leq j \leq p} \varphi_{\rho_j(z)}$, $\varphi_{\rho_j(z)}(w) = \frac{\rho_j(z)-w}{1-\overline{\rho_j(z)}w}$, and F is an outer function, Then

$$\ker T_{\phi-\phi(z)}^* = \ker T_{B_0}^* = span\{K_{\rho_j(z)} : 1 \leq j \leq p\}. \tag{3.6}$$

where K_w denote the reproducing kernel at w in $\mathcal{H}$, with $\mathcal{H} = H^2(\mathbb{D})$ or $L_a^2(\mathbb{D})$. For each $T \in \{T_\phi\}'$, T commutes with $T_{\phi-\phi(z)}$, and hence

$$T^* T_{\phi-\phi(z)}^* = T_{\phi-\phi(z)}^* T^*,$$

which implies that $\ker T_{\phi-\phi(z)}^*$ is invariant for T^*. Then by (3.6) for each $z \in Y_p \cap \Delta$, there is a sequence $\{s_j(z)\}_{j=1}^p$ of complex numbers satisfying

$$T^* K_z = \sum_j \overline{s_j(z)} K_{\rho_j(z)}. \tag{3.7}$$

Since $Tf(z) = \langle Tf, K_z \rangle = \langle f, T^* K_z \rangle, f \in \mathcal{H}$,

$$Tf(z) = \sum_j s_j(z) f(\rho_j(z)), \ z \in Y_p \cap \Delta. \tag{3.8}$$

Next we will show that both (3.7) and (3.8) hold on all of Δ. In (3.8), we get a system of equations of $s_j(z)$ by putting $f = 1, w, \cdots, w^{p-1}$. By Cramer's rule, for each $j(1 \leq j \leq p)$ we have

$$s_j(z) = \frac{\det V_j(T)(z)}{\det \left(\rho_j(z)^{i-1} \right)}, \quad z \in \Delta$$

where $V_j(T)(z)$ denotes the matrix $\left(\rho_j(z)^{i-1} \right)$ whose j-th column is replaced with $(T1(z), \cdots, Tw^{p-1}(z))$. Since $\{\rho_j\}_{j=1}^{p}$ are pairwise different, the Vandermonde determinant $\det \left(\rho_j(z)^{i-1} \right)$ has no zero, and hence s_j is holomorphic on Δ. Note that for each fixed f both sides of (3.8) are holomorphic and they equal on a set with the accumulation point $a \in \Delta$, and then (3.8) holds on Δ. Since both sides of (3.7) are co-analytic in the variable z, by the same reasoning as above (3.7) also holds on Δ.

Therefore, the proof is complete for Thomson's class.   □

Some words are in order. As in [T1] and [T2], if a local multiplier s_i in (3.7) or (3.8) is not identically zero on Δ, then the corresponding local inverse ρ_i is said to be *representing* for T in Δ. It is not necessarily that all local inverses $\rho_i(1 \leq i \leq p)$ are representing. All representing local inverses on Δ is a subset $\{\rho_{i_1}, \cdots, \rho_{i_q}\}$ of $\{\rho_i : 1 \leq i \leq p\}$. By reordering them, we may assume that $\{\rho_i : 1 \leq i \leq q\}$ is the set of all representing local inverses, and the sum in (3.7) or (3.8) are taken from 1 to q.

One can find an operator $T \in \{T_\phi\}'$ such that each representing local inverse appears in the representation of T. To see this, note that there is a finite set $\{T_1, \cdots, T_l\}$ in $\{T_\phi\}'$ such that each representing local inverse appears in the representation of one T_j. By careful choice of the coefficients of $T_i(1 \leq i \leq l)$, one can give a linear combination T of $\{T_1, \cdots, T_l\}$, all representing local inverses appear in the representation of this operator $T \in \{T_\phi\}'$. Below, T is required to be such an operator.

The next part of the proof will use the technique of analytic continuation. We require that the range of the continuation of a local inverse to be a subset of $\mathbb{D}$. It is easy to verify that the continuation of a local inverse is also a local inverse.

Step II. We will show that each representing local inverse ρ admits unrestricted continuation in some cocountable open subset of $\mathbb{D}$, say U_0, where "cocountable" means that $\mathbb{D} - U_0$ is countable. In other words, ρ admits analytic continuation for any curve in U_0.

Before continuing, an observation is in order. By the latter part of Step I, each local multiplier can be written as a formula of representing local inverses, and thus can be analytically extended to any region where all representing local inverses can be analytically extended. Then both (3.7) and (3.8) hold on that region.

Define U to be the set of all $z \in \mathbb{D}$ with the following property: there is a curve γ with $\gamma(0) = a$ and $\gamma(1) = z$ such that along γ all possible representing local inverses $\rho_j(1 \leq j \leq q)$ admit an analytic continuation.

Since analytic continuation along a curve is obtained as a sequence of direct continuations, it is clear that U is an open subset of $\mathbb{D}$.

Lemma 3.3.3 *All representing local inverses $\rho_j (1 \leq j \leq q)$ admit unrestricted continuation in U.*

Proof For a fixed curve γ in U, it will be shown that all representing local inverses $\rho_j (1 \leq j \leq q)$ admit analytic continuations along γ. Without loss of generality, we may assume that $\gamma(0) = a$. Let T be an operator in $\{T_\phi\}'$ whose representation consists of all possible representing local inverses $\rho_j (1 \leq j \leq q)$, and (3.7) holds on a neighborhood of a. By the definition of U, for each $t \in [0, 1]$ there is a disk centered at $\gamma(t)$ on which (3.7) holds. Since the image of γ is compact, applying Henie-Borel's theorem shows that there is a natural number N and a partition of $[0, 1]$:

$$0 = s_0 < s_1 < \cdots < s_N = 1 \quad \text{with} \quad s_k = \frac{k}{N},$$

and disks $D_k (0 \leq k \leq N - 1)$ such that $\gamma[s_k, s_{k+1}] \subseteq D_k$ on which a similar version of (3.7) holds:

$$T^* K_z = \sum_j s_j^k(z) K_{\rho_j^k(z)}, \quad z \in D_k. \tag{3.9}$$

On $D_k \cap D_{k+1}$, $T^* K_z$ has a unique representation since the reproducing kernels are linearly independent. Thus, for each k the family $\{\rho_j^{k+1}\}_{j=1}^q$ equals $\{\rho_j^k\}_{j=1}^q$ on $D_k \cap D_{k+1}$. After a rearrangement, ρ_j^{k+1} is a direct continuation of ρ_j^k for each j. Then it is easy to see that all representing local inverses $\rho_j (1 \leq j \leq q)$ admit analytic continuation along γ. $\square$

Write $F = \mathbb{D} - U$, which is a relatively closed subset of $\mathbb{D}$. By definition, for each point z in F and any curve γ connecting a and z, there is at least one representing local inverse not admitting analytic continuation along γ. Next we will introduce a notion called singular point [T2]. Precisely, let γ be a curve with $\gamma(0) \in U$. If there is some representing local inverse which does not admit analytic continuation along γ, let τ be the least upper bound of all t_0 for which it is possible to do the analytic continuations along γ_{t_0}, defined by $\gamma_{t_0}(t) = \gamma(t)$, $0 \leq t \leq t_0$. Then $\gamma(\tau)$ is called *a singular point*, and all singular points consists of a set, denoted by X. Note that X is a subset of F.

Proposition 3.3.4 *The singular set X is countable.*

Because of its length, the proof of Proposition 3.3.4 is placed in Sect. 3.4 of this chapter.

Lemma 3.3.5 *If X is countable, then so is F.*

Proof Suppose X is countable. Assume conversely that F is uncountable. Note that $a \in \Delta \subseteq U$, and write

$$\mathbb{C} - \{a\} = \bigsqcup_{0 \leq \theta < 2\pi} L_\theta,$$

where each L_θ denotes the open half line $\{a + re^{i\theta} : r > 0\}$. We say L_θ is good if $L_\theta \cap F$ is not empty. For a good L_θ, set

$$E_\theta = \{s > 0 : a + re^{i\theta} \in U, 0 \leq r \leq s\}.$$

Note that E_θ is bounded, and set $t(\theta) = \sup E_\theta$. If L_θ contains at least one point in F, then the point $z(\theta) = a + t(\theta)e^{i\theta}$ is in F and the segment $[a, z(\theta))$ is in U. By Lemma 3.3.3, $z(\theta)$ is a singular point, i.e. $z(\theta) \in X$. If there were uncountably many good L_θ, each containing at least one point X, then X is uncountable. This is a contradiction. Thus there are countable good half lines L_θ. At least one half line L_{θ_0} among them contains uncountably many points $\lambda \in F$. Then pick a point $a' \in \Delta$ such that the segment $\overline{aa'}$ is perpend to L_{θ_0}. For all $\lambda \in F \cap L_{\theta_0}$, the segments $(a', \lambda]$ are pairwise disjoint. By the same discussion as above, there is at least one singular point $z(\lambda) \in (a', \lambda]$, which shows that X is uncountable, a contradiction. The proof is complete.   □

Some refinement for the set U is necessary. In more detail, for each $z \in U$, there is an $r = r(z) > 0$ such that all representing local inverse ρ_j are defined on $O(z, r) \subseteq U$. Let $F_{z,r}$ denote the following set:

$$\{w \in O(z, r) : \text{there is a representing local inverse } \rho_j \text{ satisfying } \rho_j(w) \in F\}.$$

Since F is relatively closed and countable, then $F_{z,r}$ is a countable, relative closed subset of $O(z, r)$. By Proposition 3.2.1, there is a sequence $\{z_n\}$ in U and a positive sequence $\{r_n = r(z_n)\}$ such that

$$\bigcup_{z \in U} O(z, r(z)) = \bigcup_n O(z_n, r_n).$$

Thus, the union of all $F_{z,r}$ is countable, denoted by $\tilde{F}$. Since the identity map is among $\{\rho_j\}_{j=1}^q$, $\tilde{F} \supseteq F$. By definition, we have

$$\rho_j(\mathbb{D} - \tilde{F}) \subseteq \mathbb{D} - \tilde{F}.$$

Now write $U_0 \triangleq \mathbb{D} - \tilde{F}$ and replace U with U_0. Clearly, all representing local inverses ρ_j admit unrestricted continuation in U_0, and their composition are also representing local inverses.

Step III. Now we will give the construction of the finite Blaschke product B as in Theorem 3.1.1.

For each $z \in U_0$, define

$$B(z) = \prod_{j=1}^{q} \rho_j(z).$$

Firstly, B is well-defined. For this, consider a loop γ in U_0, with $\gamma(0) = \gamma(1) = z$. Each ρ_j admits an continuation $\widetilde{\rho_j}$. Since both ρ_j and $\widetilde{\rho_j}$ are presenting local inverse, $\{\widetilde{\rho_j}(z)\}_{j=1}^{q}$ is just a permutation of $\{\rho_j(z)\}_{j=1}^{q}$. Therefore, B is well-defined. Since B is a bounded holomorphic function and $\mathbb{D} - U_0 = \tilde{F}$ is a countable relatively-closed subset of $\mathbb{D}$, then by Proposition 3.2.4 B extends analytically to $\mathbb{D}$.

Below, we will show that B *is an inner function; and furthermore, B is a finite Blaschke product*. To see this, let $\mathrm{Cl}(B)$ denote the set of all possible η such that there is a sequence z_n in $\mathbb{D}$ such that $z_n \to \xi \in \mathbb{T}$ and $B(z_n) \to \eta$. Clearly, $\mathrm{Cl}(B) \subseteq \overline{\mathbb{D}}$. We claim that $\mathrm{Cl}(B) \cap \mathbb{D}$ is countable. If so, then the set of all $\xi \in \mathbb{T}$ at which B has a radial limit $w \in \mathbb{D}$ has zero measure. Thus, for almost everywhere $\xi \in \mathbb{T}$, the radial limit of B at ξ exists and lies in $\mathbb{T}$, which implies that B is an inner function. Since $\mathrm{Cl}(B)$ is contained in the union of $\mathbb{T}$ and a countable set, then by Theorem 2.1.7 $\mathrm{Cl}(B)$ must be a finite Blaschke product.

For the completeness of the proof, we must show that $\mathrm{Cl}(B) \cap \mathbb{D} \subseteq B(\tilde{F})$, forcing $\mathrm{Cl}(B) \cap \mathbb{D}$ to be countable. To see this, assume that there is a sequence $\{z_j\}$ in U_0 such that $B(z_j) \to w_0 \in \mathbb{D}$ and z_j tends to some $\xi \in \mathbb{T}$. Since B is the product of q local inverses, for each j there is a disk V_{z_j} satisfying $z_j \in V_{z_j}$ and those local inverses $\{\rho_{j,k}\}_{k=1}^{q}$ are well-defined on V_{z_j}. Since $B(z_j) \to w_0 \in \mathbb{D}$ and B is the product of q local inverses, by taking a subsequence we can assume that

$$\lim_{j \to \infty} \rho_{j,1}(z_j) = w \in \mathbb{D}.$$

Since $B(z_j) \to w_0$ and $B(\rho_{j,1}(z_j)) = B(z_j)$, we get $w_0 = B(w)$. Then it remains to show $w \in \tilde{F}$, i.e. $w \notin U_0$. Otherwise, $w \in U_0$. Then on some disk V_w centered at w, there are q representing local inverses $\sigma_k(1 \le k \le q)$. For enough large j, $\rho_{j,1}(z_j)$ lies in V_w. For such j, we have

$$\{\sigma_k \circ \rho_{j,1}\}_{k=1}^{q} = \{\rho_{j,k}\}_{k=1}^{q}.$$

In particular, there exists one integer k such that $\sigma_k \circ \rho_{j,1}(z_j) = z_j$ holds for infinitely many j. By taking a limit, $\sigma_k(w) = \xi \in \mathbb{T}$, which is a contradiction to our requirement for σ_k. Therefore, $w \notin U_0$.

Now B is a finite Blaschke product. By a bit more effort, one can prove that order $B = q$. In fact, since for a fixed point $z_0 \in U_0$, $\{\widetilde{\rho_k} \circ \rho_j(z_0)\}_{k=1}^{q}$ is a permutation of $\{\widetilde{\rho_k}(z_0)\}_{k=1}^{q}$, then $B(\rho_j(z_0)) = B(z_0)$ for $j = 1, \cdots, q$, forcing order $B \ge q$. It remains to show that order $B \le q$. For this, we first consider the case $0 \notin \phi^{-1}(\phi(Z))$.

Assume that $z_0 \in Z(B)$. Since $B(z_0) = \prod_{j=1}^{q} \rho_j(z_0) = 0$, there must be some integer i_0 such that $\rho_{i_0}(z_0) = 0$. Then

$$\{\widetilde{\rho_j} \circ \rho_{i_0}(z_0)\}_{j=1}^{q} = \{\rho_j(z_0)\}_{j=1}^{q}.$$

In particular, $z_0 \in \{\widetilde{\rho_j}(0)\}_{j=1}^{q}$. Thus $Z(B) \subseteq \{\widetilde{\rho_j}(0)\}_{j=1}^{q}$, which gives order $B \leq q$. In general, pick a sequence $\{a_n\}$ in $\mathbb{D} - \phi^{-1}(\phi(Z))$ such that $\lim_{n \to \infty} a_n = 0$, and define

$$h_n(z) = \frac{z - a_n}{1 - \overline{a_n}z}, \, z \in \mathbb{D}.$$

Consider $\phi \circ h_n$, whose local inverses are $h_n \circ \rho_j \circ h_n$. Write

$$\tilde{B}_n = \prod_{j=1}^{q} h_n \circ \rho_j \circ h_n(z),$$

a finite Blaschke product. By a similar discussion, one can get order $\tilde{B}_n \leq q$. Since h_n converges uniformly to the identity map on compact subsets of $\mathbb{D}$. $\tilde{B}_n$ converges uniformly to the finite Blashcke product B on compact subsets of $\mathbb{D}$. Note that order $\tilde{B}_n \leq q$, and an application of Rouche's theorem shows that order $B \leq q$. Therefore, order $B = q$.

Below, we will finish the proof of Theorem 3.1.1. Locally, for each $z_0 \in U_0 - Z$, there is a neighborhood V of z_0 on which all representing local inverses ρ_j are well-defined. Since $B'(z_0) \neq 0$, then $w = B(z)$ is a biholomorphic map on V, and we have two holomorphic maps on V: B and ϕ. Since $B^{-1}(B(z)) \subseteq \phi^{-1}(\phi(z))$, the map

$$\psi : B(z) \mapsto \phi(z)$$

is well defined and holomorphic on $\mathbb{D} - B(\tilde{F} \cup Z(B'))$. Note that $B(\tilde{F} \cup Z(B'))$ is a countable, relatively closed subset of $\mathbb{D}$. Then by Proposition 3.2.4 ψ extends analytically to $\mathbb{D}$, and the extension of ψ is still denoted by ψ. Therefore, $\phi = \psi \circ B$, and hence $\{T_\phi\}' \supseteq \{T_B\}'$. It remains to show that $\{T_\phi\}' \subseteq \{T_B\}'$. In fact, if $T \in \{T_\phi\}'$, then T has the form as (3.8), where ρ_j are representing local inverses of ϕ. By our construction, these ρ_j are necessarily the local inverses of B. Then it is easy to verify that $T \in \{T_B\}'$. $\square$

It is worthy to point out that in the proof of Theorem 3.1.1 the set U was not given in [T2]. This is the major difference between the proof presented here and the original one in [T2].

Now we can give the proof of Corollary 3.1.2.

Proof of Corollary 3.1.2 By Theorem 3.1.1, there exists a finite Blaschke product B and an H^∞-function ψ and such that $\phi = \psi(B)$ and $\{T_\phi\}' = \{T_B\}'$. To finish the proof, we will use the proof of Theorem 3.1.1 to give some analysis on the function ψ. First we remind a fact: for any finite Blaschke product B,

$$\frac{1}{B(z)} = \overline{B(\frac{1}{\bar{z}})}. \tag{3.10}$$

Now for each local inverse ρ, define $\hat{\rho}(z) \triangleq 1\big/\overline{\rho(\frac{1}{\bar{z}})}$. Note that $\hat{\rho}$ is defined on some domain V with $V \subseteq \{z : |z| > 1\}$. Then by (3.10) one gets $B \circ \hat{\rho}(w) = B(w)$.

Assume that ϕ is an entire function. All ρ_j extend analytically across some arc of $\mathbb{T}$ (where ρ_j are defined), denoted by $\widetilde{\rho}_j$, and clearly we have $B(\widetilde{\rho}_j) = B$. Since ρ_j has unrestricted continuation in $\mathbb{D}$, the above discussion shows that $\widetilde{\rho}_j$ admits unrestricted continuation in $\{z; |z| > 1\}$. As done in the end of the proof of Theorem 3.1.1, we will get an entire function ψ such that $\phi = \psi \circ B$. To show that B is an entire function, it suffices to show B has no pole in $\mathbb{C}$. If B had a pole w' outside $\overline{\mathbb{D}}$, then $\lim_{z \to \infty} \psi(z) = \lim_{z \to w'} \psi \circ B(z) = \phi(w') \in \mathbb{C}$, forcing ψ to be constant, which is a contradiction. Thus $B(z) = z^n$ for some integer n.

If $\phi \in Hol(\overline{\mathbb{D}})$, then by a similar discussion one can show that ψ is holomorphic on some neighborhood of $\overline{\mathbb{D}}$. The proof of Corollary 3.1.2 is complete.   □

To end this section, we will provide Cowen's proof of Lemma 3.3.2. First, we need a lemma from [Cow1].

Lemma 3.3.6 *Let U be a domain in the complex plane and $\mathcal{H} = H^2(\mathbb{D})$ or $L_a^2(\mathbb{D})$. Suppose $\mathcal{E} : U \to \mathcal{H}$ is analytic or coanalytic, and there is a point a_1 in U, a family $\{V_i : i \in I\}$ of neighborhoods of a_1 and bounded functions $H_i : V_i \to H^\infty(\mathbb{D})$ such that the following hold*

(i) *For each i,*

$$(WOT) \lim_{a \to a_1} T_{H_i(a)} = T_{H_i(a_1)};$$

(ii) *For each i, $\mathcal{E}(a) \perp H_i(a)\mathcal{H}$ for all $a \in U \cap V_i$*
(iii)

$$\bigvee_{i \in I} H_i(a_1)\mathcal{H} = \mathcal{H}.$$

Then $\mathcal{E}$ vanishes identically.

Proof Without loss of generality, assume that $\mathcal{E}$ is analytic. By (i) and (ii), $\mathcal{E}(a_1) \perp H_i(a_1)\mathcal{H}$ for each i, and then by (iii) $\mathcal{E}(a_1) = 0$. Write

$$\mathcal{E}_1(a) = \begin{cases} \dfrac{\mathcal{E}(a)}{a - a_1}, & a \in U - \{a_1\} \\ \mathcal{E}'(a_1), & a = a_1. \end{cases}$$

Clearly, $\mathcal{E}_1$ is analytic in U. Note that for $a \neq a_1$,

$$\mathcal{E}_1(a) \perp \mathcal{H}_i(a)\mathcal{H}.$$

In fact, the above identity also holds for $a = a_1$. To see this, for any $f \in \mathcal{H}$,

$$\langle \mathcal{E}_1(a), \mathcal{H}_i(a)f \rangle = 0.$$

Since $\lim\limits_{a \to a_1} \|\mathcal{E}_1(a) - \mathcal{E}_1(a_1)\| = 0$, by (i) we get

$$\langle \mathcal{E}_1(a_1), \mathcal{H}_i(a_1)f \rangle = 0, \quad f \in \mathcal{H}.$$

That is, $\mathcal{E}_1(a_1) \perp \mathcal{H}_i(a_1)\mathcal{H}$, as desired.

Replacing $\mathcal{E}$ with $\mathcal{E}_1$ and by similar reasoning, one gets $\mathcal{E}_1(a_1) = 0$. That is, $\mathcal{E}'(a_1) = 0$. By induction,

$$\mathcal{E}(a_1) = \mathcal{E}'(a_1) = \mathcal{E}''(a_1) = \cdots = 0,$$

forcing $\mathcal{E}$ to be identically zero.   □

Two notions are in order. Let λ be a point in $\mathbb{C}$. An open set W is called *a punctured neighborhood* of λ if $\lambda \notin W$ and $W \cup \{\lambda\}$ is a neighborhood of λ. A simply-connected domain V is called *a slit neighborhood* of λ if there is an analytic curve γ in $\mathbb{C} - V$ such that $V \bigsqcup \gamma$ is a neighborhood of λ and λ is one end of γ. For example, $\mathbb{D} - \{0\}$ is a punctured neighborhood of 0, and $\mathbb{D} - [0, 1)$ is a slit neighborhood of 0.

The following will be helpful to us in understanding local inverse.

Proposition 3.3.7 *Suppose*

$$f(z) = a_n z^n + a_{n+1} z^{n+1} + \cdots,$$

where $a_n \neq 0$ for $n \geq 2$. Then there exists a function

$$w = \varphi(z) = \sqrt[n]{a_n} z + b_2 z^2 + b_3 z^3 + \cdots,$$

which is biholomorphic on a neighborhood of 0 where $f(z) = \varphi(z)^n$ holds.

Let h be a bounded holomorphic function over $\mathbb{D}$ with finitely many distinct zeros:

$$b_1, b_2, \cdots, b_k,$$

and denote by $r_j(r_j \geq 1)$ the multiplicity of the zero $b_j (1 \leq j \leq k)$. Proposition 3.3.7 tells us that, by omitting a transformation, the behavior of h at b_j is like z^{r_j} at a neighborhood of 0. For each j, choose a small disk U_j centered at b_j with $U_j \subseteq \mathbb{D}$, and let ε be an enough small positive number. Consider the set

$$\bigsqcup_{1 \leq j \leq k} U_j \cap \phi^{-1}(\varepsilon)$$

For each $j(1 \leq j \leq k)$, the disk U_j contains exactly r_j distinct points around b_j, say

$$\lambda_j^1, \cdots, \lambda_j^{r_j},$$

in the anti-clockwise direction. Now rewrite $a = \lambda_1^1$. For each pair (j, m) $(1 \leq j \leq k, 1 \leq m \leq r_j)$,

$$a \mapsto \lambda_j^m$$

determines a local inverse ρ of h: ρ is locally defined on a neighborhood of a satisfying $\rho(a) = \lambda_j^m$. Clearly, different j and m gives different local inverses ρ, whose number is exactly that of the zeros of h, counting multiplicity. As a consequence of Theorem 3.3.7, if these disks U_j are chosen enough small for $1 \leq j \leq k$, then such local inverses ρ of h admit arbitrary continuation in the punctured disk $U_1 - \{b_1\}$ (remind that $a \in U_1$).

With the above comprehension of local inverse, we come to the proof of Lemma 3.3.2.

Proof of Lemma 3.3.2 The proof is due to Cowen [Cow1].

For the convenience, we rewrite $\lambda = a_1$ and assume that the inner part of $\phi - \phi(a_1)$ is a finite Blaschke product with a finite zero sequence:

$$a_1, a_2, \cdots, a_n.(\text{not necessarily distinct})$$

Below, an analytic map $\mathfrak{p}$ and a map Λ will be constructed, both are from a neighborhood of a_1 into $H^\infty(\mathbb{D})$. It will turn out that the following hold

(i) $\overline{\Lambda(a_1)\mathcal{H}} = \mathcal{H}$;

(ii)

$$(WOT) \lim_{a \to a_1} T_{\Lambda(a)} = T_{\Lambda(a_1)};$$

(iii) For each $T \in \{T_\phi\}'$, $T^*_{\mathfrak{p}(a)} T^* K_a \perp \Lambda(a)\mathcal{H}$.

For each $T \in \{T_\phi\}'$, applying Lemma 3.3.6 to $T^*_{\mathfrak{p}(a)} T^* K_a$ immediately gives $T^* K_a \perp \mathfrak{p}(a)\mathcal{H}$.

Now let $a_1, a_2, \cdots, a_n$ be the points of $\phi^{-1}(\phi(a_1))$, listed according to their multiplicities. Corresponding to these n points, there are n distinct functions $\rho_1, \rho_2, \cdots, \rho_n$ satisfying the following.

(1) For each j, ρ_j is a local inverse of ϕ;
(2) Each ρ_j is single-valued in a slit neighborhood of a_1 and ρ_j admits analytic continuation along any curve in a punctured neighborhood of a_1;
(3) Each ρ_j and its continuation map this punctured neighborhood of a_1 onto a punctured neighborhood of a_j;
(4) $\lim\limits_{w \to a_1} \rho_j(w) = a_j$.

Let Δ be an enough small disk centered at a_1, and define

$$\mathfrak{p}(a) = \prod_{j=1}^{n}(z - \rho_j(a)), \ a \in \Delta.$$

Note that for each $a \in \Delta$, $\mathfrak{p}(a) \in H^\infty(\mathbb{D})$, and clearly $\mathfrak{p}$ is uniformly bounded on Δ. Also, $\mathfrak{p}$ is well-defined in a slit neighborhood of a_1 contained in Δ. Since continuation of $\mathfrak{p}$ in $\Delta - \{a_1\}$ just reorders the factors $\{z - \rho_j(a)\}$ in the product, $\mathfrak{p}$ is thus a single-valued analytic map from Δ into $H^\infty(\mathbb{D})$.

Next, define

$$\Lambda(a) = \frac{\phi - \phi(a)}{\mathfrak{p}(a)}, \ a \in \Delta.$$

Note that $\Lambda(a_1)$ is a bounded outer function in $H^2(\mathbb{D})$. If $\mathcal{H} = H^2(\mathbb{D})$, then $\overline{\Lambda(a_1)\mathcal{H}} = \mathcal{H}$. If $\mathcal{H} = L^2_a(\mathbb{D})$, then $\overline{\Lambda(a_1)\mathcal{H}}$ contains any polynomial because

$$\|f\|_{L^2_a(\mathbb{D})} \leq \|f\|_{H^2(\mathbb{D})}.$$

Therefore, $\overline{\Lambda(a_1)\mathcal{H}} = \mathcal{H}$. This proves (i).

Next, we will show that

$$(WOT) \lim_{a \to a_1} T_{\Lambda(a)} = T_{\Lambda(a_1)}.$$

Note that for each $a \in \Delta$,

$$\Lambda(a) = \frac{\phi - \phi(a)}{\mathfrak{p}(a)} = (\phi - \phi(a))\Big(\prod_{j=1}^{n}\frac{z - \rho_j(a)}{1 - \overline{\rho_j(a)}z}\Big)^{-1}\prod_{j=1}^{n}\frac{1}{1 - \overline{\rho_j(a)}z}.$$

Then Λ is uniformly bounded in Δ since otherwise one can replace Δ with a smaller disk. Also note that $\Lambda(a)$ converges to $\Lambda(a_1)$ in measure as a tends to a_1, and thus

$$(WOT) \lim_{a \to a_1} T_{\Lambda(a)} = T_{\Lambda(a_1)},$$

completing the proof of (ii).

Finally, it remains to show (iii). To see this, for each operator $T \in \{T_\phi\}'$ define

$$\mathcal{E} : \Delta \to \mathcal{H}$$
$$a \to T_{\mathfrak{p}(a)}^* T^* K_a$$

By some computations, for each $f \in \mathcal{H}$,

$$\langle T^* K_a, (\phi - \phi(a))f \rangle = \langle K_a, T(\phi - \phi(a))f \rangle = \langle K_a, (\phi - \phi(a))Tf \rangle = 0.$$

That is, $T^* K_a \perp (\phi - \phi(a))\mathcal{H}$, which immediately gives that $\mathcal{E}(a) \perp \Lambda(a)\mathcal{H}$, as desired.

Now (i)–(iii) are proved. Since $\mathfrak{p}$ is analytic on Δ and $a \mapsto K_a$ is coanalytic, it follows that $\mathcal{E}$ is coanalytic. Then applying Lemma 3.3.6 shows that $\mathcal{E} = 0$, i.e. $T_{\mathfrak{p}(z)}^* T^* K_z = 0$, $z \in \Delta$. That is,

$$T^* K_a \perp \mathfrak{p}(a)\mathcal{H}.$$

From the definition of $\mathfrak{p}$, it follows that for each $z \in \Delta - \{a_1\}$, there exists n complex numbers $\{s_j(z)\}_{j=1}^n$ such that

$$T^* K_z = \sum_j \overline{s_j(z)} K_{\rho_j(z)}.$$

We may assume that the above identity holds for all $z \in \Delta$ because otherwise one can replace Δ with a sub-disk contained in Δ. Since for each $f \in \mathcal{H}$, $Tf(z) = \langle Tf, K_z \rangle = \langle f, T^* K_z \rangle$, we have

$$Tf(z) = \sum_j s_j(z) f(\rho_j(z)), \ z \in \Delta.$$

As done in the proof of Lemma 3.3.2, one can show that each s_j is holomorphic, thus completing the proof of Lemma 3.3.2 in the general case where ϕ lies in Cowen's class. $\qquad\square$

Remark 3.3.8 Now consider a family $\{\phi\}$ in $H^\infty(\mathbb{D})$. In the proof of Lemma 3.3.2, rewrite $\Lambda = \Lambda_\phi$, where ϕ plays the role of the index i in Lemma 3.3.6. With almost no revision, one can then get a proof of Theorem 3.1.3, both on the Hardy space and on the Bergman space. But the original proof of Theorem 3.1.3 depends heavily on the structure of the Hardy space, where Beurling's theorem plays an important role, see [Cow1].

3.4 A Proposition on Singularities

This section gives the proof of Proposition 3.3.4, which comes from [T2].

To show X is countable, it suffices to show that

$$A \triangleq X - \phi^{-1}(\phi(Z))$$

is countable. Now let $a_0 \in A$ be a singular point for some γ at τ, i.e. $\gamma(\tau) = a_0$. Then by definition, there exists a sequence $\{t_k\}$ satisfying:

(i) $0 < t_1 < t_2 < \cdots$ and $\lim_{k \to \infty} t_k = \tau$;

(ii) $\gamma(t_k) = a_k$ and by continuity we have $\lim_{k \to \infty} a_k = a_0$.

(iii) each V_k is an open disk centered at a_k and $V_k \supseteq \gamma([t_k, t_{k+1}])$ for $k = 1, 2, \cdots$;

(iv) for each k, $s_{i,k}$ and $\rho_{i,k} (1 \le i \le q)$ are holomorphic in V_k and (3.7) holds in V_k with $\rho_{i,k}$ and $s_{i,k}$ replacing ρ_i and s_i;

(v) $(\rho_{i,k+1}, V_{k+1})$ is the direct continuation of $(\rho_{i,k}, V_k)$.

Remind that q is the number of representing local inverses. Let $\{\rho_j\}_{j=1}^m (1 \le m \le q)$ be the set of representing local inverses which can be analytically continued along γ_τ. Since $a_0 = \gamma(\tau)$ is a singular point, $m < q$, Let $\sigma_j (1 \le j \le m)$ be the analytic continuation of ρ_j along γ and each σ_j is defined on a neighborhood of a_0.

Step 1. We claim that $T^* K_{a_0}$ is in the linear span of $\{K_{\sigma_j(a_0)}; j = 1, \cdots, m\}$.

To see this, we first show that for any $i > m$,

$$\lim_{n \to \infty} |\rho_{i,n}(a_n)| = 1. \tag{3.11}$$

If not, there is some subsequence n_k such that $\rho_{i,n_k}(a_{n_k})$ tends to some point $b \in \mathbb{D}$. Then

$$\phi(b) = \lim_{k \to \infty} \phi(\rho_{i,n_k}(a_{n_k})) = \lim_{k \to \infty} \phi(a_{n_k}) = \phi(a_0).$$

Since $a_0 \in A$, then $a_0 \notin \phi^{-1}(\phi(Z))$, and hence there is a local inverse σ defined on an open disk Δ containing a_0 and $\sigma(a_0) = b$. Since $\lim_{k \to \infty} \rho_{i,n_k}(a_{n_k}) = b$, for enough large k we have $\rho_{i,n_k} = \sigma$ on $V_{n_k} \cap \Delta$ and $\gamma([t_{n_k}, \tau]) \subseteq \Delta$. This is a contradiction to the assumption that ρ_i does not admit analytic continuation along γ_τ. The proof for (3.11) is complete.

By taking a subsequence, we may assume that for any $i > m$, $\{\rho_{i,n}(a_n)\}$ converges to some point on $\mathbb{T}$. For $i \le m$, we have $\lim_{n \to \infty} \rho_{i,n}(a_n) = \sigma(a_0)$. Then for any i, there is a continuous map α_i on $[0, \tau]$ such that $\alpha_i(t_n) = \rho_{i,n}(a_n)$. Set

$$P(w, t) = \prod_{i=1}^q (w - \alpha_i(t)), \quad w \in \mathbb{D}, \ t \in [0, \tau].$$

For each polynomial f, put

$$Q(f, t) = \langle P(w, t)f(w), T^* K_{\gamma(t)} \rangle, \quad t \in [0, \tau].$$

With f fixed, $Q(f, t_n) = 0$ for every n, and then by continuity we get

$$Q(f, \tau) = 0.$$

This implies that $T^* K_{\gamma(\tau)}$ is orthogonal to $P(w, \tau)f$ for any polynomial f. Since the inner part of $P(w, \tau)$ is $\prod_{i=1}^{m} \varphi_{\alpha_i(\tau)}$, then $T^* K_{a_0}$ is orthogonal to $\prod_{i=1}^{m} \varphi_{\alpha_i(\tau)} \mathcal{H}$, where $\mathcal{H} = H^2(\mathbb{D})$ (or $L_a^2(\mathbb{D})$). Thus, $T^* K_{a_0}$ is in the linear span of $\{K_{\sigma_j(a_0)}; j = 1, \cdots, m\}$. The proof of the claim is complete.

Step 2. Next we proceed to show that A is countable.

Let C_j be the set of $z \in A$ such that $T^* K_z$ can be expressed as a linear combination of exactly j reproducing kernels. Notice that C_q is empty, and thus

$$A = \cup_{j=0}^{q-1} C_j.$$

Suppose conversely that C_n is uncountable for some $n < q$. Then by Proposition 3.2.2 C_n must contain an accumulation point, say c. Let $\{\sigma_j\}_{j=1}^{n}$ be a collection of local inverses defined on some neighborhood Δ' of c, such that

$$T^* K_c \in span\{K_{\sigma_j(c)} : j = 1, \cdots, n\}.$$

Since c is an accumulation point of C_n, there is a sequence $\{c_k\}$ in C_n tending to c and for each k, there is a collection of local inverses $\{\sigma_{k,j}\}_{j=1}^{n}$ defined on some neighborhood of c_j such that

$$T^* K_{c_k} \in span\{K_{\sigma_{k,j}(c_k)} : j = 1, \cdots, n\}.$$

As done in (3.7), there are n holomorphic functions r_j that are defined on Δ'. For $z = c$,

$$T^* K_z = \sum_{j=1}^{n} \overline{r_j}(z) K_{\sigma_j(z)}. \tag{3.12}$$

We will see that (3.12) holds on Δ'. In fact, by the linear independence of the reproducing kernels, we have

$$\lim_{k \to \infty} \sigma_{k,j}(c_k) = \sigma_j(c).$$

In the above identity, we may have taken subsequence and reordered the indices j if necessary. For enough large k, we have $\sigma_{k,j}(c_k) = \sigma_j(c_k)$. Then it is not difficult to see that (3.12) holds for $z = c_k$ when k is enough large. By the uniqueness theorem, (3.12) holds on Δ' since both sides of (3.12) are coanalytic. However, since c is a singular point, a similar version of (3.8) holds along a curve approaching c:

$$TK_z = \sum_{j=1}^{q} \overline{\widetilde{s_j}(z)} K_{\widetilde{\rho_j(z)}},$$

In particular, the above holds in a small disk contained in Δ', where (3.12) also holds. Again by the linear independence of the reproducing kernels, we get $n = q$. This is a contradiction to our earlier observation that $n < q$. Therefore, A is countable. The proof of Proposition 3.3.3 is complete.   □

3.5 An Example not Satisfying Thomson's Condition

In this section, we will provide an example which does not satisfy Thomson's condition, but satisfies Cowen's condition. It is the restatement and verifying of Example 3.1.5, which comes from [GH5] that will appear elsewhere.

Example 3.5.1 There is a holomorphic function ϕ from $\mathbb{D}$ onto $\mathbb{D}$, such that the inner part of $\phi - w$ ($w \in \mathbb{D}$) is a finite Blaschke product if and only if $w = 0$.

Note that Example 3.5.1 directly shows that Thomson's class is properly contained in Cowen's class.

This section mainly furnishes the details of Example 3.1.5, where a is a point in $\mathbb{D} - \mathbb{R}$, and B denotes a Blaschke product with only simple zeros: a and $\{\lambda_n : n \geq 1\}$, where

$$\lambda_n = 1 - \prod_{j=1}^{n} \frac{1}{p_j}, \tag{3.13}$$

and $\{p_n\}$ is an increasing sequence of real numbers in $(1, +\infty)$ and

$$\lim_{n \to \infty} p_n = \infty.$$

Then one can show that $\{\lambda_n : n \geq 1\}$ is a thin Blaschke sequence, as well as

$$\{a\} \cup \{\lambda_n : n \geq 1\},$$

also see Example 5.6.3. In particular,

$$a, 1 - \frac{1}{2!}, 1 - \frac{1}{3!}, \cdots$$

is a thin Blaschke sequence as desired. As we will see below, a conformal map g will be concretely constructed from $\mathbb{D} - [0, 1)$ onto $\mathbb{D}$. Put $h = g^{-1}$, and define

$$\phi = B \circ h,$$

which proves to be the desired function as in Example 3.5.1.

Lemma 3.5.2 *With ϕ defined as above, for any $w \in \mathbb{D} - \{0\}$ the inner part of $\phi - w$ is never a finite Blaschke product.*

Proof Recall that $\phi = B \circ h$, where h is a conformal map from $\mathbb{D}$ onto $\mathbb{D} - [0, 1)$. By Proposition 2.1.8, a thin Blaschke product attains each value w in $\mathbb{D}$ for infinitely many times, and if we can show that $\{t \in [0, 1) : B(t) = w\}$ is a finite set for some $w \in \mathbb{D}$, then $B \circ h - w$ must have infinitely many zeros. The details are as follows.

To prove Lemma 3.5.2 it suffices to show that $B|_{\mathbb{D}-[0,1)}$ attains each nonzero value w in $\mathbb{D}$ for infinitely many times. For this, we will prove that $B|_{[0,1)}$ attains each nonzero value w in $B([0, 1))$ for finitely many times.

To see this, note that φ_a maps $[-1, 1]$ to a circular arc in $\overline{\mathbb{D}}$. Observe that for each fixed $r \in [-1, 1]$, the argument function $\arg \varphi_a(t)|_{[-1,1]}$ of $\varphi_a(t)$ attains the value $\arg \varphi_a(r)$ for at most k_0 times (say, $k_0 = 2$). Here the value of $\arg$ is required to be in $[0, 2\pi)$. Write

$$B = \varphi_a B_0,$$

where B_0 is a Blaschke product, and clearly $B_0(r) \in (-1, 1)$. If $B_0(r) \neq 0$, then

$$\text{either } \arg B(r) = \arg \varphi_a(r) \text{ or } \arg B(r) = \arg(\varphi_a(r)) + \pi \mod 2\pi.$$

Therefore, for each $r \in [0, 1) - Z(B)$, $\arg B$ attains the value $\arg B(r)$ for no more than $2k_0$ times on $[0, 1) - Z(B)$, which immediately implies that $B|_{[0,1)}$ attains each nonzero value w in $B([0, 1))$ for finitely many times. Since h is a conformal map from $\mathbb{D}$ onto $\mathbb{D} - [0, 1)$ and $\phi = B \circ h$, $\phi - w$ is not a finite Blaschke product for any $w \in \mathbb{D} - \{0\}$. □

The remaining part aims at showing the inner part of ϕ is a finite Blaschke product. Because of its length, it is divided into several parts.

Step 1. To begin with, we give two computational results.

Put $S_1(z) = \exp(-\frac{1+z}{1-z})$, which is continuous on the unit circle except for $z = 1$. We will see that $S_1(z)$ *has non-tangential limit 0 at $z = 1$*; that is, for each θ_0 with $0 < \theta_0 < \frac{\pi}{2}$,

$$\lim_{\varepsilon \to 0^+, |\theta| \leq \theta_0} S_1(1 - \varepsilon e^{i\theta}) = 0. \tag{3.14}$$

To show (3.14), write $z = 1 - \varepsilon e^{i\theta}$ where $\varepsilon(\varepsilon > 0)$ is enough small such that $z \in \mathbb{D}$ whenever $|\theta| \leq \theta_0$. By direct computations,

$$\text{Re}\Big(-\frac{1+z}{1-z}\Big) = -\text{Re}\Big[\frac{(1+z)(1-\bar{z})}{|1-z|^2}\Big] = -\frac{2\varepsilon\cos\theta - \varepsilon^2}{\varepsilon^2},$$

and then

$$|S_1(1 - \varepsilon e^{i\theta})| = \exp\Big(-\frac{2\varepsilon\cos\theta - \varepsilon^2}{\varepsilon^2}\Big) \leq \exp\Big(-\frac{2\cos\theta_0}{\varepsilon} + 1\Big) \to 0, \ (\varepsilon \to 0^+),$$

completing the proof of (3.14). Thus $S_1(z)$ has the non-tangential limit 0 at $z = 1$, and so does $S_1'(z) \triangleq \exp(-t\frac{1+z}{1-z})$ for any $t > 0$.

Another estimate for B is given as follows. Let θ_1 be a real number satisfying $0 < |\theta_1| < \frac{\pi}{2}$. Then we have

$$\liminf_{m\to\infty} |B(1 - \varepsilon_m e^{i\theta_1})| > 0. \tag{3.15}$$

In fact, since $B = \varphi_a B_0$, (3.15) is equivalent to

$$\liminf_{m\to\infty} |B_0(1 - \varepsilon_m e^{i\theta_1})| > 0.$$

By (3.13), we rewrite

$$\varepsilon_n = \prod_{j=1}^n \frac{1}{p_j}, \ n = 1,2,\cdots.$$

As done before, let d denote the pseudohyperbolic metric defined on $\mathbb{D}$. For an enough large integer m, $1 - \varepsilon_m e^{i\theta_1} \in \mathbb{D}$, and then

$$d(1 - \varepsilon_n, 1 - \varepsilon_m e^{i\theta_1}) = \Big|\frac{1 - \varepsilon_n - (1 - \varepsilon_m e^{i\theta_1})}{1 - (1 - \varepsilon_n)(1 - \varepsilon_m e^{-i\theta_1})}\Big|$$

$$= \Big|\frac{\varepsilon_n - \varepsilon_m e^{i\theta_1}}{\varepsilon_n + (\varepsilon_m - \varepsilon_n\varepsilon_m)e^{-i\theta_1}}\Big|$$

$$\geq \Big|\frac{\varepsilon_n - \varepsilon_m}{\varepsilon_n + (\varepsilon_m - \varepsilon_n\varepsilon_m)}\Big|$$

$$= d(1 - \varepsilon_n, 1 - \varepsilon_m). \tag{3.16}$$

Since $\{\lambda_n\}$ is a thin Blaschke sequence, then

$$\lim_{m\to\infty} \prod_{n;n\neq m} d(\lambda_n, \lambda_m) = 1.$$

That is,

$$\lim_{m\to\infty} \prod_{n;n\neq m} d(1-\varepsilon_n, 1-\varepsilon_m) = 1. \tag{3.17}$$

Also,

$$\lim_{m\to\infty} d(1-\varepsilon_m, 1-\varepsilon_m e^{i\theta_1}) = \frac{|1-e^{i\theta_1}|}{|1+e^{i\theta_1}|} > 0,$$

which, combined with (3.16) and (3.17) shows that

$$\liminf_{m\to\infty} |B_0(1-\varepsilon_m e^{i\theta_1})| > 0,$$

and hence

$$\liminf_{m\to\infty} |B(1-\varepsilon_m e^{i\theta_1})| > 0, \tag{3.18}$$

as desired.

The idea is to compare (3.14) with (3.18) to derive a contradiction.

Step 2. Below we shall give the concrete construction of h. Precisely,

$$h(z) = \left(\frac{2\lambda(z)-1}{2\lambda(z)+1}\right)^2,$$

where

$$\lambda(z) = \sqrt{-i\frac{1+z}{1-z}}.$$

Here $\sqrt{\cdot}$ denote the branch defined on $\mathbb{C} - [0, +\infty)$ satisfying $\sqrt{-1} = -i$.

However, it is not intuitive to obtain some useful information from the above formula of h. To explain the geometric property of h, we will give the detail for the construction of h. In fact, we will construct two conformal maps φ_2 and φ_1, and put

$$g \triangleq \varphi_2 \circ \varphi_1 \quad \text{and} \quad h = g^{-1}.$$

Now define

$$\varphi_1(z) = i\sqrt{-z} + 1, z \in \mathbb{D} - [0,1),$$

where $\sqrt{1} = 1$, see Fig. 3.1. Let us discuss the geometric property of φ_1. Observe that $z \mapsto -z$ is a rotation which maps $\mathbb{D} - [0,1)$ conformally onto $\mathbb{D} - (-1,0]$. A map is *conformal* if it preserves the angle between two differentiable arcs. The map

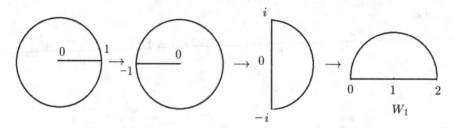

Fig. 3.1 φ_1

Fig. 3.2 φ_2

$z \mapsto \sqrt{-z}$ is conformal on $\mathbb{C} - [0, +\infty)$, and hence on $\mathbb{D} - [0, 1)$. One may imagine that the segment $[0, 1]$ is split into the upper and down parts, which are mapped onto two segments respectively: $i[-1, 0]$ and $i[0, 1]$. In particular, the point 1 is mapped to two points: $-i$ and i. Then with a rotation and a translation, φ_1 maps $\mathbb{D} - [0, 1)$ conformally onto the upper half disk

$$W_1 \triangleq \{\operatorname{Re} z > 0; |z - 1| < 1\}.$$

A biholomorphic map $\varphi_2 : W_1 \to \mathbb{D}$ will be constructed as follows. First, $z \mapsto \frac{1}{z}$ maps the upper half-disk W_1 onto a rectangular domain W_2 between two half lines: $\{\frac{1}{2} + it : t < 0\}$ and $[\frac{1}{2}, +\infty)$. Then with a rotation and a translation, W_2 is mapped onto the first quadrant W_3. Write $\varphi_3(z) = z^2$, and W_3 is mapped onto the upper half plane Π by φ_3. Then one can give a mapping which maps Π onto the unit disk, say, $z \mapsto \frac{z-i}{z+i}$. Define φ_2 to be the composition of the above maps; precisely,

$$\varphi_2(z) = \frac{(\frac{1}{z} - \frac{1}{2})^2 + i}{(\frac{1}{z} - \frac{1}{2})^2 - i}, \ z \in W_1.$$

Some words are in order. All the above maps are conformal; and except for the map $\varphi_3 : z \to z^2$ ($z \in W_3$), all maps are conformal at each point of the boundaries of their domains of definition. However, if $\theta(|\theta| < \frac{\pi}{2})$ is the angle between two differential arcs beginning at $z = 0$, then 2θ is the angle between their image-arcs under φ_3, see Fig. 3.2.

After some verification, one sees that the map $h : \mathbb{D} \to \mathbb{D} - [0, 1)$ extends continuously onto $\overline{\mathbb{D}}$, which maps exactly two points η_1 and η_2 on $\mathbb{T}$ to 1; $\mathbb{T}$ onto $\partial(\mathbb{D} - [0, 1))$; one arc $\widehat{\eta_1 \eta_2}$ onto $[0, 1)$ for twice. Precisely, by some computations we have $\eta_1 = 1$ and $\eta_2 = -1$. Note that any non-tangential domain at η_1 or η_2 will be mapped to some domain non-tangential at 1, lying either above or below the real axis, and vice versa. By the term "non-tangential", we mean the boundary of domain is not tangent to $\mathbb{T}$ nor to the segment $[0, 1]$ at 1.

Step 3. Finally, we will show that the inner part of ϕ is a finite Blaschke product by investigating its regularity.

Let S denote the inner part of $\phi = B \circ h$. Observe that $h^{-1}(0)$ contains exactly one point on $\mathbb{T}$, say η_0; that is,

$$\eta_0 = h^{-1}(0).$$

Since h is holomorphic on $\overline{\mathbb{D}}$ except for three possible points:

$$h^{-1}\{0, 1\} = \{\eta_0, \eta_1, \eta_2\},$$

and B is holomorphic on $\overline{\mathbb{D}} - \{1\}$, it follows that $\phi = B \circ h$ is holomorphic at any point $\zeta \in \mathbb{T} - \{\eta_0, \eta_1, \eta_2\}$, and hence so is S [Hof1]. As follows, one will see that none of η_0, η_1 and η_2 is a singularity of S. Let $\phi = SF$ be the inner-outer decomposition of ϕ in $H^2(\mathbb{D})$, and then $|F| = |\phi|$, a.e. on $\mathbb{T}$, forcing F to be bounded on $\mathbb{D}$ [Hof1]. Since $\phi = B \circ h$, it follows that ϕ is continuous at η_0, and $\phi(\eta_0) = B(0) \neq 0$. Then by $\phi = SF$, $|S(z)|$ is bounded below away from zero as z tends to η_0, which, combined with Theorem 2.1.7, implies that η_0 is not a singularity of S. Therefore, η_1 and η_2 are the only possible singularities of S, and thus the singular part of S is supported on $\{\eta_1, \eta_2\}$. Recall that $S_1(z) = \exp(-\frac{1+z}{1-z})$. Put $a' = h^{-1}(a)$, and write

$$S(z) = \varphi_{a'}(z) S_1^{t_1}(\overline{\eta_1} z) S_1^{t_2}(\overline{\eta_2} z),$$

where $t_1, t_2 \geq 0$. We will show that $t_1 = t_2 = 0$ to finish the proof.

For this, assume conversely that either $t_1 \neq 0$ or $t_2 \neq 0$. Without loss of generality, let $t_1 \neq 0$. Then S has non-tangential limit 0 at η_1, and by the boundedness of F, $\phi = SF$. However, with $\theta_1 = \pm\frac{\pi}{4}$, put

$$\{z_k^1\} = \{h^{-1}(1 - \varepsilon_k e^{\frac{\pi}{4}i})\} \quad \text{and} \quad \{z_k^2\} = \{h^{-1}(1 - \varepsilon_k e^{-\frac{\pi}{4}i})\},$$

where we require $k \geq n_0$ for some enough large integer n_0 such that both $\{1 - \varepsilon_k e^{\frac{\pi}{4}i}\}$ and $\{1 - \varepsilon_k e^{-\frac{\pi}{4}i}\}$ lie in $\mathbb{D}$. Since $\phi = B \circ h$, by (3.15) we get

$$\liminf_{k \to \infty} |\phi(z_k^j)| > 0, \quad j = 1, 2. \tag{3.19}$$

On the other hand, considering

$$h^{-1}(1) = \{\eta_1, \eta_2\},$$

one finds that $\{z_k^1\}$ and $\{z_k^2\}$ are two non-tangential sequences, one tending to η_1 and the other to η_2. By (3.19), this is a contradiction to the fact that ϕ has non-tangential limit 0 at η_1. Therefore, $t_1 = 0$, and similarly $t_2 = 0$. Then the inner part S of ϕ is a Möbius map, completing the proof.   □

3.6   Remarks on Chap. 3

As mentioned in [Cow1], in studying an operator T on a Hilbert space it is of interest to consider those operators commuting with T, which is of much help in understanding the structure of the operator T. Unfortunately, very little is known about this topic in general. Shields and Wallen [SWa] may be the first ones who took into account the commutants of subnormal operators. Later, Deddens and Wong noticed that their methods can be applied to prove that if ϕ is a univalent function, then $\{M_\phi\}' = \{M_z\}' = \{T_f : f \in H^\infty(\mathbb{D})\}$ [DW]. Deddens and Wong's study on this problem on $H^2(\mathbb{D})$ gave rise to six questions as follows.

Question 1 Suppose $\phi \in H^\infty(\mathbb{D})$ has inner-out factorization $\phi = \chi F$. Does it hold that $\{T_\phi\}' = \{T_\chi\}' \cap \{T_F\}'$?

Question 2 For a nonconstant function ϕ in $H^\infty(\mathbb{D})$, is zero operator the only compact operator in $\{T_\phi\}'$?

Question 3 Given $\phi \in H^\infty(\mathbb{D})$, does it hold that $\{T_\phi\}' = \{T_\psi\}'$, where ψ is an inner function and ϕ is a function of ψ?

By Frostman's theorem, for each inner function ψ there is always a member m in $\mathrm{Aut}(\mathbb{D})$ such that $m \circ \psi$ is a Blaschke product. Thus, Question 3 can be put in another way:

Suppose $\phi \in H^\infty(\mathbb{D})$. Does it hold that $\{T_\phi\}' = \{T_B\}'$ where B is some Blaschke product of which ϕ is a function?

Question 4 Suppose $\phi \in H^\infty(\mathbb{D})$. If $\{T_\phi\}' \neq \{T_z\}'$, then does $\phi = \psi \circ h$ hold, where $\psi \in H^\infty(\mathbb{D})$ and h is an inner function distinct from the Möbius map?

Question 5 Suppose $\phi \in H^\infty(\mathbb{D})$. If T commutes with T_ϕ, then does there exist an operator Y on $L^2(\mathbb{T})$ that commutes with M_ϕ satisfying $T = Y|_{H^2(\mathbb{D})}$?

As pointed out in [DW], Question 5 asks whether the commutant of T_ϕ can be lifted to the commutant of the minimal normal extension M_ϕ defined on $L^2(\mathbb{T})$.

Question 6 Suppose $\mathcal{F}$ is a family of inner functions. Is it true that

$$\{T_f : f \in \mathcal{F}\}' = \{T_B\}'$$

where B is a Blaschke product such that each $f \in \mathcal{F}$ has the form $f = g \circ B$, with $g \in H^{\infty}(\mathbb{D})$.

As mentioned in the introduction, these questions had stimulated much further work. Question 1 is negatively answered by Abrahamse [A1], who constructed a function $\phi = \chi F$ and an operator $C \in \{T_{\phi}\}'$ which does not commute with T_{χ} nor with T_F. Also, by constructing a holomorphic covering map φ from $\mathbb{D}$ onto $\{z : \frac{1}{2} < |z| < 1\}$, he showed that there is a nontrivial reducing subspace of T_{φ} and φ can not be written as $\varphi = \psi \circ h$ for some inner function h different from the Möbius map. Thus he provided a negative answer to Question 4. Cowen [Cow3] constructed a function ϕ in $H^{\infty}(\mathbb{D})$ such that T_{ϕ} commutes with a nonzero compact operator, and thereby gave a negative answer to Question 2. Also, in [Cow1] two theorems were established to give sufficient conditions on f such that $\{T_f\}'$ contains no nonzero compact operator. Concerned with Questions 2 and 5, much was done in [Cow2] on the commutants of multiplication operators. In particular, the relation between Questions 2 and 5 was shown by [Cow2, Theorem 1], which states that for a nonconstant $H^{\infty}(\mathbb{D})$-function ϕ, if $\{T_{\phi}\}'$ lifts, then T_{ϕ} does not commutes with any nonzero compact operators. As done in [Cow2], for a fixed function ϕ in $H^{\infty}(\mathbb{D})$, we say $\{T_{\phi}\}'$ *lifts* if for each A in $\{T_{\phi}\}'$, there is an operator $\tilde{A}$ on $L^2(\mathbb{T})$ such that $\tilde{A}|_{H^2(\mathbb{D})} = A$ and $\tilde{A} \in \{M_{\phi}\}'$. Concerning the problem of commutant lifting, also see [BTV]. Question 6 got a satisfactory answer in [Cow1], see Theorem 3.1.3. Furthermore, Cowen raised a more general question as follows [Cow1, Question III]: for a family $\mathcal{F}$ of $H^{\infty}(\mathbb{D})$-functions, is there a function $\psi \in H^{\infty}(\mathbb{D})$ such that

$$\{T_f : f \in \mathcal{F}\}' = \{T_{\psi}\}'?$$

For further motivations, one can refer to [Cow3] .

Question 3 is an important and interesting topic. Firstly, Baker, Deddens and Ullman [BDU] proved that for an entire function f, there is a positive integer k such that $\{T_f\}' = \{T_{z^k}\}'$. By applying function theoretic methods, Thomson gave in essence more general conditions for the commutant being equal to $\{T_B\}'$ for some finite Blaschke product B, see Sect. 3.1 in Chap. 3. In spirit, Cowen developed Thomson's techniques and gave some even more general results [Cow1]. This is the main focus of this chapter. It is an interesting question to ask what the multi-variable version of Question 3 is like and how to solve it. However, little has been done on it, see Chap. 7 for more details.

It is worthwhile to note that the above six questions can also be raised on the Bergman space. To some extent, they were partly answered by the before-mentioned literatures. Though, to the best of our knowledge, not much has been done on Question 5 in the case of the Bergman space.

For the commutants of multiplication operators, we call the reader's attention to [T1, T2, T3, T4, Cow1, Cow2, Cow3], and also [AC, ACR, AD, CDG, CGW, Cl, Cu, GW1, JL, Ro, SZ1, SZ2, Zhu1, Zhu2]. On the Hardy space, [T1, T2] gave the characterization of commutant of multiplication operator with finite

Blaschke product symbol. In [SZ2] Stessin and Zhu provided the generalized Riesz factorization of inner functions, by applying which they obtained a new description of the commutant of multiplication operators with inner symbols. Cowen [Cow1] afforded a characterization for the commutant in the case of holomorphic covering maps. Precisely, if ϕ is a bounded holomorphic covering map with the deck transformation group $G(\phi) = \{\rho_k\}$, and $S \in \{M_\phi\}'$, then there is a sequence of holomorphic functions C_k over $\mathbb{D}$ such that

$$Sh(z) = \sum_k C_k(z)h(\rho_k(z)), z \in \mathbb{D}, h \in H^2(\mathbb{D}),$$

where the above series converges uniformly on compact subsets of $\mathbb{D}$. See [Cow1, Cow2, Cow3] for related results on this line.

Example 3.1.5 is constructed by Guo and Huang [GH5].

Chapter 4
Reducing Subspaces Associated with Finite Blaschke Products

This chapter addresses on reducing subspaces associated with finite Blaschke products, which is the subject of current research receiving numerous attention. It was shown that for each finite Blaschke product B, there is always a nontrivial reducing subspace for M_B, called the distinguished reducing subspace [GSZZ, HSXY]. Therefore, the von Neumann algebra $\mathcal{V}^*(B)$ is nontrivial. Recently, Douglas, Putinar and Wang have proved that $\mathcal{V}^*(B)$ is abelian for any finite Blaschke product B. Therefore, when ϕ satisfies Cowen's condition (in particular, when $\phi \in Hol(\overline{\mathbb{D}})$), from Theorem 3.1.1, $\mathcal{V}^*(\phi) = \mathcal{V}^*(B)$ for some finite Blaschke product B. This implies that such a multiplication operator M_ϕ has at most n minimal reducing subspaces, where $n = $ order B. Analytic continuation and local inverse together provide an accessible approach to the proof of Douglas-Putinar-Wang's result. Also raised are some related problems.

4.1   The Distinguished Reducing Subspace

In the last chapter, Thomson and Cowen's results on the commutants of analytic Toeplitz operators are introduced. In particular, by Corollary 3.1.2, if $h \in Hol(\overline{\mathbb{D}})$, then there is a finite Blaschke product B satisfying $\{M_h\}' = \{M_B\}'$, which holds on both the Hardy space and the Bergman space. Therefore, the study of the reducing subspaces of M_h reduces to that of M_B, where B are finite Blaschke products and $\mathcal{V}^*(h) = \mathcal{V}^*(B)$, where $\mathcal{V}^*(B) = \{M_B, M_B^*\}'$. The following problem is naturally raised:

If B is a finite Blaschke product and B is not the Möbius map, then is there a nontrivial reducing subspace for M_B?

This is equivalent to ask whether $\mathcal{V}^*(B)$ is a nontrivial von Neumann algebra. On the Hardy space, for each closed subspace N of $H^2(\mathbb{D}) \ominus BH^2(\mathbb{D})$, then $B^m N \perp B^n N, n \neq m$, and hence the direct sum of all $B^k N (k \geq 0)$ gives a reducing

© Springer-Verlag Berlin Heidelberg 2015
K. Guo, H. Huang, *Multiplication Operators on the Bergman Space*,
Lecture Notes in Mathematics 2145, DOI 10.1007/978-3-662-46845-6_4

subspace of M_B. This shows that the answer is yes for the Hardy space. On the Bergman space, this is affirmatively answered by Hu, Sun, Xu and Yu in [HSXY]. In fact, they proved that there is always a canonical reducing subspace M_0 such that M_B, restricted on M_0, is unitarily equivalent to the Bergman shift acting on $L_a^2(\mathbb{D})$.

To see this, we need some lemmas, which comes from [GSZZ]. Let $[z−w]$ denote the closure of $(z − w)H^2(\mathbb{D}^2)$ in $H^2(\mathbb{D}^2)$.

Lemma 4.1.1 *Let $f \in H^2(\mathbb{D}^2)$, then $f \in H^2(\mathbb{D}^2) \ominus [z − w]$ iff f has the form*

$$f(z, w) = \frac{h(z) - h(w)}{z - w}$$

for some $h \in H^2(\mathbb{D})$.

Proof Given $\varphi \in H^2(\mathbb{D}^2)$ and $\varphi(0,0) = 0$, set $g = (z − w)\varphi \in H^2(\mathbb{D}^2)$. Then for any polynomial $p(z, w)$, we have

$$
\begin{aligned}
\langle \varphi, (z-w)p \rangle &= \langle g/(z-w), \ (z-w)p \rangle \\
&= \int_{\mathbb{T}^2} g(z, w) \frac{\overline{(z-w)p(z, w)}}{z - w} dm_2 \\
&= -\int_{\mathbb{T}^2} g(z, w) \overline{zwp(z, w)} dm_2 \\
&= -\langle g, zwp \rangle
\end{aligned}
\tag{4.1}
$$

Therefore, if f has the form as in Lemma 4.1.1, then (4.1) implies

$$f \in H^2(\mathbb{D}^2) \ominus [z − w].$$

Conversely, if $f \in H^2(\mathbb{D}^2) \ominus [z − w]$, write $f = g(z, w)/(z − w)$ for some $g \in H^2(\mathbb{D}^2)$. Then as done in (4.1),

$$0 = \langle f, (z-w)p \rangle = \langle g/(z-w), \ (z-w)p \rangle = -\langle g, zwp \rangle$$

for any polynomial p. Decompose $g(z, w) = g(z, 0) + g(0, w) + zw\tilde{g}$, then $\tilde{g} = 0$, and hence $g(z, w) = g(z, 0) + g(0, w)$. Since $g(z, z) = 0$, this means $g(z, 0) = -g(0, z)$. It follows that

$$g(z, w) = g(z, 0) - g(w, 0),$$

completing the proof.   □

Lemma 4.1.2 *Let $e(z, w) \in H^2(\mathbb{D}^2) \ominus [z − w]$, and ϕ be a nonconstant function in $H^\infty(\mathbb{D})$. Then we have $(\phi(z) + \phi(w))e \in H^2(\mathbb{D}^2) \ominus [z − w]$ if and only if*

$$e(z, w) = c \frac{\phi(z) - \phi(w)}{z - w}$$

for some constant c.

Proof The sufficiency follows directly from Lemma 4.1.1. For necessity, applying Lemma 4.1.1 shows that $e(z, w)$ can be represented as $e(z, w) = \frac{h(z)-h(w)}{z-w}$ for some $h \in H^2(\mathbb{D})$. Then we have

$$(\phi(z) + \phi(w))e = \frac{\phi(z)h(z) - \phi(w)h(w) + \phi(w)h(z) - \phi(z)h(w)}{z - w}.$$

Since $(\phi(z) + \phi(w))e \in H^2(\mathbb{D}^2) \ominus [z - w]$, Lemma 4.1.1 implies that there is a function $H_1 \in H^2(\mathbb{D})$ such that

$$\phi(w)h(z) - \phi(z)h(w) = H_1(z) - H_1(w).$$

Write $\phi_0 = \phi - \phi(0)$ and $h_0 = h - h(0)$, and it follows that there is a function $H_2 \in H^2(\mathbb{D})$ such that

$$\phi_0(w)h_0(z) - \phi_0(z)h_0(w) = H_2(z) - H_2(w). \tag{4.2}$$

We will show that there exists a constant c such that $h_0(z) = c\phi_0(z)$. For this, expand h_0 and ϕ_0 :

$$\phi_0(z) = \sum_{n\geq 1} a_n z^n \quad \text{and} \quad h_0(z) = \sum_{n\geq 1} b_n z^n, \ z \in \mathbb{D}.$$

Then by computations we get

$$\phi_0(w)h_0(z) - \phi_0(z)h_0(w) = \sum_{n\geq 1} (a_n h_0(z) - b_n \phi_0(z)) w^n.$$

Since $\phi_0(w)h_0(z) - \phi_0(z)h_0(w)$ has the form (4.2), then $a_n h_0(z) - b_n \phi_0(z) = 0$ holds for all $z \in \mathbb{D}$ and $n \geq 1$. Since ϕ is nonconstant, ϕ_0 is not the zero function, and then there exists a constant c such that $h_0 = c\phi_0$. Therefore,

$$e(z, w) = \frac{h(z) - h(w)}{z - w} = c\frac{\phi(z) - \phi(w)}{z - w},$$

completing the proof. $\qquad\qquad\square$

Lemma 4.1.3 *Let $\phi(z)$ be an inner function satisfying $\frac{\phi(z)-\phi(w)}{z-w} \in H^2(\mathbb{D}^2)$, then*

$$\frac{\phi(z) - \phi(w)}{z - w} \perp \phi(z)H^2(\mathbb{D}^2).$$

Proof Set $h(z, w) = \frac{\phi(z)-\phi(w)}{z-w}$. Then for $0 < r < 1$,

$$h(z, rw) = \frac{\phi(z) - \phi(rw)}{z - rw} = \bar{z}(\phi(z) - \phi(rw)) \sum_{n=0}^{\infty} r^n \bar{z}^n w^n.$$

It is easy to check that for any polynomial $p(z, w)$,

$$\langle h(z, rw), \phi(z)p(z, w) \rangle = \langle \bar{z}(\phi(z) - \phi(rw)) \sum_{n=0}^{\infty} r^n \bar{z}^n w^n, \phi(z)p(z, w) \rangle = 0.$$

This means $h(z, rw) \perp \phi(z)H^2(\mathbb{D}^2)$. Since $h(z, rw)$ converges to $h(z, w)$ in the norm of $H^2(\mathbb{D}^2)$ as $r \to 1$, we get the desired conclusion. □

Lemma 4.1.4 *If $\phi(z)$ is an inner function, then $\frac{\phi(z)-\phi(w)}{z-w} \in H^2(\mathbb{D}^2)$ if and only if $\phi(z)$ is a finite Blaschke product.*

Proof By Frostman Theorem (Theorem 2.1.4), there exist sufficiently many $\lambda \in \mathbb{D}$ such that $\frac{\phi(z)-\lambda}{1-\bar{\lambda}\phi(z)}$ are Blaschke products. Hence, we may assume that $\phi(z)$ is a Blaschke product. Write $\phi(z) = \phi_n(z)\psi_n(z)$, where $\phi_n(z)$ is a finite Blaschke product of order n. Then we have

$$\frac{\phi(z) - \phi(w)}{z - w} = \frac{\phi_n(z)\psi_n(z) - \phi_n(w)\psi_n(w)}{z - w}$$

$$= \frac{\phi_n(z) - \phi_n(w)}{z - w} \psi_n(z) + \frac{\psi_n(z) - \psi_n(w)}{z - w} \phi_n(w).$$

Set

$$G_n = \frac{\phi_n(z) - \phi_n(w)}{z - w} \psi_n(z), \quad F_n = \frac{\psi_n(z) - \psi_n(w)}{z - w} \phi_n(w).$$

In the following, we will use the fact that $M_{\phi(z)}M_{\psi(w)}^* = M_{\psi(w)}^* M_{\phi(z)}$ for $\phi, \psi \in H^\infty(\mathbb{D})$. In fact, by Lemma 4.1.3

$$\langle G_n, F_n \rangle = \langle M_{\psi_n(z)} \frac{\phi_n(z) - \phi_n(w)}{z - w}, \quad M_{\phi_n(w)} \frac{\psi_n(z) - \psi_n(w)}{z - w} \rangle$$

$$= \langle \frac{\phi_n(z) - \phi_n(w)}{z - w}, \quad M_{\psi_n(z)}^* M_{\phi_n(w)} \frac{\psi_n(z) - \psi_n(w)}{z - w} \rangle$$

$$= \langle \frac{\phi_n(z) - \phi_n(w)}{z - w}, \quad M_{\phi_n(w)} M_{\psi_n(z)}^* \frac{\psi_n(z) - \psi_n(w)}{z - w} \rangle$$

$$= 0.$$

This implies

$$\left\| \frac{\phi(z) - \phi(w)}{z - w} \right\|^2 = \left\| \frac{\phi_n(z) - \phi_n(w)}{z - w} \right\|^2 + \left\| \frac{\psi_n(z) - \psi_n(w)}{z - w} \right\|^2. \tag{4.3}$$

Let $\phi(z) = B_1(z)B_2(z) \cdots B_n(z) \cdots$, where each $B_n(z)$ is a Blaschke factor. By simple calculations, $\left\| \frac{B_n(z)-B_n(w)}{z-w} \right\|^2 = 1$. Then by iterative application of (4.3), we have

$$\left\|\frac{\phi(z)-\phi(w)}{z-w}\right\|^2 = \sum_n \left\|\frac{B_n(z)-B_n(w)}{z-w}\right\|^2 = \sum_n 1.$$

This implies the desired conclusion. □

For each integer $n \geq 0$, put $p_n(z,w) = \frac{z^{n+1}-w^{n+1}}{z-w}$. By Lemma 4.1.1 $H^2(\mathbb{D}^2) \ominus [z-w]$ is spanned by $\{p_n : n \geq 0\}$. It is easy to verify that $p_n \perp p_m, n \neq m$, and $\{\frac{p_n}{\sqrt{n+1}} : n = 0,1,\cdots\}$ consists of a orthonormal basis of $H^2(\mathbb{D}^2) \ominus [z-w]$. Rewrite $\mathcal{H}$ for $H^2(\mathbb{D}^2) \ominus [z-w]$. Then

$$H^2(\mathbb{D}^2) = \mathcal{H} \oplus [z-w].$$

Let $P_{\mathcal{H}}$ denote the projection from $H^2(\mathbb{D}^2)$ onto $\mathcal{H}$, and we have

$$P_{\mathcal{H}} M_z|_{\mathcal{H}} = P_{\mathcal{H}} M_w|_{\mathcal{H}}.$$

For $\phi(z) \in H^\infty(\mathbb{D})$, let S_ϕ denote $P_{\mathcal{H}} M_\phi|_{\mathcal{H}}$. Clearly, $S_z = S_w$, and it is easy to show that S_z is unitarily equivalent to the Bergman shift M_z on the Bergman space $L_a^2(\mathbb{D})$ [GSZZ]. In fact, this wins in general. In more detail, there is a natural unitary map between $\mathcal{H}$ and $L_a^2(\mathbb{D})$:

$$U : \mathcal{H} \to L_a^2(\mathbb{D}), \quad f \mapsto f(z,z).$$

For $\phi(z) \in H^\infty(\mathbb{D})$, we have

$$U S_\phi U^* = M_\phi, \qquad (4.4)$$

where M_ϕ is the multiplication operator on $L_a^2(\mathbb{D})$ and U^* is determined by

$$U^* : L_a^2(\mathbb{D}) \to \mathcal{H}, \quad U^* z^n = \frac{p_n(z,w)}{n+1}.$$

To see (4.4), for each $g \in \mathcal{H}$ and any $z \in \mathbb{D}$, we have

$$(U S_\phi g)(z) = (S_\phi g)(z,z) = (P_{\mathcal{H}} \phi g)(z,z) = (\phi g - P_{[z-w]}\phi g)(z,z)$$
$$= \phi(z)g(z,z) - (P_{[z-w]}\phi g)(z,z) = \phi(z)g(z,z) = (M_\phi U g)(z),$$

where $P_{[z-w]}$ denotes the projection from $H^2(\mathbb{D}^2)$ onto $[z-w]$. This leads to $U S_\phi = M_\phi U$, forcing $U S_\phi U^* = M_\phi$, as desired.

If $\phi(z)$ is a finite Blaschke product with order l, then by the proof of Lemma 4.1.4 we get

$$\left\|\frac{\phi(z)-\phi(w)}{z-w}\right\|^2 = l.$$

For $n \geq 0$, Set

$$e_n = \frac{1}{\sqrt{l(n+1)}} \frac{\phi(z)^{n+1} - \phi^{n+1}(w)}{z - w}.$$

Then we see $\|e_n\| = 1$ and $e_n \perp e_m$ if $n \neq m$. In [HSXY], Hu, Sun, Xu and Yu proved the following.

Theorem 4.1.5 (Hu-Sun-Xu-Yu) *If ϕ is a finite Blaschke product, then*

$$S_\phi e_n = \sqrt{\frac{n+1}{n+2}} e_{n+1} \quad and \quad S_\phi^* e_n = \sqrt{\frac{n}{n+1}} e_{n-1}.$$

Now write $N_0 = \overline{span}\{e_n : n \geq 0\}$, then N_0 is a reducing subspace of S_ϕ. Set $M_0 = UN_0$, where U denotes the unitary operator defined above Theorem 4.1.5. It is easy to see that $Ue_n = \sqrt{\frac{n+1}{l}} \phi' \phi^n$, and then

$$M_0 = \overline{span}\{\phi' \phi^n : n \geq 0\}$$

is a reducing subspace of M_ϕ on the Bergman space, also see [Zhu1]. Therefore, for each finite Blaschke product $\phi(z)$, by Theorem 4.1.5 M_ϕ acting on the Bergman space has a canonical reducing subspace M_0 such that $M_\phi|_{M_0}$ is unitarily equivalent to the Bergman shift M_z. In fact, for each finite Blaschke product ϕ, M_0 is unique in the following sense: if there exists another reducing subspace $\mathcal{R}$ of M_ϕ such that M_ϕ, restricted on $\mathcal{R}$, is unitarily equivalent to the Bergman shift, then $\mathcal{R} = M_0$ [GSZZ]. Thus, the canonical reducing subspace M_0 is call the *distinguished reducing subspace* of M_ϕ on $L_a^2(\mathbb{D})$.

In what follows we will present a proof of Theorem 4.1.5 which is slightly different from the original proof in [HSXY] and that in [GSZZ].

Proof of Theorem 4.1.5 We first establish two claims.

Claim 1. $f \in [z - w]$ if and only if $f(z, z) = 0, z \in \mathbb{D}$.
Claim 2. $f \in [(z - w)^2]$ if and only if $f(z, z) = 0, \frac{\partial f}{\partial w}(z, z) = 0, z \in \mathbb{D}$.

Claim 1 is obvious, and it is enough to prove Claim 2. For Claim 2, the necessity is easily verified and it remains to deal with sufficiency of Claim 2. In fact, by $f(z, z) = 0, f$ can be expressed as $f(z, w) = (z - w)h(z, w)$ for h holomorphic on the bidisk. Then we have

$$\frac{\partial f}{\partial w}(z, z) = -h(z, z) = 0,$$

and hence $h(z, w)$ has form $h(z, w) = (z - w)g(z, w)$ for g holomorphic on the bidisk. Let $g(z, w) = \sum_n g_n(z, w)$ be the homogeneous expression of g. This gives homogeneous expression of f, $f(z, w) = \sum_n (z - w)^2 g_n(z, w)$. This implies that

$\|f(z,w) - (z-w)^2 \sum_{n=0}^{m} g_n(z,w)\| \to 0$ as $m \to \infty$, forcing $f \in [(z-w)^2]$. Claim 2 is proved.

We are ready to present the proof of Theorem 4.1.5. Indeed, by a simple computation

$$\phi e_n - \sqrt{\frac{n+1}{n+2}}\, e_{n+1} = \frac{1}{(n+2)\sqrt{l(n+1)}(z-w)}$$
$$\left[(\phi^{n+2}(z) - \phi^{n+2}(w)) - (n+2)\phi^{n+1}(w)(\phi(z) - \phi(w))\right],$$

and applying claim 2, we see that

$$\left(\phi^{n+2}(z) - \phi^{n+2}(w)\right) - (n+2)\phi^{n+1}(w)\left(\phi(z) - \phi(w)\right) \in [(z-w)^2].$$

Combining this fact with the above equality ensures $\phi e_n - \sqrt{\frac{n+1}{n+2}}\, e_{n+1} \in [z-w]$, and hence

$$S_\phi e_n = \sqrt{\frac{n+1}{n+2}}\, e_{n+1}.$$

We next verify the equality $S_\phi^* e_n = \sqrt{\frac{n}{n+1}} e_{n-1}$. Firstly, by Lemma 4.1.3, it is easy to see $S_\phi^* e_0 = 0$. Since

$$S_\phi^* e_n = \frac{1}{\sqrt{l(n+1)}} M_{\phi(z)}^* \frac{\phi^{n+1}(z) - \phi^{n+1}(w)}{z-w}$$

$$= \frac{1}{\sqrt{l(n+1)}} M_{\phi(z)}^*\left[(\phi(z)^n + \phi(z)^{n-1}\phi(w) + \cdots + \phi(z)\phi^{n-1}(w) + \phi^n(w))e_0\right]$$

$$= \frac{1}{\sqrt{l(n+1)}} \left[(\phi(z)^{n-1} + \phi(z)^{n-2}\phi(w) + \cdots + \phi^{n-1}(w))e_0 + \phi^n(w)M_{\phi(z)}^* e_0\right]$$

$$= \frac{1}{\sqrt{l(n+1)}} \frac{\phi^n(z) - \phi^n(w)}{z-w}$$

$$= \sqrt{\frac{n}{n+1}} e_{n-1}.$$

The proof of Theorem 4.1.5 is complete.   □

Given $\phi \in H^\infty(\mathbb{D})$, let M be a reducing subspace of M_ϕ acting on $L_a^2(\mathbb{D})$. The reducing subspace M is call the distinguished reducing subspace of M_ϕ if M_ϕ, restricted on M is unitarily equivalent to the Bergman shift M_z on $L_a^2(\mathbb{D})$. The following theorem shows that bounded holomorphic functions enjoying this property are exactly finite Blaschke products, see [GSZZ].

Theorem 4.1.6 *Let* $\phi \in H^\infty(\mathbb{D})$. *Then* M_ϕ *has the distinguished reducing subspace in* $L_a^2(\mathbb{D})$ *if and only if* ϕ *is a finite Blaschke product.*

Proof The proof is from [GSZZ, Theorem 26].

To prove Theorem 4.1.6, it suffices to show that if M_ϕ acting on $L_a^2(\mathbb{D})$ has the distinguished reducing subspace, then ϕ is a finite Blaschke product. Now, assume M_ϕ has the distinguished reducing subspace, say M, such that $M_\phi|_M$ is unitarily equivalent to the Bergman shift M_z; that is, there exists a unitary operator $U' : M \to L_a^2(\mathbb{D})$ such that $U'^* M_z U' = M_\phi|_M$. Let K_λ^M be the reproducing kernel of M for $\lambda \in \mathbb{D}$. Then it is easy to know that $K_\lambda^M \neq 0$, except for λ in a countable set. We have

$$|\langle M_\phi K_\lambda^M, K_\lambda^M \rangle| = |\phi(\lambda)| \, \|K_\lambda^M\|^2 = |\langle M_z U K_\lambda^M, U K_\lambda^M \rangle|$$
$$\leq \|M_z\| \|U K_\lambda^M\|^2 = \|K_\lambda^M\|^2.$$

Therefore, we get $|\phi(\lambda)| \leq 1$ for all $\lambda \in \mathbb{D}$, and hence $\|\phi\|_\infty \leq 1$. Since S_ϕ acting on $\mathcal{H} = H^2(\mathbb{D}^2) \ominus [z - w]$ is unitarily equivalent to M_ϕ acting on $L_a^2(\mathbb{D})$, that is, $U : \mathcal{H} \to L_a^2(\mathbb{D})$, $f \mapsto f(z, z)$, and $U S_\phi U^* = M_\phi$. Set $N = U^* M$. This implies that S_ϕ, restricted on its corresponding reducing subspace N, is unitarily equivalent to the Bergman shift M_z acting on $L_a^2(\mathbb{D})$. Therefore, there exists a unitary operator $V : N \to L_a^2(\mathbb{D})$ such that $V^* M_z V = S_\phi|_N$. Set $e_n = V^* e_n'$, where $e_n' = \sqrt{n+1} z^n$ for $n = 0, 1, \cdots$. Then $S_\phi^* e_0 = 0$, and hence

$$M_{\phi(z)}^* e_0 = 0 \quad and \quad M_{\phi(w)}^* e_0 = 0,$$

where $M_{\phi(z)}$ and $M_{\phi(w)}$ are the operators acting on $H^2(\mathbb{D}^2)$. Noting that $S_{\phi(z)} = S_{\phi(w)}$, we have

$$\|V S_{(\phi(z)+\phi(w))} e_0\|^2 = \|z + z\|^2 = 2.$$

It is easy to verify that

$$\langle \phi(z) e_0, \phi(w) e_0 \rangle = \langle M_{\phi(w)}^* e_0, M_{\phi(z)}^* e_0 \rangle = 0.$$

Therefore,

$$\|(\phi(z) + \phi(w)) e_0\|^2 = \|\phi(z) e_0\|^2 + \|\phi(w) e_0\|^2 \leq 2$$

because $\|\phi\|_\infty \leq 1$. Since

$$2 = \|V S_{(\phi(z)+\phi(w))} e_0\|^2 = \|V P_{\mathcal{H}}(\phi(z) + \phi(w)) e_0\|^2 = \|P_{\mathcal{H}}(\phi(z) + \phi(w)) e_0\|^2,$$

we see

$$(\phi(z) + \phi(w)) e_0 \in \mathcal{H},$$

and hence by Lemma 4.1.2,

$$e_0 = c \frac{\phi(z) - \phi(w)}{z - w} \tag{4.5}$$

for some constant c. Furthermore, from

$$\|(\phi(z) + \phi(w))e_0\|^2 = \|\phi(z)e_0\|^2 + \|\phi(w)e_0\|^2 = 2$$

and $\|\phi\|_\infty \leq 1$, we have $\|\phi(z)e_0\|^2 = 1$, that is,

$$\int_{\mathbb{T}^2} (|\phi(z)|^2 - 1)|e_0|^2 dm_2 = 0.$$

Since $\|\phi\|_\infty \leq 1$, it is easy to prove that $\phi(z)$ is an inner function. From (4.5) and Lemma 4.1.4, we see that ϕ is a finite Blaschke product. The proof is complete. □

4.2 Abelian $\mathcal{V}^*(B)$

In last section it is shown that $\mathcal{V}^*(B)$ is always nontrivial for each finite Blaschke product B with order $B \geq 2$. This section will study the commutativity of $\mathcal{V}^*(B)$, which was first established in [DPW, Theorem 1.1].

Theorem 4.2.1 (Douglas-Putinar-Wang) *For each finite Blaschke product B, $\mathcal{V}^*(B)$ is abelian.*

Now let us recall some preliminary facts. Let B be a Blaschke product of order n, and set

$$F = \{z \in \mathbb{D} : \text{ there exists } w \in \mathbb{D} \text{ such that } B'(w) = 0 \text{ and } B(w) = B(z)\}$$

and it is known that F is a finite subset of $\mathbb{D}$ (for example, see Bochner's theorem [Wa1]). Suppose P is a polygon drawn through all points in F and a fixed point on the unit circle such that $\mathbb{D} - P$ is simply connected. By Theorem 12.3 in [Bli], n distinct roots $w = \rho_k(z)$ $(1 \leq k \leq n)$ of the equation

$$B(w) - B(z) = 0$$

are holomorphic functions in $\mathbb{D} - P$. Note that these ρ_k are local inverses of B.

To prove Theorem 4.2.1, we need the following lemma from [Sun1, Lemma 2], also refer to [DSZ, Theorem 4.2]. However, Sun's proof [Sun1] reads difficult. Later we will provide a different proof for completeness in Sect. 4.3 of this chapter.

Lemma 4.2.2 *Suppose B is a finite Blaschke product of order n, and S is a unitary operator which commutes with M_B, then there exist constants $c_k (1 \leq k \leq n)$*

satisfying $\sum_{k=1}^{n} |c_k|^2 = 1$ *and*

$$Sh(w) = \sum_{k=1}^{n} c_k \rho_k'(w) h \circ \rho_k(w), \quad h \in L_a^2(\mathbb{D}), \ w \in \mathbb{D}. \tag{4.6}$$

Remark 4.2.3 Let us make clear what (4.6) means. First, both sides of (4.6) are holomorphic in $\mathbb{D} - P$, and thus (4.6) holds for $w \in \mathbb{D} - P$. Since the left hand side Sh of (4.6) is holomorphic in $\mathbb{D}$, the right hand side of (4.6) can also be extend analytically to $\mathbb{D}$. Therefore (4.6) holds on $\mathbb{D}$.

Later in Theorem 5.7.2, we will get a more exact form for (4.6).

We have an immediate corollary.

Corollary 4.2.4 *For any finite Blaschke product B,* $\dim V^*(B) \le \operatorname{order} B$.

Proof Note that Lemma 4.2.2 implies all unitary operators in $V^*(B)$ span a subspace of dimension less than or equal to order B. Since $V^*(B)$ is a von Neumann algebra, and by Proposition 2.5.1 any von Neumann algebra is the finite linear span of its unitary operators, then $\dim V^*(B) \le \operatorname{order} B$, as desired. $\square$

In [DPW], it is shown that $\dim V^*(B)$ equals the number of the components of the Riemann surface

$$S_B = \{(z, w) \in \mathbb{D}^2 : B(z) = B(w), z \in \mathbb{D} - B^{-1}(\mathcal{E}_B)\},$$

where

$$\mathcal{E}_B = \{B(z) : z \in \mathbb{D} \text{ and } B'(z) = 0\}.$$

For details, also see [GSZZ, SZZ1].

The following proposition is a well-known computational result.

Proposition 4.2.5 *For any finite Blaschke product B, B' does not vanish on the unit circle* $\mathbb{T}$.

Proof Without loss of generality, assume that there are n points $a_1, \cdots, a_n$ in $\mathbb{D}$ such that

$$B(z) = \prod_{i=1}^{n} \frac{a_i - z}{1 - \overline{a_i} z}.$$

For each $\lambda \in \mathbb{D}$, we have

$$\varphi_\lambda'(z) = -\frac{1 - |\lambda|^2}{(1 - \overline{\lambda} z)^2},$$

where $\varphi_\lambda(z) = \frac{\lambda - z}{1 - \bar{\lambda} z}$. Therefore, for each $z \in \mathbb{T}$,

$$
\begin{aligned}
B'(z) &= \sum_{i=1}^{n} \left(\varphi'_{a_i}(z) \prod_{j=1, j \neq i}^{n} \frac{a_j - z}{1 - \overline{a_j} z} \right) \\
&= -B(z) \sum_{i=1}^{n} \frac{1 - |a_i|^2}{(1 - \overline{a_i} z)(a_i - z)} \\
&= B(z) \sum_{i=1}^{n} \frac{\bar{z}(1 - |a_i|^2)}{(\overline{a_i} - \bar{z})(a_i - z)} \\
&= B(z)\bar{z} \sum_{i=1}^{n} \frac{1 - |a_i|^2}{|a_i - z|^2} \neq 0.
\end{aligned}
$$

The proof is complete. $\square$

Next we will have a look at the geometric structure of finite Blaschke products. As pointed out in [DPW], a finite Blaschke product B acts on $\mathbb{T}$ just like that of z^n, where $n = \operatorname{order} B$. The following ideas are from [DPW]. Precisely, regard B as a member in $Hol(\overline{\mathbb{D}})$. That is, there is some positive constant ε such that B is holomorphic on $(1 + \varepsilon)\mathbb{D}$. Pick a point $w_0 \in \mathbb{T}$. Since Proposition 4.2.5 B' does not vanish on $\mathbb{T}$, $B^{-1}(B(w_0))$ consists of exactly n different points (in anti-clockwise direction)

$$
w_0, \cdots, w_{n-1},
$$

which lie on $\mathbb{T}$. There is a neighborhood V of w_0 on which there exists a unique local inverse $\rho_j (0 \leq j \leq n - 1)$ of B satisfying

$$
\rho_j(w_0) = w_j.
$$

All ρ_j map $V \cap \mathbb{D}$ into $\mathbb{D}$, $V \cap \mathbb{T}$ into $\mathbb{T}$, and $V - \overline{\mathbb{D}}$ into $\mathbb{C} - \overline{\mathbb{D}}$. We will extend all ρ_j analytically to some neighborhood Ω of $\mathbb{T}$, and such extensions must be local inverses of B.

For convenience, write $w_n = w_0$. For each $j(0 \leq j \leq n - 1)$, the oriented arc $\widehat{w_j w_{j+1}}$ connecting w_j with w_{j+1} must be mapped onto a connected arc of $\mathbb{T}$ by B. By the conformality of B on $\mathbb{T}$, it is not difficult to see that B maps $\widehat{w_j w_{j+1}} - \{w_{j+1}\}$ onto $\mathbb{T}$. Since $B|_\mathbb{T} : \mathbb{T} \to \mathbb{T}$ is n-to-1, the restriction of B on $\widehat{w_j w_{j+1}} - \{w_{j+1}\}$ is also injective, denoted by T_j. Then one can construct several curves γ_j whose images are $\widehat{w_j w_{j+1}}$, satisfying

$$
\gamma_j(0) = w_j \quad \text{and} \quad \gamma_j(1) = w_{j+1}.
$$

Precisely, set

$$
\gamma_j(t) = T_j^{-1}\left(e^{2\pi i t} B(w_0) \right), \quad 0 \leq t < 1,
$$

and $\gamma_j(1) = w_{j+1}$. Then γ_j is a continuous map, as desired. Moreover, it is not difficult to show that

$$\gamma_0(t) \mapsto \gamma_j(t), \; 0 \le t \le 1,$$

defines a continuous map on the image of γ_0, still denoted by ρ_j. Define $\{\gamma_k\}_{k=0}^{\infty}$ to be the sequence of curves:

$$\gamma_0, \cdots, \gamma_{n-1}; \gamma_0, \cdots, \gamma_{n-1}; \cdots,$$

and rewrite $\{w_k\}_{k=0}^{\infty}$ for the following sequence:

$$w_0, \cdots, w_{n-1}; w_0, \cdots, w_{n-1}; \cdots.$$

Then each $\rho_j (0 \le j \le n-1)$ can be extended to a continuous map as follows:

$$\gamma_k(t) \mapsto \gamma_{k+j}(t), \; 0 \le t \le 1, \; 0 \le j, k \le n-1.$$

Since

$$B(\gamma_k(t)) = B(\gamma_{k+j}(t)), 0 \le t \le 1, 0 \le j, k \le n-1,$$

then $B(\rho_j) = B$; besides, for a fixed point $w = \gamma_k(t_0)$, each ρ_j naturally induces an analytic continuation defined on a neighborhood of w. Note that ρ_j maps each arc $\overparen{w_k w_{k+1}}$ onto $\overparen{w_{k+j} w_{k+j+1}}$, and then ρ_j defines a continuous map from $\mathbb{T}$ onto $\mathbb{T}$. Since for each point $w \in \mathbb{T}$, ρ_j can be analytically extended to a small disk centered at w, and $\mathbb{T}$ can be covered by finitely many of such disks, then there is a neighborhood $\tilde{\Omega}$ of $\mathbb{T}$ such that ρ_j is a holomorphic function on $\tilde{\Omega}$. Moreover, we may require that $B^{-1}(\tilde{\Omega}) = \tilde{\Omega}$. Clearly, these n functions ρ_j are local inverses of B. By the definition of ρ_j, it is easy to verify that $\rho_j \circ \rho_k(w_0) = \rho_k \circ \rho_j(w_0)$. Since the compositions $\rho_j \circ \rho_k$ and $\rho_k \circ \rho_j$ are local inverses of B and they are equal at w_0, then they must be equal on a neighborhood of w_0. Therefore, we have $\rho_j \circ \rho_k = \rho_k \circ \rho_j$ on $\tilde{\Omega}$.

With these preparations, we are able to deal with the commutativity of $\mathcal{V}^*(B)$.

Proof of Theorem 4.2.1 Since each operator in a von Neumann algebra is a finite span of its unitary operators, then by Lemma 4.2.2 each operator S in $\mathcal{V}^*(B)$ has the following form:

$$Sh(w) = \sum_{k=1}^{n} c_k \rho_k'(w) h \circ \rho_k(w), \; h \in L_a^2(\mathbb{D}), \; w \in \mathbb{D} - P,$$

where ρ_k are local inverses of B. By the uniqueness theorem in complex analysis, we have

$$Sh(w) = \sum_{k=1}^{n} c_k \rho_k'(w) h \circ \rho_k(w), \; h \in L_a^2(\mathbb{D}), \; w \in \tilde{\Omega} \cap \mathbb{D}.$$

Now suppose T is another operator in $\mathcal{V}^*(B)$, which necessarily has a similar form as S. Since $\rho_j \circ \rho_k = \rho_k \circ \rho_j$ holds on $\tilde{\Omega}$ for all j and k, we must have

$$STh(w) = TSh(w), \ h \in L^2_a(\mathbb{D}), \ w \in \tilde{\Omega} \cap \mathbb{D},$$

and hence

$$STh(w) = TSh(w), \ h \in L^2_a(\mathbb{D}), \ w \in \mathbb{D}.$$

The proof of Theorem 4.2.1 is complete.   □

Remark 4.2.6 Let B be a finite Blaschke product with zero sequence $\{a_j\}_{j=1}^n$. The forms of local inverses ρ_j of B are given in [DPW]. Precisely, define

$$u = \frac{\sqrt[n]{(z - a_1) \cdots (z - a_n)}}{\sqrt[n]{(1 - \overline{a_1}z) \cdots (1 - \overline{a_n}z)}}, z \in \mathbb{D} - P,$$

which proves to be a well-defined holomorphic function. It is shown that there is an $r \in (0, 1)$ such that u is univalent on $u^{-1}(\{r < |z| < 1\})$. Then on the deformed annulus $\Omega \triangleq u^{-1}(\{r < |z| < 1\})$, $\rho_k(z) = u^{-1}(e^{\frac{2k\pi i}{n}} u(z))$, where $k = 1, 2, \ldots, n$ [DPW].

Note that $B^{-1}(\Omega) \subseteq \Omega$. It is shown in [DPW] that each ρ_k gives a unitary operator over $L^2_a(\Omega)$:

$$U_{\rho_k} : f \mapsto f \circ \rho_k \, \rho'_k, \quad f \in L^2_a(\Omega).$$

Let $\mathcal{W}^*(B, \Omega)$ denotes the von Neumann algebra generated by M_B defined on $L^2_a(\Omega)$, and set $\mathcal{V}^*(B, \Omega) \triangleq \mathcal{W}^*(B, \Omega)'$. Then one sees that $\mathcal{V}^*(B, \Omega)$ is exactly generated by these n unitary operators U_{ρ_k}, with $n = \text{order } B$, see [DPW].

To end this section, we shall apply Corollary 4.2.4 to give another different proof for some special cases of Theorem 4.2.1 [Zhu1, GSZZ, SZZ1].

Corollary 4.2.7 *Let B be a finite Blaschke product of order n with $1 \le n \le 4$. Then $\mathcal{V}^*(B)$ is abelian, and hence M_B has at most n minimal reducing subspaces.*

Proof Recall that $\mathcal{V}^*(B)$ is a von Neumann algebra, and by Corollary 4.2.4, $\dim \mathcal{V}^*(B) \le \text{order } B = n$. By Theorem 2.5.11, a finite dimensional von Neumann algebra is ∗-isomorphic to the direct sum of full matrix algebras

$$\bigoplus_{k=1}^r M_{n_k}(\mathbb{C}).$$

Now assume that $\mathcal{V}^*(B)$ is ∗-isomorphic to

$$\bigoplus_{k=1}^r M_{n_k}(\mathbb{C}). \tag{4.7}$$

If $1 \leq n \leq 3$, then $\dim V^*(B) \leq n \leq 3$. Since $\dim M_j(\mathbb{C}) = j^2$, clearly all n_k in (4.7) is equal to one. Thus $V^*(B)$ is abelian.

If $n = 4$, then $\dim V^*(B) \leq 4$. To reach a contradiction, assume that $V^*(B)$ is not abelian. Since $\dim V^*(B) \leq 4$, it is easy to see that

$$V^*(B) \cong M_2(\mathbb{C}),$$

whose center is trivial. But by Theorem 4.4.1 in Sect. 4.4, the orthogonal projection P_0 onto the distinguished subspace M_0 is orthogonal to any other minimal projections in $V^*(B)$. Since all minimal projections span $V^*(B)$, P_0 belongs to the center of $V^*(B)$, which is a contradiction. The proof is complete. $\square$

Remark 4.2.8 Combing Theorem 4.2.1 with Proposition 2.6.2 shows that when order $B = n$, there exist at most n minimal reducing subspaces for M_B on the Bergman space. Therefore, if ϕ satisfies the Cowen's condition (in particular, when $\phi \in Hol(\overline{\mathbb{D}})$), from Theorem 3.1.1, $V^*(\phi) = V^*(B)$ for some finite Blaschke product B, we see that a such M_ϕ has at most n minimal reducing subspaces, where $n = $ order B.

4.3   Representation for Operators in $V^*(B)$

In this section, we will established Lemma 4.2.2, which concerns with the representation for operators in $V^*(B)$. To this end, we need some preparations.

Assume that B_1 and B_2 are finite Blaschke products which have only simple zeros, with order $B_1 = $ order $B_2 = n$. Let $\beta_1, \cdots, \beta_n$ (respectively, $\gamma_1, \cdots, \gamma_n$) be n branches of B_1^{-1} (respectively, B_2^{-1}). Let Δ be some connected domain such that all β_i and γ_i are (single-valued) holomorphic on some neighborhood of $\overline{\Delta}$. In particular, since B_1 and B_2 have only simple zeros, we can choose Δ to be an open disk containing 0. Then we have the following result [GH1], which was studied in the proof of [SY, Lemma 1] in a special case.

Proposition 4.3.1 *If M_i is a closed subspace of $L_a^2(\mathbb{D})$ which is invariant under M_{B_i} ($i = 1, 2$), and if $U : M_1 \to M_2$ is a unitary operator such that $UM_{B_1} = M_{B_2}U$, then there exists an $n \times n$ numerical unitary matrix W such that*

$$W \begin{pmatrix} f(\beta_1(w))\beta_1'(w) \\ \vdots \\ f(\beta_n(w))\beta_n'(w) \end{pmatrix} = \begin{pmatrix} g(\gamma_1(w))\gamma_1'(w) \\ \vdots \\ g(\gamma_n(w))\gamma_n'(w) \end{pmatrix}, \ w \in \Delta,$$

where $f \in M_1$ and $g = Uf$.

To prove the above proposition, we need the following lemma from [GH1], which is of independent interest. Recall that given two vectors f, g in a Hilbert space H, a rank-one operator $f \otimes g$ on H is defined by

$$(f \otimes g)h = \langle h, g \rangle f.$$

Lemma 4.3.2 *Let H be a Hilbert space and suppose e_λ^k, f_μ^k ($1 \leq k \leq n$ and $\lambda, \mu \in \Lambda$) are vectors in H satisfying*

$$\sum_{k=1}^{n} e_\lambda^k \otimes e_\mu^k = \sum_{k=1}^{n} f_\lambda^k \otimes f_\mu^k, \ \lambda, \mu \in \Lambda,$$

then there is an $n \times n$ numerical unitary matrix W such that

$$W \begin{pmatrix} e_\lambda^1 \\ \vdots \\ e_\lambda^n \end{pmatrix} = \begin{pmatrix} f_\lambda^1 \\ \vdots \\ f_\lambda^n \end{pmatrix}, \ \lambda \in \Lambda.$$

In the case of Λ being a singlet, Lemma 4.3.2 was first proved by [Ar1, Proposition 5.1] and [Ar2, Proposition A.1].

Proof For each $\lambda \in \Lambda$, set

$$A_\lambda : \mathbb{C}^n \to H$$

$$(c_1, \cdots, c_n) \mapsto \sum_{k=1}^{n} c_k e_\lambda^k,$$

and

$$B_\lambda : \mathbb{C}^n \to H$$

$$(c_1, \cdots, c_n) \mapsto \sum_{k=1}^{n} c_k f_\lambda^k.$$

After simple computations, it follows that

$$A_\lambda^* : H \to \mathbb{C}^n$$

$$h \mapsto (\langle h, e_\lambda^1 \rangle, \cdots, \langle h, e_\lambda^n \rangle).$$

Besides, for any $\lambda, \mu \in \Lambda$, we have

$$A_\mu A_\lambda^* h = \sum_{k=1}^{n} \langle h, e_\lambda^k \rangle e_\mu^k = \sum_{k=1}^{n} (e_\mu^k \otimes e_\lambda^k) h$$

$$= \sum_{k=1}^{n} (f_\mu^k \otimes f_\lambda^k) h = B_\mu B_\lambda^* h, \ h \in H.$$

That is, $A_\mu A_\lambda^* = B_\mu B_\lambda^*$, which implies that

$$\sum_{i=1}^l c_i A_{\lambda_i}^* h \mapsto \sum_{i=1}^l c_i B_{\lambda_i}^* h, \ h \in H \text{ and } \lambda_i \in \Lambda$$

is a well-defined linear isometry from some subspace of $\mathbb{C}^n$ to another. This isometry can be extended to a unitary map $V^* : \mathbb{C}^n \to \mathbb{C}^n$. By the definition of V^*, we have

$$V^* A_\lambda^* = B_\lambda^*, \ \lambda \in \Lambda,$$

and hence

$$A_\lambda V = B_\lambda. \tag{4.8}$$

Let (4.8) act on $(0, \cdots, 1, 0, \cdots, 0)$ where 1 is at the k-th coordinate, and denote by $\mathbf{v}_k$ the k-th column of V, then

$$\mathbf{v}_k^T \begin{pmatrix} e_\lambda^1 \\ \vdots \\ e_\lambda^n \end{pmatrix} = f_\lambda^k, \ \lambda \in \Lambda,$$

where $\mathbf{v}_k^T$ is the transverse of $\mathbf{v}_k$. Let W be the transverse V^T of V and the above identities imply that

$$W \begin{pmatrix} e_\lambda^1 \\ \vdots \\ e_\lambda^n \end{pmatrix} = \begin{pmatrix} f_\lambda^1 \\ \vdots \\ f_\lambda^n \end{pmatrix}, \ \lambda \in \Lambda.$$

$\square$

Proof of Proposition 4.3.1 The proof is from [GH1].

Consider M_1 as a reproducing function space on $\mathbb{D}$. Let K_λ be the reproducing kernel of M_1 at $\lambda \in \mathbb{D}$, and put $L_\lambda = U K_\lambda$, where L_λ is not necessarily the reproducing kernel of M_2. Since $U M_{B_1} = M_{B_2} U$, it is easy to see that for any polynomials P and Q,

$$\langle P(B_1) K_\lambda, Q(B_1) K_\mu \rangle = \langle P(B_2) L_\lambda, Q(B_2) L_\mu \rangle, \ \lambda, \mu \in \mathbb{D}.$$

That is,

$$\int_{\mathbb{D}} \left((P\overline{Q}) \circ B_1(w) K_\lambda(w) \overline{K_\mu}(w) - (P\overline{Q}) \circ B_2(w) L_\lambda(w) \overline{L_\mu}(w) \right) dA(w) = 0. \tag{4.9}$$

Now set

$$\mathcal{Y} = span \{ P\overline{Q} : P, Q \text{ are polynomials} \}.$$

By the Stone-Weierstrass Theorem, any continuous function on $\overline{\mathbb{D}}$ can be uniformly approximated by functions in $\mathcal{Y}$. It follows from (4.9) that

$$\int_{\mathbb{D}} \Big(f(B_1(w))K_\lambda(w)\overline{K_\mu}(w) - f(B_2(w))L_\lambda(w)\overline{L_\mu}(w)\Big)dA(w) = 0, f \in C(\overline{\mathbb{D}})$$

$$(4.10)$$

By Lebesgue's Dominated Convergence Theorem, the identity (4.10) holds for any f in $L^\infty(\mathbb{D})$. In particular, for any $f \in L^\infty(\Delta)$, (4.10) gives that

$$\int_{B_1^{-1}(\Delta)} f(B_1(w))K_\lambda(w)\overline{K_\mu}(w)dA(w) = \int_{B_2^{-1}(\Delta)} f(B_2(w))L_\lambda(w)\overline{L_\mu}(w)dA(w),$$

and hence by our assumptions of Δ,

$$\int_\Delta f(z) \sum_{k=1}^n (K_\lambda \overline{K_\mu}) \circ \beta_k(z)|\beta_k'(z)|^2 dA(z) = \int_\Delta f(z) \sum_{k=1}^n (L_\lambda \overline{L_\mu}) \circ \gamma_k(z)|\gamma_k'(z)|^2 dA(z).$$

Therefore,

$$\sum_{k=1}^n (K_\lambda \overline{K_\mu}) \circ \beta_k(z)|\beta_k'(z)|^2 = \sum_{k=1}^n (L_\lambda \overline{L_\mu}) \circ \gamma_k(z)|\gamma_k'(z)|^2, \quad z \in \Delta. \qquad (4.11)$$

Next, we are to apply Lemma 4.3.2. Let H be the Bergman space over Δ and write $\Lambda = \mathbb{D}$. Set

$$e_\lambda^k = K_\lambda(\beta_k(z))\beta_k'(z), f_\mu^k = L_\mu(\gamma_k(z))\gamma_k'(z), 1 \le k \le n, \lambda, \mu \in \mathbb{D}.$$

By (4.11), the Berezin transforms of $\sum_{k=1}^n e_\lambda^k \otimes e_\mu^k$ and $\sum_{k=1}^n f_\lambda^k \otimes f_\mu^k$ are equal. Applying Theorem A.1 in Appendix A yields

$$\sum_{k=1}^n e_\lambda^k \otimes e_\mu^k = \sum_{k=1}^n f_\lambda^k \otimes f_\mu^k, \lambda, \mu \in \mathbb{D}.$$

Then by Lemma 4.3.2, there is an $n \times n$ unitary numerical matrix W satisfying

$$W \begin{pmatrix} K_\lambda(\beta_1(w))\beta_1'(w) \\ \vdots \\ K_\lambda(\beta_n(w))\beta_n'(w) \end{pmatrix} = \begin{pmatrix} L_\lambda(\gamma_1(w))\gamma_1'(w) \\ \vdots \\ L_\lambda(\gamma_n(w))\gamma_n'(w) \end{pmatrix}, w, \lambda \in \Delta,$$

This immediately leads to our conclusion. $\qquad\qquad\qquad\qquad\qquad\qquad\qquad\qquad \square$

Remark 4.3.3 In Proposition 4.3.1, special attention is paid to the case that $B_1 = B_2$ and M_1 and M_2 are reducing for M_{B_1}. In this case, $\beta_i = \gamma_i$ for all i. Also note that if $f \in M_1 \ominus B_1 M_1$, then $g = Uf \in M_2 \ominus B_1 M_2$.

If the assumption that B_1 and B_2 have only simple zeros is dropped, then Proposition 4.3.1 still holds, but Δ will be replaced with some open disk Δ' not containing 0. The proof is completely same.

Proof of Lemma 4.2.2 The proof comes from [GH1].

First we deal with the case of B having only simple zero. Let $\beta_1, \cdots, \beta_n$ be n branches of B^{-1} and they are locally holomorphic. Then regard $\rho_k(z)$ $(1 \leq k \leq n)$ as n different branches of $B^{-1} \circ B$; clearly, locally $(\rho_1(z), \cdots, \rho_n(z))$ is a permutation of $(\beta_1 \circ B, \cdots, \beta_n \circ B)$.

Now fix β_1. Without loss of generality, we may assume that the set Δ in Proposition 4.3.1 satisfies that $\beta_1(\Delta) \subseteq \mathbb{D} - P$ (otherwise we may replace Δ with some open disk Δ' and 0 is not necessarily in Δ'). Also we can assume

$$\rho_i|_{\beta_1(\Delta)} = \beta_i \circ B|_{\beta_1(\Delta)}.$$

Then we have

$$\rho_i \circ \beta_1|_\Delta = \beta_i|_\Delta, \tag{4.12}$$

and hence

$$\rho_i' \circ \beta_1 \beta_1'|_\Delta = \beta_i'|_\Delta. \tag{4.13}$$

Since S commutes with M_B, by Proposition 4.3.1 there exists an $n \times n$ numerical unitary matrix W such that

$$W \begin{pmatrix} f(\beta_1(w))\beta_1'(w) \\ \vdots \\ f(\beta_n(w))\beta_n'(w) \end{pmatrix} = \begin{pmatrix} Sf(\beta_1(w))\beta_1'(w) \\ \vdots \\ Sf(\beta_n(w))\beta_n'(w) \end{pmatrix}, f \in L_a^2(\mathbb{D}) \text{ and } w \in \Delta.$$

Clearly, there are n constants $c_1, \cdots, c_n$ satisfying $\sum_{i=1}^n |c_i|^2 = 1$ and

$$Sf(\beta_1(w))\beta_1'(w) = \sum_{i=1}^n c_i f(\beta_i(w))\beta_i'(w), \ f \in L_a^2(\mathbb{D}) \text{ and } w \in \Delta.$$

Then by (4.12) and (4.13), we have for each $f \in L_a^2(\mathbb{D})$,

$$Sf(\beta_1(w))\beta_1'(w) = \sum_{i=1}^n c_i f(\rho_i \circ \beta_1(w))\rho_i' \circ \beta_1(w)\beta_1'(w), \ w \in \Delta,$$

and thus

$$Sf(\beta_1(w)) = \sum_{i=1}^{n} c_i f(\rho_i \circ \beta_1(w))\rho_i' \circ \beta_1(w), \ w \in \Delta.$$

That is, (4.6) holds on $\beta_1(\Delta)$. Moreover, since both sides of (4.6) are holomorphic on the connected set $\mathbb{D} - P$, then (4.6) holds on $\mathbb{D} - P$. By Remark 4.2.3, we have (4.6) on $\mathbb{D}$. Then the proof is complete in the case that B has only simple zeros.

In general, by Remark 4.3.3 Proposition 4.3.1 still holds, but Δ will be replaced with some other disk. By the same discussion as above, (4.6) holds. $\square$

4.4 Further Consideration on Reducing Subspaces

Though some results in this section are covered by Douglas, Putinar and Wang's theorem (Theorem 4.2.1), we intend to provide a different method, which will inspire some more general results.

First we state the following result, which comes from [GH1].

Theorem 4.4.1 *Let B be a finite Blaschke product and M, N be two different minimal reducing subspaces of M_B on the Bergman space. If one of the following holds:*

(1) $\dim M \ominus BM = \dim N \ominus BN = 1$;
(2) $\dim M \ominus BM \neq \dim N \ominus BN$,

then $M \perp N$, and M is not unitarily equivalent to N.

Clearly, Theorem 4.4.1 is a special case of Theorem 4.2.1. However, a different approach will be provided later. By using Theorem 4.4.1, one can give the following corollary.

Corollary 4.4.2 *If M is a reducing subspace of M_B satisfying $\dim M \ominus BM = 1$, then P_M is in the center of $V^*(B)$.*

Proof The idea of the proof comes from that of [GH1, Theorem 3.3].

Suppose M is a reducing subspace of M_B satisfying $\dim M \ominus BM = 1$. Clearly, M is minimal. If there is another reducing subspace N of M_B, we must show that P_M commutes with P_N. To see this, we may assume that $P_N P_M \neq 0$. Rewrite p for $P_N P_M$, and $p \in V^*(B)$. Let $p = u|p|$ be the polar decomposition. Then u is a partial isometry in $V^*(B)$. By minimality of P_M,

$$u^*u = P_M \quad \text{and} \quad uu^* = Q,$$

where Q denotes the orthogonal projection onto the range of u. Clearly, $Q \leq P_N$. Now write $N_0 = Range\ Q$. Since the restriction $u|_M : M \to N_0$ is a unitary operator satisfying $uM_B = M_B u$ and $\dim M \ominus BM = 1$, then $\dim N_0 \ominus BN_0 = 1$. By Theorem 4.4.1 either $M = N_0$ or $M \perp N_0$. Since $P_{N_0} = Q \leq P_N$,

$$P_{N_0} P_M = P_{N_0} P_N P_M = P_{N_0} p = u u^* u |p| = p \neq 0,$$

forcing $M = N_0$. Then by $P_N \geq Q = P_{N_0}$, P_N commutes with P_M. Since by Proposition 2.5.1(e) each von Neumann algebra is the norm-closure of the linear span of its projections, P_M commutes with any member in $V^*(B)$, i.e. $P_M \in V^*(B)'$. Therefore, P_M is in the center of $V^*(B)$ because $P_M \in V^*(B)$. The proof is complete.

$\square$

Also note that Corollary 4.4.2 is an easy application of Corollary 2.5.7, combined with Theorem 4.4.1.

It is well known that there is a unique distinguished minimal reducing subspace M_0 satisfying $\dim M_0 \ominus BM_0 = 1$, see Theorem 4.1.5. Precisely, M_0 is the Bergman subspace spanned by $\{B'B^m : m \in \mathbb{Z}_+\}$ [SZ2]. Theorem 4.4.1 implies that any minimal reducing subspace different from M_0 must be orthogonal to M_0, also see [GSZZ, Theorem 27].

Before proving Theorem 4.4.1, let us state a result from Theorem 3 in [GSZZ], which is a special case of Corollary 2.5.6 and Proposition 2.6.4.

Proposition 4.4.3 *Let M and N be two minimal reducing subspaces of M_B. If M and N are not orthogonal, then M is unitarily equivalent to N.*

Furthermore, let us make an observation. It is easy to see that the reducing subspaces of M_B are the same as those of $M_{\varphi_\lambda(B)}(\lambda \in \mathbb{D})$, where

$$\varphi_\lambda(z) = \frac{\lambda - z}{1 - \bar{\lambda}z}, \quad z \in \mathbb{D}.$$

If λ is not in the following finite set

$$\{z : \text{ there exists } z' \text{ such that } B'(z') = 0 \text{ and } B(z') = B(z)\},$$

then $\varphi_\lambda(B)$ has only simple zeros. Thus, *when studying the reducing subspaces of M_B, we can reduce to the case that B has only simple zeros.* As mentioned in Sect. 4.3 of this chapter, the set Δ is fixed to be a disk containing 0. Recall that there are n branches of B^{-1}; $\beta_1, \cdots, \beta_n$ which are (single-valued) holomorphic on some neighborhood of $\overline{\Delta}$. For each reducing subspace N of M_B, put

$$\mathcal{L}_{N,\Delta} = span\left\{ \begin{pmatrix} h(\beta_1(w))\beta_1'(w) \\ \vdots \\ h(\beta_n(w))\beta_n'(w) \end{pmatrix} : h \in N, w \in \Delta \right\} \subseteq \mathbb{C}^n.$$

Now let us make an observation. Given two orthogonal reducing subspaces M and M' of M_B, take any f in M and g in M'. In the proof of Proposition 4.3.1, replace K_λ

with f and K_μ with g, and then we get

$$\langle P(B)f, Q(B)g \rangle = 0.$$

As done in the proof of Proposition 4.3.1, we have on the Bergman space $L_a^2(\Delta)$

$$\sum_{k=1}^{n} \left(f(\beta_k)\beta_k'\right) \otimes \left(g(\beta_k)\beta_k'\right) = 0, f \in M, g \in M'.$$

Let G_z denote the reproducing kernel at z in $L^2(\Delta)$, and thus

$$\langle \sum_{k=1}^{n} [(f(\beta_k)\beta_k') \otimes (g(\beta_k)\beta_k')]G_z, G_z \rangle = 0, z \in \Delta, f \in M, g \in M'.$$

That is,

$$\sum_{k=1}^{n} f(\beta_k(z))\beta_k'(z)\overline{g(\beta_k(z))\beta_k'(z)} = 0, z \in \Delta, f \in M, g \in M',$$

from which we see $\mathcal{L}_{M,\Delta} \perp \mathcal{L}_{M',\Delta}$. Thus, *if M and M' are two orthogonal reducing subspaces, then $\mathcal{L}_{M,\Delta} \perp \mathcal{L}_{M',\Delta}$.* Furthermore, we have the following lemma.

Lemma 4.4.4 *For each reducing subspace M of M_B, we have*

$$\dim \mathcal{L}_{M \ominus BM, \Delta} = \dim \mathcal{L}_{M,\Delta} = \dim M \ominus BM.$$

Proof Rewrite M_1 for M and put $M_2 = M^\perp$. From the above observation $\mathcal{L}_{M_1,\Delta} \perp \mathcal{L}_{M_2,\Delta}$. Therefore $\dim \mathcal{L}_{M_1,\Delta} + \dim \mathcal{L}_{M_2,\Delta} \leq n$, and hence

$$\dim \mathcal{L}_{M_1 \ominus BM_1, \Delta} + \dim \mathcal{L}_{M_2 \ominus BM_2, \Delta} \leq n.$$

Also noting that $\dim M_1 \ominus BM_1 + \dim M_2 \ominus BM_2 = n$, it suffices to show that

$$\dim \mathcal{L}_{M_i \ominus BM_i, \Delta} \geq \dim M_i \ominus BM_i, \quad i = 1, 2.$$

To see this, consider M_1 and set $r = \dim M_1 \ominus BM_1$. Pick r linearly independent functions $h_1, \cdots, h_r$ in $M_1 \ominus BM_1$, and we will show that there is a $w \in \Delta$ such that the matrix

$$\begin{pmatrix} h_1(\beta_1(w))\beta_1'(w) & \cdots & h_r(\beta_1(w))\beta_1'(w) \\ \vdots & \ddots & \vdots \\ h_1(\beta_n(w))\beta_n'(w) & \cdots & h_r(\beta_n(w))\beta_n'(w) \end{pmatrix}$$

has rank r. This is equivalent to prove that

$$\begin{pmatrix} h_1(\beta_1(w)) & \cdots & h_r(\beta_1(w)) \\ \vdots & \ddots & \vdots \\ h_1(\beta_n(w)) & \cdots & h_r(\beta_n(w)) \end{pmatrix}$$

has rank r. For simplicity, we denote the above matrix by $H(w)$.

In fact, the matrix $H(w)$ has rank r when $w = 0$. The reasoning is as follows. Assume conversely that $H(0)$ has rank less than r. Then the columns of $H(0)$ span a subspace in $\mathbb{C}^n$ with dimension less than r, and hence there is a nonzero vector $\mathbf{c} = (c_1, \cdots, c_r)$ in $\mathbb{C}^r$ satisfying

$$\sum_{i=1}^{r} c_i h_i(\beta_j(0)) = 0 \ (1 \leq j \leq n).$$

That is,

$$\langle \sum_{i=1}^{r} c_i h_i, K_{\beta_j(0)} \rangle = 0 \ (1 \leq j \leq n). \tag{4.14}$$

By our assumption, B has only simple zeros $\{\beta_j(0)\}_{j=1}^{n}$, and thus the set $\{K_{\beta_j(0)} : 1 \leq j \leq n\}$ spans $L_a^2(\mathbb{D}) \ominus BL_a^2(\mathbb{D})$. Then (4.14) gives that

$$\sum_{i=1}^{r} c_i h_i \in BL_a^2(\mathbb{D}).$$

On the other hand,

$$\sum_{i=1}^{r} c_i h_i \in M_1 \ominus BM_1 \subseteq L_a^2(\mathbb{D}) \ominus BL_a^2(\mathbb{D}),$$

and hence $\sum_{i=1}^{r} c_i h_i = 0$, which is a contradiction to the linear independence of $h_1, \cdots, h_r$. Therefore,

$$\dim \mathcal{L}_{M_1 \ominus BM_1, \Delta} \geq \dim M_1 \ominus BM_1.$$

Similarly, we have $\dim \mathcal{L}_{M_2 \ominus BM_2, \Delta} \geq \dim M_2 \ominus BM_2$ as desired. The proof is complete.   □

Lemma 4.4.5 *Let B be a finite Blaschke product and M is a reducing subspace of M_B. Then for any Möbius map ϕ,*

$$\dim M \ominus BM = \dim M \ominus \phi(B)M.$$

Proof Fix $\lambda \in \mathbb{D}$, and consider the Blaschke product $\varphi_\lambda(B)$. Since $M_{\varphi_\lambda(B)}$ is a Fredholm operator on the Bergman space and M is a reducing subspace of $M_{\varphi_\lambda(B)}$, it is easy to verify that

$$M_{\varphi_\lambda(B)}|_M : M \to M$$

is also a Fredholm operator. Moreover, since $\lambda \to M_{\varphi_\lambda(B)}|_M$ is a continuous map from $\mathbb{D}$ to bounded operators on M, *Index* $M_{\varphi_\lambda(B)}|_M$ is a continuous integer-valued function in λ. From

$$Index\ M_{\varphi_\lambda(B)}|_M = -\dim M \ominus \varphi_\lambda(B)M,$$

and $\varphi_0(z) = -z$, we get

$$\dim M \ominus BM = \dim M \ominus \varphi_0(B)M = \dim M \ominus \varphi_\lambda(B)M, \ \lambda \in \mathbb{D}.$$

This immediately leads to our conclusion. $\square$

Now we are prepared to prove Theorem 4.4.1.

Proof of Theorem 4.4.1

(1) As mentioned above, for a finite Blaschke product B, there is always a Möbius map ϕ such that $\phi(B)$ has only simple zeros. Notice also that M_B and $M_{\phi(B)}$ has the same reducing subspaces. Then by Lemma 4.4.5, it suffices to deal with the case of B being a finite Blaschke product with only simple zeros.

Now suppose M and N are two distinct minimal reducing subspaces of M_B, and

$$\dim M \ominus BM = \dim N \ominus BN = 1.$$

Assume conversely that M is unitarily equivalent to N; that is, there exists a unitary operator $U : M \to N$ such that $UM_B = M_B U$. Then by Proposition 4.3.1 and Remark 4.3.3, there exists an open disk Δ and an $n \times n$ numerical unitary matrix W such that

$$W \begin{pmatrix} f(\beta_1(w))\beta_1'(w) \\ \vdots \\ f(\beta_n(w))\beta_n'(w) \end{pmatrix} = \begin{pmatrix} Uf(\beta_1(w))\beta_1'(w) \\ \vdots \\ Uf(\beta_n(w))\beta_n'(w) \end{pmatrix}, \ w \in \Delta, \quad (4.15)$$

where $f \in M$.

Now take $f = f_0$, a nonzero function in M and put $g_0 = Uf_0 \in N$. Since by Lemma 4.4.4

$$\dim \mathcal{L}_{M,\Delta} = \dim \mathcal{L}_{N,\Delta} = 1,$$

there exist two nonzero vectors $\mathbf{c}$ and $\mathbf{d}$ in $\mathbb{C}^n$ such that $\mathbf{c} \in \mathcal{L}_{M,\Delta}$ and $\mathbf{d} \in \mathcal{L}_{N,\Delta}$. In addition, $\mathbf{c}$ and $\mathbf{d}$ can be required to satisfy the following:

$$\begin{pmatrix} f_0(\beta_1(w))\beta_1'(w) \\ \vdots \\ f_0(\beta_n(w))\beta_n'(w) \end{pmatrix} = f_0(\beta_1(w))\beta_1'(w)\mathbf{c}, \ w \in \Delta,$$

and

$$\begin{pmatrix} g_0(\beta_1(w))\beta_1'(w) \\ \vdots \\ g_0(\beta_n(w))\beta_n'(w) \end{pmatrix} = g_0(\beta_1(w))\beta_1'(w)\mathbf{d}, \ w \in \Delta.$$

Then by (4.15), on Δ we have

$$f_0(\beta_1)\beta_1'W\mathbf{c} = g_0(\beta_1)\beta_1'\mathbf{d},$$

and hence $f_0 = cg_0$ for some constant $c \neq 0$. Thus the reducing subspace $[f_0]$ generated by f_0 equals $[g_0]$. Since M is minimal, $M = [f_0]$ and similarly $N = [g_0]$, and hence $M = N$, which is a contradiction. So M is not unitarily equivalent to N. Applying Proposition 4.4.3 yields $M \perp N$, as desired.

(2) If $\dim M \ominus BM \neq \dim N \ominus BN$, then obviously M is not unitarily equivalent to N. Therefore by Proposition 4.4.3, $M \perp N$. The proof of Theorem 4.4.1 is complete.   □

In fact, by similar methods Guo and Huang proved the following proposition. Because of its length, the proof is placed in the next section.

Proposition 4.4.6 *Let B be an interpolating Blaschke product. Then all minimal reducing subspaces M (if exist) satisfying $\dim M \ominus BM = 1$ are pairwise orthogonal, and P_M is in the center of $\mathcal{V}^*(B)$.*

Inspired by Corollary 4.4.2 and Proposition 4.4.6, we make the following

Conjecture 4.4.7 *Suppose $\phi \in H^\infty$ and $Z(\phi) \cap \mathbb{D} \neq \emptyset$. If M is a reducing subspace of M_ϕ satisfying $\dim M \ominus \phi M < \infty$, then P_M lies in the center of $\mathcal{V}^*(\phi)$.*

Here, ϕM is not necessarily closed. Note that if Conjecture 4.4.7 holds, then under the same assumption the von Neumann algebra $P_M \mathcal{V}^*(\phi)|_M$ is abelian. This follows from an easy observation: any orthogonal projection in $P_M \mathcal{V}^*(\phi)|_M$ has the form P_N, where N is a reducing subspace satisfying $N \subseteq M$, and thus

$$\dim N \ominus \phi N \leq \dim M \ominus \phi M < \infty.$$

By Conjecture 4.4.7 all $P_N \in Z(\mathcal{V}^*(\phi))$. Then $P_M \mathcal{V}^*(\phi)|_M$ is a $*$-subalgebra of $Z(\mathcal{V}^*(\phi))$, and hence it is abelian.

Special interest is focused on the case when M equals the whole space $L^2_a(\mathbb{D})$. Theorem 4.2.1 can be regarded as a good example for Conjecture 4.4.7. We also give the following.

Conjecture 4.4.8 *If* $\dim V^*(\phi) < \infty$, *then* $V^*(\phi)$ *is abelian.*

If Conjecture 4.4.8 holds, then a follow-up problem on $L^2_a(\mathbb{D})$ is posed as follows: *If* $V^*(\phi)$ *is a type I von Neumann algebra, then is* $V^*(\phi)$ *abelian?*

Conjectures 4.4.7 and 4.4.8 fail in multivariable case, see Example 8.4.3. Also, they fail on the Hardy space $H^2(\mathbb{D})$. On the Dirichlet space, little is known, even in the case of ϕ being a finite Blaschke product. Precisely, there are several distinct norms on Dirichlet spaces $\mathcal{D}$: one is adapted with the norm

$$\|f\| \triangleq \left(\|f\|^2_{H^2(\mathbb{D})} + \int_{\mathbb{D}} |f'(z)|^2 dA(z)\right)^{\frac{1}{2}}, \, f \in \mathcal{D}.$$

and the other is with the norm

$$\|f\|_0 = \left(|f(0)|^2 + \int_{\mathbb{D}} |f'(z)|^2 dA(z)\right)^{\frac{1}{2}}, \, f \in \mathcal{D}.$$

Zhao [Zhao] showed that for a Blaschke product B of order 2, the von Neumann algebra $V^*(B)$ on $(\mathcal{D}, \|\cdot\|_0)$ is nontrivial if and only if $B = \varphi(z^2)$ for some $\varphi \in \mathrm{Aut}(\mathbb{D})$. In this case, $V^*(B)$ is abelian and M_B has exactly two minimal reducing subspaces. Later, Chen et al. [CLY] obtained a similar result on $(\mathcal{D}, \|\cdot\|)$. Also, they gave a characterization for the reducibility of M_B on $(\mathcal{D}, \|\cdot\|)$. However, when order $B \geq 3$, it remains open when M_B is reducible on either Dirichlet space. This is equivalent to ask when $V^*(B)$ is nontrivial.

4.5 Proof of Proposition 4.4.6

This section mainly provide the proof of Proposition 4.4.6, which is restated as follows.

Proposition 4.5.1 *Let* B *be an interpolating Blaschke product. Then all minimal reducing subspaces* M *(if exist) satisfying* $\dim M \ominus BM = 1$ *are pairwise orthogonal, and in this case* P_M *is in the center of* $V^*(B)$.

Two lemmas are needed. One is the following, also see [Ga, p. 395, Lemma 1.4]. A different proof will be provided at the end of this section.

Lemma 4.5.2 *[Ga] Let* B *be an interpolating Blaschke product with the zero sequence* $\{z_k\}$. *Then there is an open disk* Δ *containing* 0 *such that there is a* *sequence of functions $\{\beta_k\}$ holomorphic on a neighborhood of $\overline{\Delta}$, satisfying*

$$\beta_k(0) = z_k \text{ and } B \circ \beta_k(z) = z, \ z \in \Delta, \ k \in \mathbb{Z}_+.$$

Moreover, $\{\beta_k(\overline{\Delta})\}$ are pairwise disjoint.

Now fix an open disk Δ as in Lemma 4.5.2. For each $f \in L_a^2(\mathbb{D})$,

$$\sum_{k=1}^{\infty} \int_{\Delta} |f(\beta_k(z))\beta_k'(z)|^2 dA(z) = \int_{\bigsqcup_{k=1}^{\infty} \beta_k(\Delta)} |f(z)|^2 dA(z) \le \int_{\mathbb{D}} |f(z)|^2 dA(z).$$

Thus, the sequence $\{f(\beta_k)\beta_k'\}_k$ in $L_a^2(\Delta)$ satisfies

$$\sum_{k=1}^{\infty} \|f(\beta_k)\beta_k'\|^2 < \infty,$$

and then

$$\{\langle f(\beta_k)\beta_k', G_w\rangle\}_k \in l^2, \ w \in \Delta,$$

where G_w denotes the reproducing kernel at w in $L_a^2(\Delta)$. That is, the sequence $\{f(\beta_k(w))\beta_k'(w)\}_k$ is in l^2 for any $w \in \Delta$. Then for each $w \in \Delta$, define

$$\widetilde{\pi_w} : L_a^2(\mathbb{D}) \to l^2$$

$$f \mapsto \{f(\beta_k(w))\beta_k'(w)\}_k,$$

which is a bounded operator. For each closed subspace M of $L_a^2(\mathbb{D})$, set

$$\mathcal{L}_{M,E} = \overline{\operatorname{span}}\{\widetilde{\pi_w}f \mid f \in M, w \in E\},$$

where E is a subset of Δ. Following Sects. 4.3 and 4.4 of Chap. 4, one can show that for any two orthogonal reducing subspaces M and N of M_B, $\mathcal{L}_{M,\Delta} \perp \mathcal{L}_{N,\Delta}$. Furthermore, we have the following.

Lemma 4.5.3 *If M is a reducing subspace of M_B, then we have*

$$\dim \mathcal{L}_{M,\Delta} = \dim \mathcal{L}_{M \ominus BM,0} = \dim M \ominus BM.$$

Proof First, we show that

$$\mathcal{L}_{L_a^2(\mathbb{D}),0} = \mathcal{L}_{L_a^2(\mathbb{D}) \ominus BL_a^2(\mathbb{D}),0} = l^2. \tag{4.16}$$

The reasoning is as follows. Since the sequence $\{z_k\}$ is uniformly separated, then by Theorem 2.4.3 M_B is bounded below, and hence $BL_a^2(\mathbb{D})$ is closed. Therefore,

$$L_a^2(\mathbb{D}) = BL_a^2(\mathbb{D}) \oplus (L_a^2(\mathbb{D}) \ominus BL_a^2(\mathbb{D})).$$

Note that $\mathcal{L}_{BL_a^2(\mathbb{D}),0} = 0$, and then $\mathcal{L}_{L_a^2(\mathbb{D}),0} = \mathcal{L}_{L_a^2(\mathbb{D})\ominus BL_a^2(\mathbb{D}),0}$. Thus it is enough to show that $\mathcal{L}_{L_a^2(\mathbb{D}),0} = l^2$. Since $\{z_k\}$ is an interpolating sequence, there is a constant $\delta > 0$ such that

$$\prod_{j,j\neq k} d(z_j, z_k) \geq \delta, \ k = 1, 2, \cdots,$$

where

$$d(z_j, z_k) = \frac{z_j - z_k}{1 - \overline{z}_j z_k}.$$

Write $B(z) = \varphi_{z_k}(z) B_k(z)$ for $1 \leq k < \infty$. Since $|B_k(z_k)| \geq \delta$, and

$$\varphi_\lambda'(\lambda) = -\frac{1}{1 - |\lambda|^2},$$

it follows that

$$\delta \frac{1}{1 - |z_k|^2} \leq |B'(z_k)| \leq \frac{1}{1 - |z_k|^2}, \ k = 1, 2, \cdots \tag{4.17}$$

On the other hand, since $\beta_k(0) = z_k$ and

$$B \circ \beta_k(z) = z, \ z \in \Delta, \ k \in \mathbb{Z}_+,$$

then

$$\beta_k'(0) = \frac{1}{B'(z_k)}. \tag{4.18}$$

As mentioned above, for each $f \in L_a^2(\mathbb{D})$ and $w \in \Delta$, the sequence

$$\widetilde{\pi}_w(f) = \{f(\beta_k(w))\beta_k'(w)\}_k$$

is in l^2. In particular, $\{f(z_k)\beta_k'(0)\}_k$ is in l^2 for any $f \in L_a^2(\mathbb{D})$. Thus, by (4.17) we get two linear maps

$$\widetilde{\pi}_0 : L_a^2(\mathbb{D}) \to l^2$$

$$f \mapsto \{f(z_k)\beta_k'(0)\}_k,$$

and

$$\widetilde{\pi}_1 : L_a^2(\mathbb{D}) \to l^2$$

$$f \mapsto \{f(z_k)(1 - |z_k|^2)\}_k.$$

Since $\{z_k\}$ is uniformly separated, then by Theorem 2.4.1 the sequence $\{z_k\}$ is an interpolating sequence for $L_a^2(\mathbb{D})$, which implies that $\widetilde{\pi}_1$ is onto. From (4.17) and (4.18), it is easy to see that $\widetilde{\pi}_0$ is also onto. That is, $\mathcal{L}_{L_a^2(\mathbb{D}),0} = l^2$, as desired. The proof of (4.16) is complete.

Next we claim that *for each reducing subspace M of M_B, we have*

$$\dim \mathcal{L}_{M \ominus BM,0} = \dim M \ominus BM.$$

If this claim is proved, the remaining part of Lemma 4.5.3 can be given as follows. Put $N = M^{\perp}$, and by the observation preceding Lemma 4.5.3,

$$\mathcal{L}_{M,\Delta} \perp \mathcal{L}_{N,\Delta}. \tag{4.19}$$

But by (4.16), $\mathcal{L}_{M \ominus BM,0} \oplus \mathcal{L}_{N \ominus BN,0} = \mathcal{L}_{L_a^2(\mathbb{D}),0} = l^2$. Combining this identity with (4.19) implies that

$$\mathcal{L}_{M \ominus BM,0} = \mathcal{L}_{M,\Delta} \text{ and } \mathcal{L}_{N,\Delta} = \mathcal{L}_{N \ominus BN,0}.$$

Then by the claim we get $\dim \mathcal{L}_{M,\Delta} = \dim M \ominus BM$, as desired.

For the completeness of the proof, we must prove the claim. Clearly,

$$\dim \mathcal{L}_{M \ominus BM,0} \le \dim M \ominus BM.$$

Then it suffices to show that $\dim \mathcal{L}_{M \ominus BM,0} \ge \dim M \ominus BM$. For this, write $r = \dim M \ominus BM$ and first assume that $r < \infty$. Pick r linearly independent functions in $M \ominus BM$: $f_1, \cdots, f_r$, and we will show that these r vectors in l^2

$$(f_i(\beta_1(0))\beta_1'(0), f_i(\beta_2(0))\beta_2'(0), \cdots) \ (1 \le i \le r)$$

span a subspace $\mathcal{L}$ of dimension r. Assume conversely that the vectors

$$(f_i(\beta_1(0))\beta_1'(0), f_i(\beta_2(0))\beta_2'(0), \cdots) \ (1 \le i \le r)$$

span a subspace of dimension less than r. Then there is a nonzero vector $\mathbf{c} = (c_1, \cdots, c_r)$ in $\mathbb{C}^r$ satisfying

$$\sum_{i=1}^{r} c_i f_i(\beta_k(0))\beta_k'(0) = 0, \ k = 1, 2, \cdots$$

Since $\beta_k'(0) \ne 0$ and $\beta_k(0) = z_k$, the function $\sum_{i=1}^{r} c_i f_i$ vanishes at $z_1, z_2, \cdots$. Now denote by $\mathcal{N}$ the subspace of all functions in $L_a^2(\mathbb{D})$ that vanish on the zero set of B. By Theorem 2.4.3(1) and (4),

$$\mathcal{N} = BL_a^2(\mathbb{D}),$$

and hence $\sum_{i=1}^{r} c_i f_i$ is in $BL_a^2(\mathbb{D})$. Since M is a reducing subspace,

$$\sum_{i=1}^{r} c_i f_i \in M \ominus BM \subseteq L_a^2(\mathbb{D}) \ominus BL_a^2(\mathbb{D}).$$

Therefore $\sum_{i=1}^{r} c_i f_i = 0$, which is a contradiction.

For the case when $r = \infty$, consider m linearly independent functions in $M \ominus BM$. By a similar discussion as above, we have $\dim \mathcal{L}_{M \ominus BM,0} \geq m$. By the arbitrariness of m,

$$\dim \mathcal{L}_{M \ominus BM,0} = \dim M \ominus BM = \infty,$$

completing the proof of the claim.

Thus, the proof of Lemma 4.5.3 is complete.   □

With Lemmas 4.5.2 and 4.5.3, applying the proofs of Theorem 4.4.1 and Corollary 4.4.2 gives Proposition 4.5.1.

To end this section, we will provide an alternative proof for Lemma 4.5.2. For this, two lemmas will be established.

Lemma 4.5.4 *Suppose h_n are holomorphic functions over $\mathbb{D}$ for $n \geq 1$ and there is a constant $C > 1$ satisfying $\frac{1}{C} \leq |h_n| \leq C$, then there is an open disk Δ centered at 0 such that all $zh_n(z)|_\Delta$ are univalent.*

Proof To prove this lemma, assume conversely that there is a subsequence $\{n_k\}$ such that none of $zh_{n_k}(z)|_{\frac{1}{k}\mathbb{D}}$ is univalent.

Since h_n are uniformly bounded, by the normal family method, there is a subsequence of $\{h_n\}$, say $\{h_{n_k}\}$, converging to some holomorphic function h uniformly on compact sets in $\mathbb{D}$. Put $g(z) = zh(z)$, and we have $g'(0) \neq 0$ since $h(0) \neq 0$. Before continuing, we prove the following.

Claim There are two constants ε_0 and ε_1 in $(0, 1)$ such that for any $w \in \varepsilon_1 \mathbb{D}$, $zh_{n_k}(z)|_{\varepsilon_0 \mathbb{D}}$ takes the value w at most once for $k \geq K$, where K is a positive integer not depending on w.

Since $g'(0) \neq 0$, there is a constant $\varepsilon_0 \in (0, 1)$ such that g is univalent on $\varepsilon_0 \mathbb{D}$. Let $\varepsilon_1 = \frac{\varepsilon_0}{2C}$ and set

$$c_0 = \min \{ |zh(z) - w| : z \in \varepsilon_0 \mathbb{T} \text{ and } w \in \varepsilon_1 \overline{\mathbb{D}} \}.$$

Since $\frac{1}{C} \leq |h| \leq C$, $c_0 > 0$.

Since $\{h_{n_k}\}$ converges to h uniformly on compact sets in $\mathbb{D}$, there is an enough large integer K such that

$$\|h - h_{n_k}\|_{\varepsilon_0 \overline{\mathbb{D}}, \infty} < \frac{c_0}{\varepsilon_0}, \ k \geq K.$$

Thus for each $w_0 \in \varepsilon_1 \mathbb{D}$,

$$|(zh(z) - w_0) - (zh_{n_k}(z) - w_0)| = \varepsilon_0 |h(z) - h_{n_k}(z)|$$

$$< c_0 \leq |zh(z) - w_0|, \ z \in \varepsilon_0 \mathbb{T}.$$

That is,

$$|(zh(z) - w_0) - (zh_{n_k}(z) - w_0)| < |zh(z) - w_0|, \ z \in \varepsilon_0 \mathbb{T}.$$

Therefore, by Rouché's theorem the numbers of zeros of $zh(z) - w_0$ and $zh_{n_k}(z) - w_0$ in $\varepsilon_0 \mathbb{D}$ are equal for $k \geq K$ and $w_0 \in \varepsilon_1 \mathbb{D}$. Notice also that $zh(z) - w_0$ is injective in $\varepsilon_0 \mathbb{D}$, and thus for any $w_0 \in \varepsilon_1 \mathbb{D}$, $zh_{n_k}(z)|_{\varepsilon_0 \mathbb{D}}$ takes the value w_0 at most once for $k \geq K$. The proof of the claim is complete.

Now take $\eta_0 = \dfrac{\varepsilon_1}{C}$, and by $\dfrac{1}{C} \leq |h_n| \leq C$, it follows that the images of all $zh_n(z)|_{\eta_0 \mathbb{D}}$ are contained in $\varepsilon_1 \mathbb{D}$. By the claim, $zh_{n_k}(z)|_{\varepsilon_0 \mathbb{D}}$ takes the values in $\varepsilon_1 \mathbb{D}$ at most once for $k \geq K$. Clearly this is also true for $zh_{n_k}(z)|_{\eta_0 \mathbb{D}}$. Therefore, $zh_{n_k}(z)|_{\eta_0 \mathbb{D}}$ are univalent for $k \geq K$. In particular, take a $k_0 \geq K$ such that $\dfrac{1}{k_0} \leq \eta_0$ and then $zh_{n_{k_0}}(z)|_{\frac{1}{k_0} \mathbb{D}}$ is univalent, which is a contradiction to our assumption. The proof is complete. $\square$

Remark 4.5.5 In fact, Lemma 4.5.4 is a direct consequence of a theorem in [Ne, p. 171, Exercise 5], which states that: if f is a holomorphic function from $\mathbb{D}$ to $\mathbb{D}$ such that

$$f(0) = 0, |f'(0)| = a,$$

then f is univalent in $t\mathbb{D}$ where $t = \dfrac{a}{1 + \sqrt{1 - a^2}}$.

In this book, denote by $\Delta(z, r)$ the *pseudohyperbolic disk* centered at z with radius r, i.e.

$$\Delta(z, r) = \{w \in \mathbb{D} : |\frac{z - w}{1 - \bar{z}w}| < r\}.$$

When $z = 0$, $\Delta(0, r)$ is nothing but the disk $r\mathbb{D}$. The following is covered by [Ga, p. 395, Lemma 1.4], but we include an alternative proof.

Lemma 4.5.6 *Let B be an interpolating Blaschke product with the zero sequence $\{z_k\}$. Then there is an $\varepsilon > 0$ such that all $\Delta(z_k, \varepsilon)$ are pairwise disjoint and $B|_{\Delta(z_k, \varepsilon)}$ are univalent.*

Proof By assumption, $\{z_k\}$ is an interpolating sequence for $H^\infty(\mathbb{D})$; and hence $\{z_k\}$ is uniformly separated. That is, there is a constant $\delta > 0$ satisfying

$$\prod_{j \geq 1, j \neq k} \rho(z_j, z_k) \geq \delta, \ k = 1, 2, \cdots. \tag{4.20}$$

Note that the pseudohyperbolic metric ρ is a metric and $\rho(z_j, z_k) \geq \delta$ for all $j \neq k$. Then it follows that if $\varepsilon' < \frac{\delta}{2}$, the pseudohyperbolic disks $\Delta(z_k, \varepsilon')$ are pairwise disjoint.

For each $n \in \mathbb{Z}_+$, there is a Blaschke product B_n such that $B = \varphi_{z_n} B_n$. Put $g_n(z) = B \circ \varphi_{z_n}(z)$ and it is easy to see that there are holomorphic functions h_n bounded by one such that $g_n(z) = z h_n(z)$. By the derivation formula of holomorphic functions,

$$h'_n(z) = \frac{1}{2\pi i} \int_{|\zeta|=\frac{3}{4}} \frac{h_n(\zeta)}{(\zeta - z)^2} d\zeta, \quad |z| \leq \frac{1}{2}.$$

Since h_n are uniformly bounded on $\mathbb{D}$, there is an $M_0 > 0$ such that $h'_n|_{\frac{1}{2}\mathbb{D}}$ are bounded by M_0. Let $[0, w]$ denote the segment joining 0 with w, and we have

$$h_n(w) - h_n(0) = \int_{[0,w]} h'_n(z) dz, \quad w \in \mathbb{D}.$$

Thus, there is an $r(0 < r < \frac{1}{2})$ satisfying

$$|h_n(w) - h_n(0)| < \frac{\delta}{2}, \quad n \in \mathbb{Z}_+ \text{ and } w \in r\mathbb{D}.$$

Then by $|h_n(0)| \geq \delta$,

$$\frac{\delta}{2} \leq |h_n(z)| \leq 1, \, z \in r\mathbb{D}, \, n = 1, 2, \cdots \tag{4.21}$$

Then applying Lemma 4.5.4 shows that there is an $\varepsilon(0 < \varepsilon < \frac{\delta}{2})$ such that all $g_n(z)|_{\varepsilon\mathbb{D}}$ are univalent. Since $g_n(z) = B \circ \varphi_{z_n}(z)$, $B|_{\Delta(z_k,\varepsilon)}$ are univalent. In addition, $\Delta(z_k, \varepsilon)$ are pairwise disjoint because $\varepsilon < \frac{\delta}{2}$. The proof is complete. $\qquad\Box$

Below, we present the proof of Lemma 4.5.2.

Proof of Lemma 4.5.2 Take an ε as in Lemma 4.5.6, and $B|_{\Delta(z_n,\varepsilon)}$ are univalent. Thus, $B|_{\overline{\Delta(z_n,\frac{\varepsilon}{2})}}$ is injective. Let h_n be the function as in the proof of Lemma 4.5.6, and the image of $B|_{\Delta(z_n,\frac{\varepsilon}{2})}$ is the same as that of $z h_n(z)|_{\Delta(0,\frac{\varepsilon}{2})}$.

Let Θ_n denote the image of $z h_n(z)|_{\Delta(0,\frac{\varepsilon}{2})}$ for each $n \geq 1$. Since $z h_n(z)|_{\Delta(0,\varepsilon)}$ is univalent, the boundary $\partial\Theta_n$ of Θ_n is

$$\{z h_n(z) : z \in \partial\Delta(0, \frac{\varepsilon}{2})\},$$

and by (4.21) $\partial\Theta_n$ is contained in $\{z : \frac{\varepsilon\delta}{4} \leq |z| \leq 1\}$. Therefore it is easy to see that

$$\frac{\varepsilon\delta}{4}\mathbb{D} \subseteq \Theta_n, \, n = 1, 2, \cdots.$$

Write $\Delta = \frac{\varepsilon\delta}{4}\mathbb{D}$. Then by the above arguments, there is a domain $\Omega_n \subseteq \Delta(z_n, \frac{\varepsilon}{2})$ such that

$$B|_{\Omega_n} : \Omega_n \to \Delta$$

is a biholomorphic map. Note that by Lemma 4.5.6, $\overline{\Delta(z_n, \frac{\varepsilon}{2})}$ are pairwise disjoint. Then let β_n be the inverse of $B|_{\Omega_n}$, and the proof is complete. $\qquad\square$

4.6   Abelian $\mathcal{V}^*(B)$ for Order $B = 5, 6$

For the cases of order $B = 5, 6$, Guo and Huang proved that $\mathcal{V}^*(B)$ is abelian, see [GH1]. In this section we will present the original proof given by Guo and Huang.

Theorem 4.6.1 *Let B be a Blaschke product of order n with $n = 5$ or 6, then M_B has at most n minimal reducing subspaces. In this case, $\mathcal{V}^*(B)$ is abelian.*

Proposition 2.6.8 will be needed in the proof of Theorem 4.6.1 later. For convenience, we restate it here with an alternative proof.

Proposition 4.6.2 *Let B be a finite Blaschke product of order n. Then the following are equivalent.*

(1) M_B has at most n minimal reducing subspaces;
(2) $\mathcal{V}^*(B)$ is abelian;
(3) All minimal projections in $\mathcal{V}^*(B)$ are orthogonal.

Proof (3)$\Rightarrow$ (2). Lemma 4.2.2 shows that $\mathcal{V}^*(B)$ finite dimensional von Neumann algebra. Recall that a finite dimensional von Neumann algebra is spanned by its minimal projections. Then (3)$\Rightarrow$ (2) follows directly.

(2)$\Rightarrow$ (3) follows from a simple fact: if $\mathcal{A}$ is an abelian von Neumann algebra and $\mathcal{A}$ has a minimal projection, then all minimal projections in $\mathcal{A}$ are orthogonal.

(3) $\Rightarrow$ (1). Recall that every minimal reducing subspace is exactly the range of a minimal projection in $\mathcal{V}^*(B)$. By (3), all minimal reducing subspaces are orthogonal. Thus there are only finitely many minimal reducing subspaces, denoted by $M_i(1 \leq i \leq t)$, whose direct sum is the Bergman space. Note that

$$L_a^2(\mathbb{D}) \ominus BL_a^2(\mathbb{D}) = \bigoplus_{i=1}^{t} M_i \ominus BM_i,$$

and hence $t \leq n$, where n is the order of B, as desired. One can also refer to [GSZZ] for (3) $\Rightarrow$ (1).

(1) $\Rightarrow$ (3) follows from [GSZZ, Theorem 31]. Here, we include a similar but shorter proof.

Suppose conversely that (3) does not hold, and then there exist two distinct minimal reducing subspaces M and N, which are not orthogonal. By Proposition 4.4.3, M is unitarily equivalent to N; that is, there is a unitary operator U from M onto N commuting with M_B. Now for each $0 < a < 1$, set

$$M_a = \{f + aUf : f \in M\}.$$

It is easy to see that each M_a is closed and a reducing subspace of M_B, and is minimal since M is minimal. Moreover, it is not difficult to verify that if $0 < a < a' < 1$, then $M_a \neq M_{a'}$ because $M \cap N = \{0\}$. This is a contradiction to (1).   □

Let P and ρ_k be defined as in Sect. 4.2 of this chapter. As before, P is a polygon in $\overline{\mathbb{D}}$ such that $\mathbb{D} - P$ is simply connected. It is mentioned that $w = \rho_k(z)$ $(1 \leq k \leq n)$ are local inverses of B, holomorphic on $\mathbb{D} - P$, and satisfying

$$B(\rho_k(z)) = B(z), \ z \in \mathbb{D} - P, \ k = 1, \cdots, n.$$

Lemma 4.6.3 *Suppose $B = z^2 \varphi_\alpha \varphi_\beta \varphi_\gamma$ with $\alpha, \beta, \gamma \in \mathbb{D}$, then all ρ_k can be extended analytically to the unit disk if and only if $\alpha = \beta = \gamma = 0$.*

Proof If $\alpha = \beta = \gamma = 0$, then ρ_k has the form $\omega^k z$, where $\omega = e^{i\frac{2\pi}{5}}$.

Now write $B = z^2 \varphi_\alpha \varphi_\beta \varphi_\gamma$ and all ρ_k extended analytically to the unit disk. We will show $\alpha = \beta = \gamma = 0$. Since each ρ_k satisfies

$$B(\rho_k(z)) - B(z) = 0, \tag{4.22}$$

then $|\rho_k(z)| \to 1(|z| \to 1)$ and hence all ρ_k are Blaschke products, of order one. Among ρ_k, one is the identity map. There are two cases to distinguish:

Case I: there is an integer k such that $\rho_k(z) = \overline{c}z$, where $|c| = 1$ and $c \neq 1$. Then by (4.22) we have

$$B(\overline{c}z) = B(z). \tag{4.23}$$

First we show that either $\alpha = \beta = \gamma = 0$ or $\alpha\beta\gamma \neq 0$. If not, then there are essentially two cases to discuss. If $\alpha \neq 0$ and $\beta = \gamma = 0$, then by (4.23), $c\alpha = \alpha$. Thus $\alpha = 0$, which is a contradiction. If $\alpha\beta \neq 0$ and $\gamma = 0$, then by comparing the zeros of both sides of (4.23) we get $c = -1$ and $\beta = -\alpha$. Then $B = z^3 \varphi_{\alpha^2}(z^2)$ and $B(\overline{c}z) \neq B(z)$, which is a contradiction.

Next we will exclude the case $\alpha\beta\gamma \neq 0$ to finish the proof. Again by (4.23), $(c\alpha, c\beta, c\gamma)$ is a permutation of (α, β, γ). Without loss of generality, $\beta = ca$ and then $(c^2\alpha, c\gamma)$ is a permutation of (α, γ). Since $c \neq 1$, $c\gamma \neq \gamma$, so $c^2\alpha = \gamma$ and $c\gamma = \alpha$. Therefore $c^3 = 1$. By a simple computation, we get

$$B(\overline{c}z) = \overline{c}^2 B(z) \neq B(z),$$

which is a contradiction.

Case II: there is an integer k such that $\rho_k(z)$ is a Möbius map different from the identity map and $\rho_k(0) \neq 0$. Write $\rho_k = \phi$. By $B \circ \phi = B$, z^2 is a factor of $B \circ \phi$. Then we can assume that $\alpha = \beta = \phi(0)$. Now $\alpha \neq 0$ is a zero of B with multiplicity ≥ 2, and by $B \circ \phi = B$, α is also a zero of $B \circ \phi$ with multiplicity ≥ 2. This implies $\phi(\alpha) = 0$, and thus there exists a constant $\xi (|\xi| = 1)$ satisfying $\phi = \xi \varphi_\alpha$. From $\alpha = \phi(0)$, we get $\phi = \varphi_\alpha$. By (4.22),

$$z^2 \varphi_\alpha^2 \varphi_\gamma \circ \varphi_\alpha = z^2 \varphi_\alpha^2 \varphi_\gamma.$$

Thus $\varphi_\gamma \circ \varphi_\alpha = \varphi_\gamma$, which is impossible. $\qquad\qquad\square$

Now we are ready for the proof of Theorem 4.6.1.

Proof of Theorem 4.6.1 The proof is from [GH1].

Let B be a finite Blaschke product of order n. By Proposition 4.6.2, it suffices to show that M_B has at most n minimal reducing subspaces. There are two cases under consideration: $n = 5$ and $n = 6$.

First assume that $n = 5$. Recall that for any Blaschke product B of order 5, there always exists an a and c in $\mathbb{D}$ such that $\varphi_a \circ B \circ \varphi_c = z^2 \varphi_\alpha \varphi_\beta \varphi_\gamma$ with $\alpha, \beta, \gamma \in \mathbb{D}$. Without loss of generality, assume that $B = z^2 \varphi_\alpha \varphi_\beta \varphi_\gamma$.

Observe that there is always an orthogonal decomposition of $L_a^2(\mathbb{D})$:

$$L_a^2(\mathbb{D}) = \bigoplus_{i=0}^{t} M_i, \qquad\qquad (4.24)$$

where each $M_i (0 \leq i \leq t)$ is a minimal reducing subspace. From now on, M_0 always denotes the distinguished reducing subspace. Since (4.24) gives

$$L_a^2(\mathbb{D}) \ominus BL_a^2(\mathbb{D}) = \bigoplus_{i=0}^{t} M_i \ominus BM_i, \qquad\qquad (4.25)$$

we get

$$\sum_{i=0}^{t} \dim M_i \ominus BM_i = \text{order } B = 5. \qquad\qquad (4.26)$$

The next discussion is based on (4.25) and (4.26). In fact, noting that $\dim M_0 \ominus BM_0 = 1$, it suffices to consider the following cases:

(i) $t = 4$ and $\dim M_i \ominus BM_i = 1 (1 \leq i \leq 4)$;
(ii) $t = 3$, $\dim M_i \ominus BM_i = 1 (1 \leq i \leq 2)$ and $\dim M_3 \ominus BM_3 = 2$;
(iii) $t = 1$ and $\dim M_1 \ominus BM_1 = 4$;
(iv) $t = 2$, $\dim M_1 \ominus BM_1 = 1$ and $\dim M_2 \ominus BM_2 = 3$;
(v) $t = 2$ and $\dim M_1 \ominus BM_1 = \dim M_2 \ominus BM_2 = 2$.

Case (i)–(iv) can be done by using Theorem 4.4.1. For example, let us deal with case (ii). Suppose conversely there is some other minimal reducing subspace other than $M_i(0 \le i \le 3)$, say N. Since $\dim M_i \ominus BM_i = 1 (0 \le i \le 2)$, by Theorem 4.4.1 N is orthogonal to $M_i(0 \le i \le 2)$. Therefore $N \subseteq M_3$, and hence by minimality, $N = M_3$. This is a contradiction. Case (i), (iii) and (iv) can be done similarly. The difficulty lies in case (v), which will be discussed in details. Below, we will show that in case (v) there is no minimal reducing subspace other than M_0, M_1 and M_2.

To see this, assume conversely that M is a minimal reducing subspace other than M_0, M_1 and M_2. Then M is orthogonal to M_0, and hence $M \subseteq M_1 \oplus M_2$. A simple application of Proposition 4.4.3 shows that M_1 and M_2 are unitarily equivalent. That is, there is a unitary U from M_1 onto M_2 which commutes with M_B. Now extend U to $\tilde{U}$ such that $\tilde{U}|_{M_1} = U$ and $\tilde{U}|_{M_1^\perp} = 0$. Let P_j denote the orthogonal projection from $L_a^2(\mathbb{D})$ onto $M_j (j = 0, 1, 2)$, and it is easy to verify that

$$P_0, P_1, P_2, \tilde{U} \text{ and } \tilde{U}^*$$

are linearly independent. Since for any pair $(c_1, c_2) \in \mathbb{T}^2$,

$$P_0 + \sum_{i=1}^{2} c_i P_i \text{ and } P_0 + c_1 \tilde{U} + c_2 \tilde{U}^*$$

are unitary operators which commute with M_B, and also with M_B^*, then all five operators $P_0, P_1, P_2, \tilde{U}$ and $\tilde{U}^*$ lie in $\mathcal{V}^*(B)$. Now denote them by $A_1, \cdots, A_5$, and by Lemma 4.2.2 and Proposition 2.5.1(b), there is a 5×5 matrix $(c_{i,j})$ such that

$$A_i h(w) = \sum_{j=1}^{n} c_{i,j} \rho_j'(w) h \circ \rho_j(w), \ h \in L_a^2(\mathbb{D}), \ w \in \mathbb{D}, \ 1 \le i \le 5. \qquad (4.27)$$

Since $A_1, \cdots, A_5$ are linearly independent, we get $\det(c_{i,j}) \ne 0$, and hence

$$\det(c_{i,j})^T \ne 0. \quad (T \text{ represents the transverse})$$

For each k $(1 \le k \le 5)$, there is a vector $\mathbf{d} = (d_1, \cdots, d_5)^T$ such that

$$(c_{i,j})^T \mathbf{d} = \mathbf{e}_k,$$

where $\mathbf{e}_k = (0, \cdots, 1 \cdots, 0)^T$ with 1 on the k-th coordinate. This, combined with (4.27), shows that $\sum_{i=1}^{5} d_i A_i$ is a well-defined map from $L_a^2(\mathbb{D})$ to $L_a^2(\mathbb{D})$, which turns out to be

$$h \mapsto \rho_k'(w) h \circ \rho_k(w), \ h \in L_a^2(\mathbb{D}).$$

Writing $h = 1$, one sees that $\rho'_k(w)$ is holomorphic in $\mathbb{D}$ and hence ρ_k extends to an holomorphic function over $\mathbb{D}$. Then by Lemma 4.6.3, we get $B(z) = -z^5$. In this case, it is well-known that M_B has exactly 5 minimal reducing subspace and each one of them, say N, satisfies $\dim N \ominus BN = 1$, which is a contradiction.

It remains to deal with the case of $n = 6$. Similarly, we can assume that $B = z^2 \varphi_\alpha \varphi_\beta \varphi_\gamma \varphi_\delta$ with $\alpha, \beta, \gamma, \delta \in \mathbb{D}$. By (4.25) and similar arguments as in the case of $n = 5$, it suffices to deal with this case:

$$t = 3, \ \dim M_1 \ominus BM_1 = 1 \text{ and } \dim M_i \ominus BM_i = 2 (i = 2, 3).$$

After careful verifications, one can established a similar version of Lemma 4.6.3, which will derive a contradiction if we assume that there is some other minimal reducing subspace different from $M_i (0 \le i \le 3)$. The proof is just like that of case (v). Thus Theorem 4.6.1 also holds in this case. The proof of Theorem 4.6.1 is complete. $\square$

4.7 Remarks on Chap. 4

It has been a long time since the study of commutants of a given operator. A lot of work has been done on multiplication operators, see [T1, T2, T3, T4, Cow1, Cow2, Cow3]. In Chap. 3, some theorems due to Thomson and Cowen are presented. In particular, Theorem 3.1.1 implies that if ϕ is an H^∞-function satisfying some mild conditions, then there is some finite Blaschke product B such that two von Neumann algebras are equal: $\mathcal{V}^*(\phi) = \mathcal{V}^*(B)$. On the Hardy space, the structure of $\mathcal{V}^*(B)$ is clear. However, when it passed to the Bergman space, the case is different. Hu et al. [HSXY] first showed that $\mathcal{V}^*(B)$ is nontrivial by establishing the existence of distinguished reducing subspace of M_B. This work stimulated further work on the structure of $\mathcal{V}^*(B)$. For order $B = 2$, it was shown that M_B has exactly 2 reducing subspaces in [SW] and [Zhu1] independently. In that case, $\mathcal{V}^*(B)$ is abelian. The similar results for order $B = 3, 4$ were shown in [GSZZ] and [SZZ1], respectively; and for order $B = 5, 6$, see [GH1]. Affirmative results have been recently attained in [DSZ] by Douglas, Sun and Zheng in the case of order $B = 7, 8$. Very recently, Douglas et al. [DPW] shows that for all finite Blaschke products B, $\mathcal{V}^*(B)$ are abelian. Therefore, for a large class of multiplication operators M_ϕ (for example, ϕ satisfies the Cowen's condition; in particular, when $\phi \in Hol(\overline{\mathbb{D}})$), $\mathcal{V}^*(\phi)$ is abelian since $\mathcal{V}^*(\phi) = \mathcal{V}^*(B)$ for some finite Blaschke product B. This implies that such a M_ϕ has at most finitely many minimal reducing subspaces. However, any infinite Blaschke product is not contained in the above class, and this will be the focus of the next chapter.

Xu and Yan [XY] extended the main result in [HSXY]; they proved that $\mathcal{V}^*(B)$ is nontrivial on those weighted Bergman space $L^2_{a,\alpha}(\mathbb{D})$ where the parameter $\alpha = 0, 1, \cdots$.

As mentioned in Sect. 4.1 of this chapter, on the Hardy space the structure of $\mathcal{V}^*(B)$ is clear for all Blaschke products B due to the fact the multiplication operator M_B is an isometric operator. However, on the Dirichlet space $\mathcal{D}$ little is known about $\mathcal{V}^*(B)$ even if B is a finite Blaschke product. One can refer to [Zhao] and [CLY].

Most of Sect. 4.1 in this chapter is from Sects. 2 and 4 in the paper [GSZZ]. The remaining mostly comes from [GH1], and the exceptions have been mentioned in the context. It is notable that Proposition 4.4.6 is new.

Chapter 5
Reducing Subspaces Associated with Thin Blaschke Products

Last chapter mainly concerns with reducing subspace problem of multiplication operators M_B induced by finite Blaschke products B. This chapter still focuses on the same theme, whereas the symbol B is replaced with a thin Blaschke product. In Chap. 3 it was shown that the geometric property of this symbol B is a key to the study of the abelian property of $\mathcal{V}^*(B)$. However, the geometry of thin Blaschke products is far more complicated than that of finite Blaschke products.

Geometry of thin Blaschke products will be the main focus of this chapter. The techniques of local inverse and analytic continuation prove useful, along with a solid treatise of function theory. Finite Blaschke products and thin Blaschke products are a curious pair in that both the thin Blaschke product B and its related Riemann surface S_B have nice geometric structures. In this book, a Riemann surface is a complex manifold of complex dimension 1, not necessarily connected. Also, we give a representation of those operators in $\mathcal{V}^*(B)$ for such thin Blaschke products B. Based on this representation, a geometric characterization is obtained for M_B having a nontrivial reducing subspace. Furthermore, it is shown that $\mathcal{V}^*(B)$ is always abelian. In addition, for "most" thin Blaschke products B, M_B has no nontrivial reducing subspace and such a function B is constructed. It is remarkable that this phenomenon never happens in the case of a finite Blaschke product. This reveals some differences between the geometries of thin and finite Blaschke products. As a further application of the methods, it is shown that for finite Blaschke products B, M_B usually has exactly two minimal reducing subspaces.

This chapter is mainly based on Guo and Huang's paper [GH4].

5.1 Properties of Thin Blaschke Products

In this section, it is shown that each thin Blaschke product is a branched covering map, and some properties are obtained.

© Springer-Verlag Berlin Heidelberg 2015
K. Guo, H. Huang, *Multiplication Operators on the Bergman Space*,
Lecture Notes in Mathematics 2145, DOI 10.1007/978-3-662-46845-6_5

Recall that a thin Blaschke product is a Blaschke product whose zero sequence $\{z_n\}$ of B satisfies

$$\lim_{k \to \infty} \prod_{j \geq 1, j \neq k} d(z_j, z_k) = 1.$$

Below, we denote by $\Delta(z, r)$ the pseudohyperbolic disk centered at z with radius r, that is

$$\Delta(z, r) = \{w \in \mathbb{D} : |\frac{z - w}{1 - \bar{z}w}| < r\}.$$

As done in Chap. 4, for a Blaschke product B we denote the critical value set

$$\mathcal{E}_B = \{B(z) : z \in \mathbb{D} \text{ and } B'(z) = 0\},$$

where B' denotes the derivative of B. Let $Z(B)$ and $Z(B')$ denote the zero sets of B and B', respectively. We have the following proposition.

Proposition 5.1.1 *Let B be a thin Blaschke product, then both $\mathcal{E}_B$ and $B^{-1}(\mathcal{E}_B)$ are discrete in $\mathbb{D}$.*

Proof The proof is divided into two parts.

First we will prove that the critical value set $\mathcal{E}_B$ is discrete in $\mathbb{D}$.

It suffices to show that for any $r \in (0, 1)$, each $w \in r\mathbb{D}$ is not an accumulation point of $\mathcal{E}_B$. To see this, we first show that $\forall \varepsilon > 0$, *except for finitely many n*, $B|_{\Delta(z_n, \varepsilon)}$ *is univalent*, and hence B' has no zero point on $\Delta(z_n, \varepsilon)$. The reasoning is as follows. Put

$$g_n = B \circ \varphi_{z_n},$$

and it is easy to verify that

$$\lim_{n \to \infty} |g_n'(0)| = \lim_{n \to \infty} \prod_{j \geq 1, j \neq n} d(z_j, z_n) = 1. \tag{5.1}$$

A theorem in [Ne, p. 171, Exercise 5] states that, if f is a holomorphic function from $\mathbb{D}$ to $\mathbb{D}$ such that

$$f(0) = 0 \text{ and } |f'(0)| = a,$$

then f is univalent in $t\mathbb{D}$ with $t \equiv t(a) = \frac{a}{1 + \sqrt{1 - a^2}}$. Thus, each g_n is univalent on $t(|g_n'(0)|)\mathbb{D}$. By (5.1) there is some natural number N_0 such that

$$t(|g_n'(0)|) > \varepsilon, \ n \geq N_0.$$

Thus, for all $n \geq N_0$, g_n are univalent on $\varepsilon\mathbb{D}$; that is, B is univalent on $\Delta(z_n, \varepsilon)$.

Now write $\delta = \frac{1+r}{2}$, and by Proposition 2.1.9 there exists an $\varepsilon \in (0,1)$ such that $|B(z)| \geq \delta > r$ whenever $d(z, z_n) \geq \varepsilon$ for all n. Thus $|B(z)| \leq r$ only if $d(z, z_n) \leq \varepsilon$ for some n. Also note that B is univalent on $\Delta(z_n, \varepsilon)$ for $n \geq N_0$, and then $Z(B') \cap B^{-1}(r\overline{\mathbb{D}})$ is necessarily contained in a compact subset K of $\mathbb{D}$, where

$$K \triangleq \bigcup_{1 \leq n < N_0} \overline{\Delta(z_n, \varepsilon)}.$$

For any $w \in r\mathbb{D}$, if w were an accumulation point of $\mathcal{E}_B$, then there would be a sequence $\{\lambda_n\}$ in K satisfying $B'(\lambda_n) = 0$ and $\lim_{n \to \infty} B(\lambda_n) = w$. Without loss of generality, we may assume that $\{\lambda_n\}$ converges to some point in K. Since $B'(\lambda_n) = 0$, B' was identically zero on $\mathbb{D}$, and then B was a constant. This is a contradiction. Therefore, $\mathcal{E}_B \cap r\mathbb{D}$ has no accumulation point. By the arbitrariness of r, $\mathcal{E}_B$ is discrete in $\mathbb{D}$.

Next we show that $B^{-1}(\mathcal{E}_B)$ is discrete in the open unit disk; equivalently, $B^{-1}(\mathcal{E}_B) \cap r\mathbb{D}$ is finite for any $0 < r < 1$.

Now fix such an r. The discrete property of $\mathcal{E}_B$ shows that $B(r\mathbb{D}) \cap \mathcal{E}_B$ is finite, and hence $B^{-1}(\mathcal{E}_B) \cap r\mathbb{D}$ is discrete. By arbitrariness of r, $B^{-1}(\mathcal{E}_B)$ is discrete in $\mathbb{D}$, as desired. The proof is complete. $\qquad\square$

Later, one will see that $Z(B')$ always contains infinite points, say $w_1, \cdots, w_n, \cdots$. Since $\mathcal{E}_B$ is discrete in $\mathbb{D}$, $\lim_{n \to \infty} |B(w_n)| = 1$.

Below, it will be shown that each thin Blaschke product is a branched covering map. First we need a lemma [GH4] of independent interest.

Lemma 5.1.2 *Suppose ϕ is a holomorphic function over $\mathbb{D}$ and Δ is a domain in $\mathbb{C}$. If Δ_0 is a component of $\phi^{-1}(\Delta)$ satisfying $\overline{\Delta_0} \subseteq \mathbb{D}$, then $\phi|_{\Delta_0} : \Delta_0 \to \Delta$ is a proper map onto Δ.*

Proof First we show that $\phi|_{\Delta_0} : \Delta_0 \to \Delta$ is a proper map. To see this, assume conversely that there is a sequence $\{z_n\}$ in Δ_0 with no limit point in Δ_0, such that $\phi(z_n)$ has a limit point $w_0 \in \Delta$. Without loss of generality, assume that $\{z_n\}$ converges to some $z_0 \in \partial \Delta_0$ and

$$\lim_{n \to \infty} \phi(z_n) = w_0.$$

Since $\overline{\Delta_0} \subseteq \mathbb{D}$, $z_0 \in \mathbb{D}$. Then by continuity of ϕ, $\phi(z_0) = w_0$, which shows that there is a connected neighborhood Δ_1 of z_0 such that $\Delta_1 \subseteq \phi^{-1}(\Delta)$. Note that $\Delta_0 \subsetneqq \Delta_0 \cup \Delta_1$, which is a contradiction to the assumption that Δ_0 is a component of $\phi^{-1}(\Delta)$. Therefore, $\phi|_{\Delta_0} : \Delta_0 \to \Delta$ is a proper map.

Next we show that $\phi|_{\Delta_0} : \Delta_0 \to \Delta$ is onto. Otherwise, there is a point $w_1 \in \Delta - \phi(\Delta_0)$. Since Δ is connected, then there is a path $l : [0, 1] \to \Delta$ such that $l(0) \in \phi(\Delta_0)$ and $l(1) = w_1$. Let t_0 be the supremum of

$$\{s \in [0, 1]; l(t) \in \phi(\Delta_0) \text{ for all } t \in [0, s]\}.$$

Write $w_2 = l(t_0)$. By an argument of analysis, we have $w_2 \in \partial\phi(\Delta_0)$. Thus, there is a sequence $\{z_n'\}$ in Δ_0 such that

$$\lim_{n\to\infty} \phi(z_n') = w_2.$$

Since $w_2 \in \partial\phi(\Delta_0)$, it follows that $\{z_n'\}$ has no limit point in Δ_0. But $\{\phi(z_n')\}$ has a limit point $w_2 \in \Delta$, which is a contradiction to the statement that $\phi|_{\Delta_0} : \Delta_0 \to \Delta$ is a proper map. Therefore, $\phi|_{\Delta_0} : \Delta_0 \to \Delta$ is onto.   □

Proposition 5.1.3 *All thin Blaschke products are branched covering maps.*

Proof The following proof is from [GH4].

Suppose B is a thin Blaschke product with its zero sequence $\{z_n\}$. By Proposition 2.1.8, $B(\mathbb{D}) = \mathbb{D}$. Now we will show that $B : \mathbb{D} \to \mathbb{D}$ is a branched covering map. For each $z \in \mathbb{D}$, set $\delta = \dfrac{1 + |z|}{2}$. Clearly, $z \in \Delta(0, \delta)$. To finish the proof, we must show that each connected component $\mathcal{J}$ of $B^{-1}(\Delta(0, \delta))$ maps onto $\Delta(0, \delta)$ by a proper map, namely, by the restriction $B|_{\mathcal{J}}$ of B on $\mathcal{J}$.

To see this, note that by Proposition 2.1.9, there is an $\varepsilon \in (0, 1)$ such that

$$B^{-1}(\Delta(0, \delta)) \subseteq \bigcup_n \Delta(z_n, \varepsilon). \tag{5.2}$$

We first study when two pseudohyperbolic disks $\Delta(z_n, \varepsilon)$ and $\Delta(z_k, \varepsilon)$ have a common point. If $\Delta(z_n, \varepsilon) \cap \Delta(z_k, \varepsilon)$ contains a point, say z', then by Garnett [Ga, p. 4, Lemma 1.4]

$$d(z_n, z_k) \le \frac{d(z_n, z') + d(z', z_k)}{1 + d(z_n, z')d(z', z_k)} \le d_\varepsilon, \tag{5.3}$$

where

$$d_\varepsilon = \max\{\frac{x + y}{1 + xy} : 0 \le x, y \le \varepsilon\} < 1.$$

By the definition of thin Blaschke product, it is clear that

$$\lim_{\substack{n\to\infty \\ k\neq n}} \inf d(z_n, z_k) = 1.$$

This, combined with (5.3), shows that there is a positive integer N_0 such that $\Delta(z_n, \varepsilon) \cap \Delta(z_k, \varepsilon)$ is empty whenever $n \ge N_0$ and $k \neq n$. Therefore, by (5.2) the closure of each component of $B^{-1}(\Delta(0, \delta))$ is contained in $\mathbb{D}$. Applying Lemma 5.1.2 shows that the restriction of B on each connected component of $B^{-1}(\Delta(0, \delta))$ is a proper map onto $\Delta(0, \delta)$. The proof is complete.   □

Thin Blaschke products share some good properties. For example, Lemma 5.1.2 states that the inverse image of the critical value set of a thin Blaschke product is discrete. Under some mild condition, the product of two thin Blaschke products is also a thin Blaschke product. Furthermore, we restate Proposition 2.1.8 as follows.

Proposition 5.1.4 *Let B be a thin Blaschke product. Then the following hold:*

(1) *Each value in $\mathbb{D}$ can be achieved for infinitely many times by B.*
(2) *For every w in $\mathbb{D}$, $\varphi_w \circ B$ is a thin Blaschke product.*

5.2 Representation for Operators in $\mathcal{V}^*(B)$

This section will apply the method of Sect. 4.3 in Chap. 4 to give a representation for those operators in $\mathcal{V}^*(B)$, where B is a thin Blaschke product. By using the techniques of analytic continuation and local inverse (see Sect. 2.3), we provide a sufficient condition for when $\mathcal{V}^*(B)$ is trivial.

For a thin or finite Blaschke product B, $\mathcal{E}_B$ always denotes its critical value set. As done in [T1, T2] and [DSZ], we set

$$E = \mathbb{D} - B^{-1}(\mathcal{E}_B).$$

By Proposition 5.1.1, E is a connected open set.

Before continuing, let us make an observation. For an interpolating Blaschke product ϕ, by Garnett [Ga, p. 395, Lemma 1.4] there is an $r \in (0, 1)$ such that $\phi^{-1}(r\mathbb{D})$ equals the disjoint union of domains $V_i(i = 0, 1, \cdots)$, and $\phi|_{V_i} : V_i \to r\mathbb{D}$ is biholomorphic for each i. Define biholomorphic maps $\rho_i : V_0 \to V_i$ by setting

$$\rho_i = \phi|_{V_i}^{-1} \circ \phi|_{V_0},$$

and clearly $\phi \circ \rho_i|_{V_0} = \phi|_{V_0}$. That is, ρ_i are local inverses of ϕ on V_0. Now for any fixed point $z_0 \in E$, write $\lambda = B(z_0)$ and consider $\phi = \varphi_\lambda \circ B$. Proposition 5.1.4 implies that ϕ is an interpolating Blaschke product. This observation soon leads to the following.

Lemma 5.2.1 *Let B be a thin Blaschke product. For each $z_0 \in E$, there always exists a neighborhood V of z_0 and a sequence of holomorphic functions $\{\rho_i\}_{i=0}^{\infty}$ on V satisfying the following:*

1. *$B^{-1} \circ B(z) = \{\rho_i(z) : i = 0, 1, \cdots\}$ holds for each $z \in V$.*
2. *ρ_i are local inverses of B on V, i.e. $B \circ \rho_i(z) = B(z)$, $z \in V$;*
3. *Each $\rho_i : V \to \rho_i(V)$ is biholomorphic.*

In this case, we say that V admits a complete local inverse $\{\rho_i\}_{i=0}^{\infty}$ of B.

Note that in Lemma 5.2.1, $\rho_i(V) \cap \rho_j(V)$ is empty if $i \neq j$. Throughout this chapter, ρ_0 always denotes the identity map, i.e. $\rho_0(z) = z, z \in V$.

By applying Lemma 5.2.1, Guo and Huang [GH4] obtained a local representation of operators S in $\mathcal{V}^*(B)$ as follows.

Lemma 5.2.2 *Let B be a thin Blaschke product. For each $z_0 \in E$, there always exists a neighborhood V of z_0 and a sequence of holomorphic functions $\{\rho_i\}_{i=0}^{\infty}$ on V satisfying the conditions in Lemma 5.2.1. Furthermore, for each operator S in $\mathcal{V}^*(B)$, there is a unique vector $\{c_k\}$ in l^2 satisfying*

$$Sh(z) = \sum_{k=0}^{\infty} c_k h \circ \rho_k(z)\rho_k'(z), \ h \in L_a^2(\mathbb{D}), \ z \in V. \tag{5.4}$$

Proof The idea of the proof comes from [GH2]. For each $z_0 \in E$, let V be a neighborhood of z_0 as in Lemma 5.2.1. Without loss of generality, we may assume that V is a disk and rewrite Δ for V. Then by Lemma 5.2.1,

$$B^{-1}(B(\Delta)) = \bigsqcup_k \rho_k(\Delta).$$

By Proposition 5.1.1 $B^{-1}(\mathcal{E}_B)$ is discrete in $\mathbb{D}$, and then we may require that $\Delta \cap B^{-1}(\mathcal{E}_B)$ is empty (otherwise, one can shrink Δ).

For each local inverse ρ_k of B, define $U_{\rho_k} : L_a^2(\mathbb{D}) \to L_a^2(\Delta)$ by

$$U_{\rho_k}h(z) = h \circ \rho_k(z) \, \rho_k'(z), z \in \Delta, \ h \in L_a^2(\mathbb{D}).$$

For each $w \in \Delta$, set

$$\pi_w : h \mapsto \{U_{\rho_k}h(w)\}, \ h \in L_a^2(\mathbb{D}).$$

As will be soon seen, each π_w is a bounded operator from the Bergman space to l^2. In fact, we have the following.

Claim The sequence $\{U_{\rho_k}h\}$ is a square-summable sequence of $L_a^2(\Delta)$-functions for any $h \in L_a^2(\mathbb{D})$. Consequently, for any $\{c_k\} \in l^2$,

$$\sum_{k=0}^{\infty} c_k U_{\rho_k}h(z), z \in \mathbb{D}$$

converges uniformly on each compact subset of Δ, and hence it is holomorphic in Δ.

To see this, note that

$$\int_{\Delta} |U_{\rho_k}h(z)|^2 dA(z) = \int_{\Delta} |h \circ \rho_k(z) \, \rho_k'(z)|^2 dA(z) = \int_{\rho_k(\Delta)} |h(z)|^2 dA(z),$$

which implies that

$$\sum_{k=0}^{\infty} \int_{\Delta} |U_{\rho_k} h(z)|^2 dA(z) = \int_{\bigsqcup_{k \geq 0} \rho_k(\Delta)} |h(z)|^2 dA(z) \leq \int_{\mathbb{D}} |h(z)|^2 dA(z) < \infty.$$

Therefore, $\{U_{\rho_k} h\}$ is a square-summable sequence of $L_a^2(\Delta)$. Also note that

$$\{U_{\rho_k} h(w)\} = \{\langle U_{\rho_k} h, G_w \rangle\}, \ w \in \Delta,$$

where G_w denotes the reproducing kernel at w in $L_a^2(\Delta)$. Then for each $w \in \Delta$, π_w defines a bounded operator from the Bergman space to l^2. The proof of the claim is complete.

To show Lemma 5.2.2, we first consider the case of S being a unitary operator in $\mathcal{V}^*(B)$. In this case, we will show that there is a unique operator $W : l^2 \to l^2$ such that

$$W\pi_w(h) = \pi_w(Sh), \ h \in L_a^2(\mathbb{D}), \ w \in \Delta. \tag{5.5}$$

This W is necessarily unitary.

Once (5.5) is proved, we can prove (5.4) as follows. In fact, recall that $\rho_0(z) = z$ and $U_{\rho_0} = I$. Denote the first row of the matrix W by $\{c_k\}$ and expanding (5.5) yields

$$Sh(z) = \sum_{k=0}^{\infty} c_k U_{\rho_k} h(z), \ h \in L_a^2(\mathbb{D}), \ z \in \Delta. \tag{5.6}$$

By Proposition 2.5.1, a von Neumann algebra is the finite linear span of its unitary operators, and then each operator in $\mathcal{V}^*(B)$ also has the same representation as (5.6). That is, (5.4) holds.

For the uniqueness of the coefficients c_k, just note that for any fixed w_0 in $\mathbb{D} - B^{-1}(\mathcal{E}_B)$, $\{\rho_k(w_0)\}$ is an interpolating sequence, see Proposition 5.1.4. Pick such a $w_0 \in \Delta$. For each $i \geq 0$, there is a bounded holomorphic function g satisfying $g(\rho_k(w_0)) = \delta_k^i$. If there is another sequence $\{c_k'\} \in l^2$ such that

$$Sh(z) = \sum_{k=0}^{\infty} c_k' U_{\rho_k} h(z), \ h \in L_a^2(\mathbb{D}), \ z \in \Delta,$$

then

$$Sg(w_0) = \sum_{k=0}^{\infty} c_k' g(\rho_k(w_0)) \rho_k'(w_0) = \sum_{k=0}^{\infty} c_k g(\rho_k(w_0)) \rho_k'(w_0),$$

which immediately gives $c_i = c_i'$.

Below we will derive (5.5) to finish the proof of Lemma 5.2.1.

Proof of (5.5) Now write $H = L_a^2(\Delta)$ and $\Lambda = L_a^2(\mathbb{D})$. For each $h \in \Lambda$, set

$$e_h^k(z) = U_{\rho_k}h(z);\ f_h^k(z) = U_{\rho_k}(Sh)(z), z \in \Delta,\ k = 0, 1, \cdots .$$

Clearly e_h^k and f_h^k are in H. Recall that for f_1, f_2 in H, a rank-one operator $f_1 \otimes f_2$ on H is defined by

$$(f_1 \otimes f_2)f = \langle f, f_2 \rangle f_1,\ f \in H.$$

By the claim, for any $h \in L_a^2(\mathbb{D})$ $\{e_h^k\}$ is a square-summable sequence of H, and hence for any g and h in Λ, both $\sum_{k=0}^{\infty} e_g^k \otimes e_h^k$ and $\sum_{k=0}^{\infty} f_g^k \otimes f_h^k$ converge in the operator norm on $B(H)$. Soon we will see that

$$\sum_{k=0}^{\infty} e_g^k \otimes e_h^k = \sum_{k=0}^{\infty} f_g^k \otimes f_h^k.$$

In fact, since $SM_B = M_B S$, then for any polynomials P and Q

$$\langle P(B)g, Q(B)h \rangle = \langle P(B)Sg, Q(B)Sh \rangle.$$

Writing $\tilde{g}$ for Sg and $\tilde{h}$ for Sh, we get

$$\int_{\mathbb{D}} \left((P\overline{Q}) \circ B(w) g(w)\overline{h(w)} - (P\overline{Q}) \circ B(w)\tilde{g}(w)\overline{\tilde{h}(w)} \right) dA(w) = 0.$$

Write $\Omega \triangleq B(\mathbb{D})$, and $\Omega = \mathbb{D}$. By Theorem 2.1.21, any continuous function on $\overline{\Omega}$ can be uniformly approximated by functions in the linear span of

$$\{P\overline{Q} : P, Q \text{ are polynomials}\}.$$

Therefore,

$$\int_{\mathbb{D}} v(B(w)) \left(g(w)\overline{h(w)} - \tilde{g}(w)\overline{\tilde{h}(w)} \right) dA(w) = 0,\ v \in C(\overline{\Omega}). \tag{5.7}$$

Lusin's Theorem (Theorem 2.1.23) shows that for each $v \in L^{\infty}(\Omega)$, there is a uniformly bounded sequence $\{v_n\}$ in $C(\overline{\Omega})$ such that $\{v_n\}$ converges in measure to v. Thus by the Dominated Convergence Theorem, (5.7) holds for any $v \in L^{\infty}(\Omega)$. In particular, for any $v \in L^{\infty}(B(\Delta))$, (5.7) gives that

$$\int_{B^{-1}(B(\Delta))} v(B(w))g(w)\overline{h(w)}dA(w) = \int_{B^{-1}(B(\Delta))} v(B(w))\tilde{g}(w)\overline{\tilde{h}(w)}dA(w).$$

Note that $B^{-1}(B(\Delta)) = \bigsqcup_k \rho_k(\Delta)$ and $B|_{\rho_k(\Delta)} : \rho_k(\Delta) \to B(\Delta)$ are biholomorphic. By a simple calculation one can show that for each $u \in L^\infty(\Delta)$,

$$\int_\Delta u(z) \sum_{k=0}^\infty (g\overline{h}) \circ \rho_k(z) |\rho_k'(z)|^2 dA(z) = \int_\Delta u(z) \sum_{k=0}^\infty (\widetilde{g}\overline{h}) \circ \rho_k(z) |\rho_k'(z)|^2 dA(z).$$

Therefore,

$$\sum_{k=0}^\infty (g\overline{h}) \circ \rho_k(z) |\rho_k'(z)|^2 = \sum_{k=0}^\infty (\widetilde{g}\overline{h}) \circ \rho_k(z) |\rho_k'(z)|^2, \quad z \in \Delta.$$

That is,

$$\langle \sum_{k=0}^\infty (e_g^k \otimes e_h^k) G_z, G_z \rangle = \langle \sum_{k=0}^\infty (f_g^k \otimes f_h^k) G_z, G_z \rangle, z \in \Delta,$$

where G_z is the reproducing kernel of $L_a^2(\Delta)$ at z. By the property of Berezin transform (see Appendix A), we get

$$\sum_{k=0}^\infty e_g^k \otimes e_h^k = \sum_{k=0}^\infty f_g^k \otimes f_h^k. \tag{5.8}$$

Next we show the existence of W. For each $g \in \Lambda$, we set

$$A_g : l^2 \to H$$

$$\{c_k\} \mapsto \sum_{k=0}^\infty c_k e_g^k,$$

and

$$B_g : l^2 \to H$$

$$\{c_k\} \mapsto \sum_{k=0}^\infty c_k f_g^k,$$

It is easy to check that A_g and B_g are well-defined since both $\{e_g^k\}$ and $\{f_g^k\}$ are square-summable sequences from H. Furthermore, we have

$$A_g^* : H \to l^2$$

$$p \mapsto \{\langle p, e_g^k \rangle\},$$

and

$$B_g^* : H \to l^2$$

$$p \mapsto \{\langle p, f_g^k \rangle\}.$$

Then for any $g, h \in \Lambda$,

$$A_h A_g^* p = \sum_{k=0}^{\infty} \langle p, e_g^k \rangle e_h^k = \sum_{k=0}^{\infty} (e_h^k \otimes e_g^k) p$$

$$= \sum_{k=0}^{\infty} (f_h^k \otimes f_g^k) p = B_h B_g^* p, \quad p \in H.$$

The third equality follows from (5.8). So $A_h A_g^* = B_h B_g^*$, $g, h \in \Lambda$. Then it is easy to verify that

$$\sum_{i=1}^{l} A_{g_i}^* p_i \mapsto \sum_{i=1}^{l} B_{g_i}^* p_i, \quad p_i \in H \text{ and } g_i \in \Lambda$$

is a well-defined isometry from some subspace of l^2 to another, and by a careful observation, we see that the initial space and the final space are the same. Therefore, this isometry can be extended to a unitary operator from l^2 onto l^2, say U. Specifically, we have

$$U A_g^* = B_g^*, \quad g \in \Lambda.$$

Recall that G_w is the reproducing kernel at w of H ($w \in \Delta$), and

$$U A_g^* G_w = B_g^* G_w, \quad g \in \Lambda \text{ and } w \in \Delta.$$

That is,

$$U\{\overline{e_g^k(w)}\} = \{\overline{f_g^k(w)}\}. \tag{5.9}$$

Now write the infinite matrix form of U as

$$(a_{i,j})_{i,j=1}^{\infty},$$

and put

$$W = (\overline{a_{i,j}})_{i,j=1}^{\infty}.$$

Also denote the corresponding operator by W, and clearly W is unitary since U is unitary. Then by (5.9), we get

$$W\{e_g^k(w)\} = \{f_g^k(w)\}, \ g \in \Lambda, w \in \Delta. \tag{5.10}$$

This is nothing but (5.5).

The proof of (5.5) is now complete. However, we would like to show something else: *(5.10) uniquely determines the operator W.* To see this, it suffices to show that for each fixed $w_0 \in \Delta$, the span of all $\{e_g^k(w_0)\}$ $(g \in \Lambda)$ is dense in l^2. Suppose conversely that there is a nonzero vector $\{d_i\}$ in l^2 such that $\langle\{e_g^k(w_0)\}, \{d_i\}\rangle = 0$ for all $g \in \Lambda$. That is,

$$\sum_{k=0}^{\infty} \overline{d_k}\rho_k'(w_0)g(\rho_k(w_0)) = 0. \tag{5.11}$$

On the other hand, for each $w_0 \in \mathbb{D} - B^{-1}(\mathcal{E}_B)$, $\{\rho_k(w_0)\}$ is an interpolating sequence. In particular, for each i there is a bounded holomorphic function g satisfying

$$g(\rho_k(w_0)) = \delta_k^i,$$

where δ_k^i denotes the *Kronecker delta*. That is,

$$\delta_k^i = \begin{cases} 0, k \neq i, \\ 1, k = i. \end{cases}$$

By (5.11), we have $\overline{d_i}\rho_i'(w_0) = 0$, forcing $d_i = 0$ $(\forall i)$. This is a contradiction. Thus the operator W in (5.5) is unique.   □

Note that the identity (5.4) holds on V, locally. The notion π_w and matrix W are also locally defined. Below, we will try to get a global version for them.

In Sect. 2.3 in Chap. 2, we have introduced the definitions of direct continuations and analytic continuation. By convention, *throughout this chapter all curves and loops concerned are required to be contained in E, where $E = \mathbb{D} - B^{-1}(\mathcal{E}_B)$.*

If ρ is a local inverse of B and γ is a curve in E such that $\gamma(0)$ belongs to $D(\rho)$, the definition domain of ρ, then ρ always admits an analytic continuation $\tilde{\rho}$ along γ. By Theorem 2.3.1, such an analytic continuation is unique, and one can show that this $\tilde{\rho}$ is necessarily a local inverse of B. Then applying the monodromy theorem (Theorem 2.3.4) shows that each local inverse ρ of B can be extended analytically to any simply connected subdomain of E. Below we will seek a curve L which passes all points in $B^{-1}(\mathcal{E}_B)$, such that $\mathbb{D} - L$ is simply connected. In this way, one can get a global version of (5.4). The following is from [GH4].

Lemma 5.2.3 *Suppose $\mathcal{E}$ is a discrete subset in $\mathbb{D}$. Then there is a curve L containing $\mathcal{E}$ such that $\mathbb{D} - L$ is simply connected.*

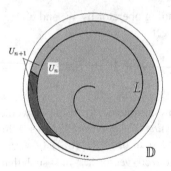

Fig. 5.1 Snail

Proof Let $\mathcal{E}$ be a discrete set in $\mathbb{D}$. Without loss of generality, we assume that $\mathcal{E}$ consists of countably infinitely many points. Since $\mathcal{E}$ is discrete, there is a strictly increasing sequence $\{r_n\}$ satisfying

$$\lim_{n \to \infty} r_n = 1 \text{ and } \mathcal{E} \subseteq \bigcup r_n \mathbb{T}.$$

Note that each circle $r_n \mathbb{T}$ contains only finite points in $\mathcal{E}$, and then we can draw an infinite polygon L passing along these $r_n \mathbb{T}$ and through all points in $\mathcal{E} \cap (r_n \mathbb{T})$. Since L tends to $\mathbb{T}$ and has no end point on $\mathbb{T}$, $\mathbb{D} - L$ looks like a snail. See Fig. 5.1, which is provided by Dr. Xie. It is easy to construct a sequence of simply connected open sets U_n whose union is the snail $\mathbb{D} - L$ and

$$U_n \subseteq U_{n+1}, \ n = 1, 2, \cdots.$$

Therefore, $\mathbb{D} - L$ is simply connected.   □

In the proof Lemma 5.2.3, one sees that the curve L consists of at most countably many arcs and L is relatively closed in $\mathbb{D}$. Here, by an arc we mean a subset in $\mathbb{C}$ which is C^1-homeomorphic to some segment.

By Proposition 5.1.1 and Lemma 5.2.3, there is a curve L containing $B^{-1}(\mathcal{E}_B)$ such that $\mathbb{D} - L$ is simply connected. Note that all ρ_k are local inverse of B, and by the monodromy theorem all ρ_k are holomorphic in $\mathbb{D} - L$. For each w in $\mathbb{D} - L$, define

$$\pi_w(h) = \{h \circ \rho_k(w)\rho_k'(w)\}, \ h \in L_a^2(\mathbb{D}).$$

By the claim below Lemma 5.2.2, $\pi_w(h) \in l^2$. Furthermore, Guo and Huang [GH4] obtained the following.

Theorem 5.2.4 *Suppose B is a thin Blaschke product. If S is a unitary operator on the Bergman space which commutes with M_B, then there is a unique operator $W : l^2 \to l^2$ such that*

$$W\pi_w(h) = \pi_w(Sh), \ h \in L_a^2(\mathbb{D}), \ w \in \mathbb{D} - L. \tag{5.12}$$

This W is necessarily a unitary operator. Thus, there is a unique vector $\{c_k\}$ in l^2 satisfying

$$Sh(z) = \sum_{k=0}^{\infty} c_k h \circ \rho_k(z)\rho_k'(z), \ h \in L_a^2(\mathbb{D}), \ z \in \mathbb{D} - L. \tag{5.13}$$

where all holomorphic functions ρ_k satisfy $B \circ \rho_k|_{\mathbb{D}-L} = B$, and the right side of (5.13) converges uniformly on compact subsets of $\mathbb{D} - L$.

The operator W is exactly the one that appeared in the proof of Lemma 5.2.2. If S is any operator in $\mathcal{V}^*(B)$, then both (5.12) and (5.13) still hold. In this case, W is not necessarily unitary.

In fact, for a finite Blaschke product B, such an operator W in (5.12) was also obtained by Douglas, Sun and Zheng in [DSZ]. Motivated by a result in [DSZ], we can show that W has a very restricted form. To see this, let V be chosen as in Lemma 5.2.1 and put $\mathcal{E} = B^{-1}(\mathcal{E}_B)$ in Lemma 5.2.3. Since V can be shrunk such that $\rho_i(V) \cap \rho_j(V)$ are empty whenever $i \neq j$, we may carefully choose a curve L with all $L \cap \rho_i(V)$ being empty.

Note that each ρ_k maps $\rho_j(V)$ biholomorphically onto some $\rho_i(V)$, and hence $\rho_k \circ \rho_j|_V$ makes sense; locally $\rho_k \circ \rho_j$ equals ρ_i. Now consider the $(n+1)$-th row of both sides of (5.12). Let $\{d_k\} \in l^2$ be the $(n+1)$-th row of W, then by (5.12),

$$Sh(\rho_n(w))\rho_n'(w) = \sum_{k=0}^{\infty} d_k h(\rho_k(w))\rho_k'(w), \ h \in L_a^2(\mathbb{D}) \text{ and } w \in V. \tag{5.14}$$

By the above discussion, there is a permutation π_n such that

$$\rho_k \circ \rho_n|_V = \rho_{\pi_n(k)}|_V, k = 1, 2, \cdots, \tag{5.15}$$

and hence

$$\rho_k' \circ \rho_n \rho_n'|_V = \rho_{\pi_n(k)}'|_V. \tag{5.16}$$

Combining (5.14), (5.15) with (5.16) shows that for each $h \in L_a^2(\mathbb{D})$,

$$Sh(\rho_n(w))\rho_n'(w) = \sum_{k=0}^{\infty} d_{\pi_n(k)}h(\rho_k \circ \rho_n(w))\rho_k' \circ \rho_n(w)\rho_n'(w), \ w \in V.$$

Therefore,

$$Sh(\rho_n(w)) = \sum_{k=0}^{\infty} d_{\pi_n(k)}h(\rho_k \circ \rho_n(w))\rho_k' \circ \rho_n(w), \ w \in V,$$

and hence

$$Sh(z) = \sum_{k=0}^{\infty} d_{\pi_n(k)} h \circ \rho_k(z) \rho_k'(z), \ z \in \mathbb{D} - L.$$

Therefore by (5.13)

$$\sum_{k=0}^{\infty} d_{\pi_n(k)} h \circ \rho_k(z) \rho_k'(z) = \sum_{k=0}^{\infty} c_k h \circ \rho_k(z) \rho_k'(z), \ h \in L_a^2(\mathbb{D}), \ z \in \mathbb{D} - L.$$

By uniqueness of the coefficients, we get

$$d_{\pi_n(k)} = c_k, k = 1, 2, \cdots.$$

Put $\sigma_n = \pi_n^{-1}(n \geq 1)$, and then $d_k = c_{\sigma_n(k)}, k = 1, 2, \cdots$. Therefore, the form of W is presented as follows, see [GH4].

Theorem 5.2.5 *The infinite unitary matrix W in Theorem 5.2.4 has a very restricted form. Precisely, there is a unit vector $\{c_k\} \in l^2$ and permutations $\sigma_j(j \geq 1)$ of positive integers such that*

$$W = \begin{pmatrix} c_1 & c_2 & \cdots & c_n & \cdots \\ c_{\sigma_1(1)} & c_{\sigma_1(2)} & \cdots & c_{\sigma_1(n)} & \cdots \\ \vdots & \vdots & \ddots & \vdots & \vdots \\ c_{\sigma_m(1)} & c_{\sigma_m(2)} & \cdots & c_{\sigma_m(n)} & \cdots \\ \vdots & \vdots & \ddots & \vdots & \ddots \end{pmatrix}.$$

Below, we will use Theorem 5.2.4 to give a geometric characterization for when $V^*(B)$ is trivial. First, we need the following result from [GH4]. In the case of B being a finite Blaschke product, it is given by Douglas et al. [DSZ, Lemma 5.1].

Lemma 5.2.6 *With the notations in Theorem 5.2.4, suppose S is an operator in $V^*(B)$ with the form:*

$$Sh(z) = \sum_{k=0}^{\infty} c_k h \circ \rho_k(z) \rho_k'(z), \ h \in L_a^2(\mathbb{D}), \ z \in \mathbb{D} - L,$$

where $\{c_k\} \in l^2$. Then $c_i = c_j$ if $[\rho_i] = [\rho_j]$. In particular, for each $c_i \neq 0$, $\sharp\{k : [\rho_k] = [\rho_i]\} < \infty$.

Recall that for each local inverse ρ of B, $[\rho]$ denotes the equivalent class of all analytic continuation σ of ρ such that $D(\sigma) \subseteq E$.

Proof Suppose S is an operator in $V^*(B)$ given by

$$Sh(z) = \sum_{k=0}^{\infty} c_k h \circ \rho_k(z) \rho_k'(z), \quad h \in L_a^2(\mathbb{D}), \ z \in \mathbb{D} - L,$$

where $\{c_k\} \in l^2$. Since $\{c_k\}$ is in l^2, to prove Lemma 5.2.6 it is enough to show that
if ρ_j and $\rho_{j'}$ are equivalent, i.e. $[\rho_j] = [\rho_{j'}]$, then $c_j = c_{j'}$. To see this, fix a point
z_0 in $\mathbb{D} - L$ and assume that ρ_j and $\rho_{j'}$ are equivalent. Then there is a loop γ in E
based at z_0 such that $\rho_{j'}$ is the analytic continuation of ρ_j along γ. By (5.4), for each
$f \in L_a^2(\mathbb{D})$, the sum

$$\sum_{k=0}^{\infty} c_k f \circ \rho_k(z) \rho_k'(z)$$

makes sense for each $z \in \gamma$ and it is locally holomorphic. As done in the proof
of [DSZ, Lemma 5.1], suppose along the loop γ each ρ in $\{\rho_i\}_{i=0}^{\infty}$ is extended
analytically to a holomorphic function denoted by $\hat{\rho}$, defined on some neighborhood
of z_0. Then we have

$$\sum_{k=0}^{\infty} c_k f \circ \rho_k(z_0) \rho_k'(z_0) = \sum_{k=0}^{\infty} c_k f \circ \hat{\rho}_k(z_0) \hat{\rho}_k'(z_0). \tag{5.17}$$

Since $\{\rho_k(z_0)\}$ is an interpolating sequence, there exists a bounded holomorphic
function f over $\mathbb{D}$ such that

$$f(\rho_{j'}(z_0)) = 1 \quad \text{and} \quad f(\rho_k(z_0)) = 0, \ k \neq j'.$$

Since $\rho_{j'} = \hat{\rho}_j$, then by (5.17) $c_j = c_{j'}$.   □

Denote by $D(f)$ the domain of definition for any function f. Now define $G[\rho]$ to
be the union of all graphs $G(\sigma)$ of members σ in the equivalent class $[\rho]$ of ρ, with
$D(\sigma) \subseteq E$; that is,

$$G[\rho] = \bigcup_{\sigma \in [\rho]} \{(z, \sigma(z)) : z \in D(\sigma) \subseteq E\}.$$

Clearly, $G[\rho]$ is a subset of $E \times E$. In fact, $G[\rho]$ is a connected Riemann surface
since it contains only points which locally can be written as $(z, \sigma(z))$, where σ are
some local inverses defined on a neighborhood of some $z_0 \in E$. Also, observe that
for any local inverses ρ and σ of B, $G[\rho] = G[\sigma]$ if and only if $[\rho] = [\sigma]$; and either
$G[\rho] = G[\sigma]$ or $G[\rho] \cap G[\sigma] = \emptyset$. Following [DSZ] and [GSZZ], define

$$S_B \triangleq \{(z, w) \in \mathbb{D}^2; z \in E \text{ and } B(z) = B(w)\},$$

equipped with the topology inherited from the natural topology of $\mathbb{D}^2$. It is not difficult to see that S_B *is a Riemann surface* whose components are $G[\rho]$, where ρ are local inverses of B. The Riemann surface

$$G[z] = \{(z, z) : z \in E\}$$

is called *the identity component*, or *the trivial component* of S_B.

Define π_1 and π_2 by setting

$$\pi_1(z, w) = z,$$

and

$$\pi_2(z, w) = w$$

for all $(z, w) \in \mathbb{C}^2$. Rewrite π_ρ for the restriction $\pi_1 : G[\rho] \to E$. By the discussion above Lemma 5.2.3, for any local inverse ρ of B and any curve $\gamma \in E$, there must exist a unique analytic continuation $\tilde{\rho}$ of ρ along γ. This implies that π_ρ is onto. Then for each $z \in E$, one can put

$$[\rho](z) = \{w : (z, w) \in G[\rho]\},$$

which is nonempty. Furthermore, by using Lemma 5.2.1 one can show the following [GH4].

Proposition 5.2.7 *For each local inverse ρ of B, $\pi_\rho : G[\rho] \to E$ is a covering map.*

A standard result in topology says that for a covering map $p : Y \to X$ with X arc-connected, the number $\sharp p^{-1}(x)$ does not depend on the choice of $x \in X$, called the multiplicity of p, or the number of sheets of p. Therefore,

$$\sharp(\pi_\rho)^{-1}(z) = \sharp[\rho](z)$$

is either a constant integer or ∞, called *the multiplicity* of $G[\rho]$, or *the number of sheets* of $G[\rho]$. For simplicity, we rewrite $\sharp[\rho]$ for $\sharp[\rho](z)$ to emphasize that it does not depend on z. If $\sharp[\rho] < \infty$, then the component $G[\rho]$ is called *finite*; otherwise, $G[\rho]$ is called *infinite*.

Combining Theorem 5.2.4, Lemma 5.2.6 with Proposition 5.2.7 yields the following, due to Guo and Huang [GH4].

Proposition 5.2.8 *For a thin Blaschke product B, if there is no nontrivial component $G[\rho]$ of S_B such that $\sharp[\rho] < \infty$, then $\mathcal{V}^*(B)$ is trivial; equivalently, M_B has no nontrivial reducing subspace.*

The existence of thin Blaschke products with the above property will be carried out in Sect. 5.4 of Chap. 5. Also, the sufficient condition in Proposition 5.2.8 is also necessary for $\mathcal{V}^*(B)$ to be trivial, as will be seen in Sect. 5.5 in Chap. 5.

Remark 5.2.9 All definitions appearing in this section, such as local inverse, analytic continuation, can also be carried to the case of finite Blaschke products ϕ. However, since $\sharp\phi^{-1}(\phi(z)) = \text{order}\,\phi < \infty$ for each $z \in E$, any component $G[\rho]$ of S_ϕ must have finite multiplicity where ρ is a local inverse of ϕ [GH4].

5.3 Geometric Characterization for $\mathcal{V}^*(B)$

Last section presents the form of operators in $\mathcal{V}^*(B)$ where B is a thin Blaschke product. Based on this, this section will provide a complete geometric characterization for the nontriviality of $\mathcal{V}^*(B)$.

The following, due to Guo and Huang [GH4], is the main result of this section. Its proof will be placed at the end of this section.

Theorem 5.3.1 *Suppose B is a thin Blaschke product. Then $\mathcal{V}^*(B)$ is nontrivial if and only if the Riemann surface S_B has a nontrivial component with finite multiplicity.*

Some notations are in order. For any local inverse ρ of B, the inverse map ρ^{-1} of ρ is also a local inverse of B, and we rewrite ρ^- for ρ^{-1}. For any subset F of $E \times E$, write $F^- = \{(w, z) : (z, w) \in F\}$. Thus,

$$G^-[\rho] = \{(w, z) : (z, w) \in G[\rho]\}.$$

In [GH4], a geometric characterization for $G[\rho^-]$ was obtained as follows.

Proposition 5.3.2 *For each local inverse ρ, we have $G[\rho^-] = G^-[\rho]$.*

Proof Suppose $z_0 \in D(\rho)$. Clearly, $\rho^-(\rho(z_0)) = z_0$. Let $z \in E$, and γ be any curve that connects z_0 with z. Denote by $\tilde\rho$ the analytic continuation of ρ along γ, and then $\tilde\rho(\gamma)$ is a curve connecting $\rho(z_0)$ and $\tilde\rho(z)$. Let $\widetilde{\rho^-}$ be the analytic continuation of ρ^- along $\tilde\rho(\gamma)$. Since $\rho^-(\rho(z_0)) = z_0$, then

$$\widetilde{\rho^-}(\tilde\rho) = id$$

holds on the curve γ. In particular, $\widetilde{\rho^-}(\tilde\rho(z)) = z$, forcing

$$(\tilde\rho(z), z) \in G[\widetilde{\rho^-}] = G[\rho^-].$$

The above means that

$$G^-[\rho] \subseteq G[\rho^-].$$

Therefore,

$$\{G^-[\rho]\}^- \subseteq G^-[\rho^-] \subseteq G[\{\rho^-\}^-],$$

Clearly,

$$G[\rho] \subseteq G^-[\rho^-] \subseteq G[\rho],$$

forcing $G^-[\rho^-] = G[\rho]$, and hence $G[\rho^-] = G^-[\rho]$.   □

Since $\pi_1 : G[\rho^-] \to E$ is a covering map, by Proposition 5.3.2 $\pi_2 : G[\rho] \to E, (z, w) \to w$ is also a covering map, and the multiplicity $\sharp[\rho^-]$ of $\pi_1|_{G[\rho^-]}$ equals that of $\pi_2|_{G[\rho]}$. If both $\sharp[\rho]$ and $\sharp[\rho^-]$ are finite, then we say that the component $G[\rho]$ of S_B is *bi-finite*. Later, it turns out that $G[\rho]$ is bi-finite if and only if it is finite, see Proposition 5.3.4. Note that S_B has a nontrivial bi-finite component if and only if there exists a nontrivial local inverse ρ of B, such that one of the covering maps $\pi_\rho : G[\rho] \to E$ and $\pi_2 : G[\rho] \to E$ has finite multiplicity.

Before continuing, we need some preparations. If $\sharp[\rho] = n < \infty$, then $\pi_1 : G[\rho] \to E$ is a covering map of finite sheets. Then for each $z \in E$, there is a small disk $\Delta(z \in \Delta)$ and n local inverse $\rho_i(1 \le i \le n)$ defined on Δ such that

$$\pi_1|_{G[\rho]}^{-1}(\Delta) = \bigsqcup_{i=1}^{n} \{(w, \rho_i(w)) : w \in \Delta\}.$$

In this case, define

$$\sum_{\sigma \in [\rho]} h \circ \sigma(w) \, \sigma'(w) \triangleq \sum_{i=1}^{n} h \circ \rho_i(w) \, \rho_i'(w), w \in \Delta, \tag{5.18}$$

where h is any function over E or $\mathbb{D}$. Then we get a linear map $\mathcal{E}_{[\rho]}$ defined by

$$\mathcal{E}_{[\rho]} h(z) \equiv \sum_{\sigma \in [\rho]} h \circ \sigma(z) \, \sigma'(z), z \in E.$$

One can verify that the map $\mathcal{E}_{[\rho]}$ is well-defined; that is, the value $\mathcal{E}_{[\rho]} h(z)$ does not depend on the choice of disks Δ which contains z. The functions h over E or $\mathbb{D}$ are not necessarily holomorphic. Sometimes we write

$$\mathcal{E}_{[\rho]} h = \sum_{\rho_i \in [\rho]} h \circ \rho_i \rho_i'$$

to emphasize that this $\mathcal{E}_{[\rho]}$ maps holomorphic functions to holomorphic functions. In particular, the trivial component $[z]$ gives the identity operator I, i.e. $\mathcal{E}_{[z]} = I$. It is worthwhile to mention that those operators $\mathcal{E}_{[\rho]}$ were first considered in [DSZ]. Similar as (5.18), define

$$\mathcal{E}_{[|\rho|]} h(z) = \sum_{\rho_i \in [\rho]} h \circ \rho_i(z) |\rho_i'(z)|.$$

It is well-known that a square-summable holomorphic function on a punctured disk can always be extended analytically to the whole disk. Since $\mathbb{D} - E$ is discrete in $\mathbb{D}$, then each function f in $L_a^2(E)$ can be extended analytically to the unit disk $\mathbb{D}$; that is, f can be regarded as a member in $L_a^2(\mathbb{D})$. Thus, it is reasonable to write

$$L_a^2(E) = L_a^2(\mathbb{D}).$$

As mentioned above, $\mathcal{E}_{[\rho]}$ maps holomorphic functions to holomorphic functions. Soon we will see that $\mathcal{E}_{[\rho]}$ defines a linear map from $L_a^2(\mathbb{D})$ to $L_a^2(E)$. Furthermore, $\mathcal{E}_{[\rho]} : L_a^2(\mathbb{D}) \to L_a^2(\mathbb{D})$ is a bounded operator, see as follows [GH4].

Lemma 5.3.3 *If $G[\rho]$ is a bi-finite component of S_B, then both $\mathcal{E}_{[\rho]}$ and $\mathcal{E}_{[\rho^-]}$ are in $\mathcal{V}^*(B)$, and $\mathcal{E}_{[\rho]}^* = \mathcal{E}_{[\rho^-]}$.*

Proof Assume that $G[\rho]$ is a bi-finite component of S_B. First we will show that $\mathcal{E}_{[\rho]}$ is bounded. To see this, let $h \in C_c(E)$; that is, h is a compactly supported continuous function over E. For each open set U that admits a complete local inverse $\{\rho_i\}_{i=0}^\infty$, we have

$$\int_U |\mathcal{E}_{[\rho]} h(z)|^2 dA(z) \leq C \int_U \sum_{\rho_i \in [\rho]} |h \circ \rho_i(z)|^2 |\rho_i'(z)|^2 dA(z)$$

$$= C \sum_{\rho_i \in [\rho]} \int_{\rho_i(U)} |h(z)|^2 dA(z), \tag{5.19}$$

where C is a numerical constant that depends only on $\sharp[\rho]$.

Since E is an open set which can be covered by open disks with each admitting a complete local inverse, E can be written as the union of countably many of them. The reasoning is as follows. For each $z \in E$, there is an $r = r(z) > 0$ such that the disk $O(z, r(z))$ admits a complete local inverse. Then by Proposition 3.2.1, one can pick countable disks $O(z, r(z))$ whose union equals E.

Now assume that $\{U_j\}$ is such a covering of E and each U_j admits a complete local inverse $\{\rho_i\}_{i=0}^\infty$. We emphasize that the family $\{\rho_i\}_{i=0}^\infty$ is fixed once their common definition domain U_j is fixed. Set

$$E_1 = U_1 \quad \text{and} \quad E_j = U_j - \bigcup_{1 \leq i < j} U_i \ (j \geq 2).$$

As (5.19), we also have

$$\int_{E_j} \left| \mathcal{E}_{[\|\rho\|]} |h|(z) \right|^2 dA(z) \leq C \sum_{\rho_i \in [\rho]} \int_{\rho_i(E_j)} |h(z)|^2 dA(z) \equiv C \sum_{\rho_i \in [\rho]} \mu(\rho_i(E_j)),$$

where μ is a measure defined by

$$\mu(F) = \int_F |h(z)|^2 dA(z)$$

for each Lebesgue measurable subset F of $\mathbb{D}$. Since $E = \cup U_j = \sqcup E_j$, then

$$\int_E |\mathcal{E}_{[|\rho|]}|h|(z)|^2 dA(z) \leq C \sum_{j=1}^{\infty} \sum_{\rho_i \in [\rho]} \mu(\rho_i(E_j)). \qquad (5.20)$$

Write $m = \sharp[\rho^-] < \infty$. Let us recall a fact in measure theory: for measurable sets F and F_j, if $\cup_{j=1}^k F_j \subseteq F$, and for each point $z \in F$, there are at most m sets F_j containing z, then

$$\sum_{j=1}^{k} \chi_{F_j} \leq m \chi_F,$$

and hence $\sum_{j=1}^{k} \mu(F_j) \leq m\mu(F)$. Below, we will apply this result. By our choice of E_j ($E_j \subseteq U_j$) and Proposition 5.3.2, it follows from that for any $w \in E$, there are at most m sets E_j such that $w \in \rho_i(E_j)$ for some $\rho_i \in [\rho]$. Therefore,

$$\sum_{\rho_i \in [\rho]} \sum_{j=1}^{k} \mu(\rho_i(E_j)) \leq m\mu(E).$$

Letting k tends to infinity shows that

$$\sum_{\rho_i \in [\rho]} \sum_{j=1}^{\infty} \mu(\rho_i(E_j)) \leq m\mu(E);$$

and thus by (5.20)

$$\int_{\mathbb{D}} \left| \mathcal{E}_{[|\rho|]}|h|(z) \right|^2 dA(z) \leq mC\|h\|^2.$$

By a limit argument, it is easy to see that the above identity also holds for each $h \in C(\mathbb{D})$, which immediately gives that

$$\int_{\mathbb{D}} |\mathcal{E}_{[\rho]}f(z)|^2 dA(z) \leq \int_{\mathbb{D}} \left| \mathcal{E}_{[|\rho|]}|f|(z) \right|^2 dA(z) \leq mC\|f\|^2, f \in L_a^2(\mathbb{D}).$$

That is, $\mathcal{E}_{[\rho]} : L_a^2(\mathbb{D}) \rightarrow L_a^2(E)$ is a bounded operator. By the comments above Lemma 5.3.3, $\mathcal{E}_{[\rho]}$ can be regarded as a bounded operator from $L_a^2(\mathbb{D})$ to $L_a^2(\mathbb{D})$.

Also, $\mathcal{E}_{[\rho^-]}$ is a bounded operator from $L_a^2(\mathbb{D})$ to $L_a^2(\mathbb{D})$. Below we will show that $\mathcal{E}_{[\rho]}^* = \mathcal{E}_{[\rho^-]}$.

To see this, let U be a fixed open subset of E admitting a complete local inverse, h be a bounded measurable function whose support is contained in $\rho_{i_0}(U)$ for some

i_0, and $g \in C_c(E)$. Since $m = \sharp[\rho^-]$, we may write

$$[\rho^-]\rho_{i_0}(U) = \bigsqcup_{j=1}^{m} \check{\rho}_j(U), \tag{5.21}$$

where $\check{\rho}_j$ are local inverses of B on U. That is, on $\rho_{i_0}(U)$ there are m members $\rho_{k_j}(1 \le j \le m)$ in $[\rho^-]$, and there are m integers $k'_j(1 \le j \le m)$ such that

$$\rho_{k_j} \circ \rho_{i_0}(z) = \rho_{k'_j}(z), \ z \in U.$$

Rewrite $\check{\rho}_j$ for $\rho_{k'_j}(1 \le j \le m)$, and then we have (5.21).

Then for each $j(1 \le j \le m)$, there is a local inverse $\sigma_j : \check{\rho}_j(U) \to \rho_{i_0}(U)$ and σ_j is onto. Since $\sigma_j \in [\rho]$ and h is supported on $\rho_{i_0}(U)$, we have

$$\int_{\check{\rho}_j(U)} \sum_{\rho_k \in [\rho]} h \circ \rho_k(z)\rho'_k(z)\overline{g(z)}dA(z) = \int_{\check{\rho}_j(U)} h \circ \sigma_j(z)\sigma'_j(z)\overline{g(z)}dA(z)$$

$$= \int_{\rho_{i_0}(U)} h \circ \sigma_j(\sigma_j^-(w))$$

$$\sigma'_j(\sigma_j^-(w))|\sigma_j^{-'}(w)|^2 \overline{g(\sigma_j^-(w))}dA(w)$$

$$= \int_{\rho_{i_0}(U)} h(w)\overline{g(\sigma_j^-(w))\sigma_j^{-'}(w)}dA(w).$$

That is,

$$\int_{\check{\rho}_j(U)} \sum_{\rho_k \in [\rho]} h \circ \rho_k(z)\rho'_k(z)\overline{g(z)}dA(z) = \int_{\rho_{i_0}(U)} h(w)\overline{g(\sigma_j^-(w))\sigma_j^{-'}(w)}dA(w).$$

Since h is supported in $\rho_{i_0}(U)$, by (5.21) one can verify that

$$\mathcal{E}_{[\rho]}h = \sum_{\rho_i \in [\rho]} \sum_{j=1}^{m} \chi_{\check{\rho}_j(U)} h \circ \rho_i \rho'_i,$$

and hence

$$\langle \mathcal{E}_{[\rho]}h, g \rangle = \sum_{j=1}^{m} \int_{\check{\rho}_j(U)} \sum_{\rho_i \in [\rho]} h \circ \rho_i(z)\rho'_i(z)\overline{g(z)}dA(z)$$

$$= \sum_{j=1}^{m} \int_{\rho_{i_0}(U)} h(w)\overline{g(\sigma_j^-(w))\sigma_j^{-'}(w)}dA(w).$$

$$= \int_{\mathbb{D}} h(w)\chi_{\rho_{i_0}(U)} \overline{\sum_{\sigma \in [\rho^-]} g(\sigma(w))\sigma'(w)} dA(w)$$

$$= \langle h, \mathcal{E}_{[\rho^-]}g \rangle.$$

That is,

$$\langle \mathcal{E}_{[\rho]}h, g \rangle = \langle h, \mathcal{E}_{[\rho^-]}g \rangle.$$

Since each compact subset of E is contained in the union of finite open sets like $\rho_{i_0}(U)$, the above identity also holds for any $h \in C_c(E)$. Then by a limit argument, the above holds for any $g, h \in L_a^2(\mathbb{D})$. Therefore,

$$\mathcal{E}_{[\rho]}^* = \mathcal{E}_{[\rho^-]}.$$

Besides, since both $\mathcal{E}_{[\rho^-]}$ and $\mathcal{E}_{[\rho]}$ commute with M_B, $\mathcal{E}_{[\rho]} \in \mathcal{V}^*(B)$. The proof is complete.   □

Below, we need a property of finite Blaschke products, as mentioned in [DSZ] and implied in [DPW], but without proof. This property is also shared by thin Blaschke products.

Proposition 5.3.4 (DPW) *For a finite or thin Blaschke product B, let ρ be one of its local inverse. Then $\sharp[\rho] = \sharp[\rho^-]$, allowed to be ∞.*

Proof First, assume that B is a finite Blaschke product. Let us have a sketch of the geometric analysis of B, see the paragraph below Corollary 4.2.4. For a fixed point $w_0 \in \mathbb{T}$, $B^{-1}(B(w_0))$ consists of exactly n different points (in anti-clockwise direction)

$$w_0, \cdots, w_{n-1},$$

which lie on $\mathbb{T}$. For each $j(0 \le j \le n-1)$, there is a unique local inverse ρ_j of B defined on a neighborhood of w_0 such that

$$\rho_j(w_0) = w_j,$$

and all ρ_j extend analytically to a neighborhood Ω of $\mathbb{T}$. Let $\widehat{w_0} \in \mathbb{D}$ be enough close to w_0, and then $B^{-1}(B(\widehat{w_0}))$ consists of exactly n different points $\widehat{w_0}, \cdots, \widehat{w_{n-1}}$, in anti-clockwise direction. Furthermore, we have $\rho_j \circ \rho_k = \rho_k \circ \rho_j$ on Ω.

Below, we will derive a simple formula for $\sharp[\rho]$. For simplicity, we define the sequence $\{\widehat{w_n}\}$ to be

$$\widehat{w_0}, \cdots, \widehat{w_{n-1}}, \widehat{w_0}, \cdots, \widehat{w_{n-1}}, \cdots.$$

Then those ρ_j satisfies

$$\rho_j(\widehat{w_i}) = \widehat{w_{i+j}}, 0 \le i \le n-1.$$

Since $[(\widehat{w_0}, \widehat{w_j})] = [(\widehat{w_i}, \widehat{w_{i+j}})]$, for any $\rho \in \{\rho_j\}$ we get

$$\sharp[\rho] = \frac{\sharp\{(w, z) \in G[\rho] : w = \widehat{w_0}, \cdots, \widehat{w_{n-1}}\}}{n}.$$

Note that

$$\sharp\{(w, z) \in G[\rho] : w = \widehat{w_0}, \cdots, \widehat{w_{n-1}}\} = \sharp\{(w, z) \in G[\rho] : z = \widehat{w_0}, \cdots, \widehat{w_{n-1}}\},$$

and by Proposition 5.3.2 we get $\sharp[\rho] = \sharp[\rho^-]$.

Now assume that B is a thin Blaschke product and $[\rho]$ is one component of S_B. As we will see, B restricted on an appropriate domain $\Omega_t \subseteq \mathbb{D}$ behaves like a finite Blaschke product, and local inverses ρ^* of $B|_{\Omega_t}$ are well-defined. In fact, let r be any fixed number in $(0, 1)$ and put

$$t' = \max\{|B(z)| : |z| \le r\}.$$

By Proposition 5.1.1, $\{B(z) : z \in Z(B')\}$ is a discrete subset of $\mathbb{D}$, and then we may take a number t such that

$$t \notin \{|B(z)| : z \in Z(B')\} \quad \text{and} \quad t' < t < 1.$$

Let Ω_t denote the component of $\{z : |B(z)| < t\}$ containing $r\mathbb{D}$. By the maximum modulus principle, it is not difficult to see that Ω_t is a simply-connected domain; and by the choice of t, the boundary of Ω_t is analytic, and hence a Jordan curve. Then by Riemann's mapping theorem, there is a biholomorphic map $f : \mathbb{D} \to \Omega_t$, which extends to a homeomorphism from $\overline{\mathbb{D}}$ onto $\overline{\Omega_t}$. By Lemma 5.1.2, $B|_{\Omega_t} : \Omega_t \to t\mathbb{D}$ is a proper map, and then $B \circ f : \mathbb{D} \to t\mathbb{D}$ is also a proper map. By Theorem 2.1.3, there is a finite Blaschke product B_0 such that $B \circ f = tB_0$.

This observation shows that if $\sharp[\rho] < \infty$, then by choosing enough large Ω_t we get $\sharp[\rho^*] = \sharp[\rho]$. Then by the above result for finite Blaschke products,

$$\sharp[\rho^-] \ge \sharp[\rho^{-*}] = \sharp[\rho^*] = \sharp[\rho].$$

If $\sharp[\rho] = \infty$, then the above reasoning shows that for any integer k, $\sharp[\rho^-] \ge k$, and hence $\sharp[\rho^-] \ge \sharp[\rho]$. Therefore, in either case $\sharp[\rho^-] \ge \sharp[\rho]$, and similarly $\sharp[\rho] \ge \sharp[\rho^-]$, forcing $\sharp[\rho] = \sharp[\rho^-]$. The proof of Proposition 5.3.4 is complete. $\square$

Proof of Theorem 5.3.1 Suppose B is a thin Blaschke product. It suffices to show that $\mathcal{V}^*(B)$ is nontrivial if and only if there is a nontrivial local inverse ρ such that $\sharp[\rho] < \infty$. First we deal with the "if" part. If there is a nontrivial local inverse ρ

such that $\sharp[\rho] < \infty$, then by Proposition 5.3.4

$$\sharp[\rho^-] = \sharp[\rho] < \infty,$$

which, combined with Lemma 5.3.3, shows that $\mathcal{E}_{[\rho]} \in \mathcal{V}^*(B)$. By the local representation of $\mathcal{E}_{[\rho]}$, we have $\mathcal{E}_{[\rho]} \neq cI$ for any constant c since $\rho \neq id$. Therefore, $\mathcal{V}^*(B)$ is nontrivial.

The converse direction follows directly from Proposition 5.2.8. The proof is complete. $\qquad\qquad\qquad\qquad\qquad\qquad\qquad\qquad\qquad\qquad\qquad\qquad\quad\square$

It is natural to ask for which thin Blaschke product B, M_B has a nontrivial reducing subspace. The following provides such an example.

Example 5.3.5 Write $\phi = h \circ \varphi$, where $h \in H^\infty(\mathbb{D})$ and φ is a finite Blaschke product with order $\varphi \geq 2$. Then we have $\mathcal{V}^*(\phi) \supseteq \mathcal{V}^*(\varphi)$. Since $\mathcal{V}^*(\varphi)$ is nontrivial [HSXY], then $\mathcal{V}^*(\phi)$ is nontrivial. For example, Proposition 2.1.11 shows that for each thin Blaschke product B, $B \circ \varphi$ is still a thin Blaschke product, and $M_{B\circ\varphi}$ has nontrivial reducing subspaces.

For a thin Blaschke product B, if there is a member ρ in $\mathrm{Aut}(\mathbb{D})(\rho \neq id)$ satisfying $B \circ \rho = B$, then

$$U_\rho : f \rightarrow f \circ \rho\,\rho', f \in L_a^2(\mathbb{D})$$

defines a unitary operator in $\mathcal{V}^*(B)$, and thus $\mathcal{V}^*(B)$ is nontrivial. However, such a B is strongly restricted. This is characterized by the following, which is of independent interest, see [GH4].

Proposition 5.3.6 *Given a thin Blaschke product B, if there is a member ρ in $\mathrm{Aut}(\mathbb{D})$ ($\rho \neq id$) satisfying $B \circ \rho = B$, then there exists a thin Blaschke product B_1, an integer $n(n \geq 2)$ and a point $\lambda \in \mathbb{D}$ such that*

$$B = B_1(\varphi_\lambda^n).$$

The finite-Blaschke-product version of the above first appeared in [DSZ, Lemma 8.1]. Proposition 5.3.6 is sharp in the sense that it fails for interpolating Blaschke products, especially for those holomorphic covering maps $B : \mathbb{D} \rightarrow \mathbb{D}-F$, with F a discrete subset of $\mathbb{D}$, see Example 6.3.9 in Chap. 6.

Proof The proof is from [GH4].

Assume that B is a thin Blaschke product, and there is a member ρ in $\mathrm{Aut}(\mathbb{D})(\rho \neq id)$ satisfying $B \circ \rho = B$. Without loss of generality, we may assume that B has only simple zeros; otherwise we shall replace B with $\varphi_w \circ B$, and Proposition 5.1.4(2) shows that $\varphi_w \circ B$ is also a thin Blaschke product. First, we will show that *the group G generated by ρ is finite*. To see this, assume conversely that G is not finite. Under this assumption, we first show that there is a $z_0 \in Z(B)$ such that for all $i, j \in \mathbb{Z}$ with $i \neq j$,

$$\rho^i(z_0) \neq \rho^j(z_0), \tag{5.22}$$

where ρ^j denotes the j-th iterate of ρ. If (5.22) were not true, then there would be a positive integer n_1 satisfying

$$\rho^{n_1}(z_0) = z_0.$$

Now take some $z_1 \in Z(B)$ such that $z_1 \neq z_0$. If $\{\rho^j(z_1)\}$ are pairwise different, then we are done; otherwise, there would be a positive integer n_2 satisfying

$$\rho^{n_2}(z_1) = z_1.$$

Clearly, $\rho^{n_1 n_2}$ is a member in $\mathrm{Aut}(\mathbb{D})$ satisfying

$$\rho^{n_1 n_2}(z_0) = z_0 \quad \text{and} \quad \rho^{n_1 n_2}(z_1) = z_1.$$

That is, $\rho^{n_1 n_2}$ has two different fixed points z_0 and z_1 in $\mathbb{D}$. However, by Milnor [Mi, Theorem 1.12] each member in $\mathrm{Aut}(\mathbb{D})$ admitting at least two different fixed points must be the identity map, and hence $\rho^{n_1 n_2} = id$. This is a contradiction. Therefore, there exists some $z_0 \in Z(B)$ such that $\{\rho^j(z_0)\}$ are pairwise different, as desired.

Thus, if G were not infinite, then $\{\rho^j(z_0)\}_{j \in \mathbb{Z}}$ were a subsequence in $Z(B)$, and hence a thin Blaschke sequence. Thus we would have

$$\lim_{j \to \infty} \prod_{k \in \mathbb{Z}, k \neq j} d(\rho^j(z_0), \rho^k(z_0)) = 1.$$

However, since the hyperbolic metric d is invariant under Mobius maps,

$$\prod_{k \in \mathbb{Z}, k \neq j} d(\rho^j(z_0), \rho^k(z_0)) \equiv \prod_{k \in \mathbb{Z} - \{0\}} d(z_0, \rho^k(z_0)) < 1, \quad j \in \mathbb{Z}.$$

This is a contradiction. Therefore, G is finite.

By the theory of complex dynamics (see [Mi, Appendix E]), the finite group G gives a holomorphic regular branched covering map, ψ. Since G is a cyclic finite group, ψ has the following form: $\psi = h(\varphi_\lambda^n)$ for some $h \in \mathrm{Aut}(\mathbb{D})$ (also see Theorem 6.6.2). Write $\rho^* \triangleq \varphi_\lambda \circ \rho \circ \varphi_\lambda$, and then $z^n \circ \rho^* = z^n$ since $\psi \circ \rho = \psi$. Then there is an n-th root ξ of unity such that $\rho^*(z) = \xi z$, and hence

$$\rho(z) = \varphi_\lambda(\xi \varphi_\lambda(z)).$$

The above representation for ρ was also obtained in [DSZ, Lemma 8.1]. Note that $B \circ \varphi_\lambda$ is also a thin Blaschke product satisfying $(B \circ \varphi_\lambda) \circ \rho^* = B \circ \varphi_\lambda$, and $B = (B \circ \varphi_\lambda) \circ \varphi_\lambda$. To finish the proof, it suffices to discuss the case of ρ being a rational rotation, i.e. $\rho(z) = \xi z$. In this case, we will show that $B(z) = B_1(z^n)$ for some thin Blaschke product B_1.

For this, observe that

$$B(\rho^j) = B, \ 1 \le j < n,$$

which implies that for any $z_0 \in Z(B)$, z_0 and $\rho^j(z_0)$ are zeros of B, with the same multiplicity. Therefore, there is a subsequence $\{z_j\}$ in $Z(B)$ such that the zero sequence of $Z(B)$ is as follows:

$$z_0, \rho(z_0), \cdots, \rho^{n-1}(z_0); z_1, \rho(z_1), \cdots, \rho^{n-1}(z_1); \cdots,$$

where $\rho(z) = \xi z$. Note that the zero sequence of $\varphi_{z_k^n}(z^n)$ is exactly

$$z_k, \rho(z_k), \cdots, \rho^{n-1}(z_k).$$

On the other hand, $\{z_j\}$ satisfies the Blaschke condition:

$$\sum_{j=1}^{\infty} (1 - |z_j|^2) < \infty.$$

Since $1 - |z_j^n|^2 = 1 - (|z_j|^2)^n \le n(1 - |z_j|^2)$, then

$$\sum_{j=1}^{\infty} (1 - |z_j^n|^2) < \infty.$$

That is, $\{z_j^n\}$ also gives a Blaschke product, denoted by B_1. By omitting a unimodular constant, we have $B_1(z^n) = B(z)$. Since B is a thin Blaschke product, it's easy to verify that B_1 is also a thin Blaschke product. The proof is complete. □

Note that those thin Blaschke products B in Proposition 5.3.6 have rather special form and for such B, $V^*(B)$ is clearly nontrivial. In next section, it will be shown that for most thin Blaschke product B, $V^*(B)$ is trivial. Equivalently, most M_B are irreducible.

5.4   Most M_B Are Irreducible

In last section, a complete characterization is presented for when $V^*(B)$ is nontrivial in terms of Riemann surface associated with this thin Blaschke product B. Based on this, this section develops techniques of analytic continuation for local inverses and then show that $V^*(B)$ is trivial under a mild condition. That is, for most thin Blaschke products B, M_B has no nontrivial reducing subspace. This is equivalent to say that most M_B are irreducible. As we will see, this fact depends heavily on the geometric property of thin Blaschke products.

The main result of this section, due to Guo and Huang [GH4], is stated as follows.

Theorem 5.4.1 *Suppose B is a thin Blaschke product, and there is a finite subset F of $Z(B')$ such that:*

(1) $B|_{Z(B')-F}$ *is injective;*
(2) *Each point in $Z(B') - F$ is a simple zero of B'.*

Then M_B has no nontrivial reducing subspace. Equivalently, $V^*(B)$ is trivial.

It seems that either (1) or (2) rarely fails for a thin Blaschke product B. If B has the form $B = B_1 \circ B_2$ for two Blaschke products B_1, B_2 and $2 \leq \mathrm{order}\, B_2 < \infty$, then (1) fails. In this case, $V^*(B)$ is nontrivial since $V^*(B) \supseteq V^*(B_2)$, see Example 5.3.5.

In what follows, we will introduce a notion called *gluable*, which comes from [GH4] and proves to play an important role in the proof of Theorem 5.4.1.

Now given a holomorphic function ψ defined on $\mathbb{D}$, set

$$\mathfrak{F} \triangleq \psi^{-1} \circ \psi\big(Z(\psi')\big),$$

i.e.

$$\mathfrak{F} = \{z \in \mathbb{D};\ \text{there is a } w \in \mathbb{D} \text{ such that } \psi(w) = \psi(z) \text{ and } \psi'(w) = 0\}.$$

We assume that $\mathfrak{F}$ is relatively closed in $\mathbb{D}$, and its complement set $E \triangleq \mathbb{D} - \mathfrak{F}$ is connected. Special interest is focused on the case when ψ is a thin (or finite) Blaschke product. In this case, Proposition 5.1.1 says that Q is a discrete subset of $\mathbb{D}$, and hence the open set E is connected.

As follows, we introduce some notions from [GH4].

Definition 5.4.2 Given a function ψ as above and two points $z_1, z_2 \in E$ satisfying $\psi(z_1) = \psi(z_2)(z_1 \neq z_2)$, if there are two paths σ_1 and σ_2 satisfying the following

(a) $\sigma_i(0) = z_i$ for $i = 1, 2, \sigma_1(1) = \sigma_2(1)$.
(b) $\psi(\sigma_1(t)) = \psi(\sigma_2(t)), 0 \leq t \leq 1$;
(c) For $0 \leq t < 1, \sigma_i(t) \in E, i = 1, 2,$

then z_1 and z_2 are directly ψ-glued, denoted by $z_1 \overset{\psi}{\simeq} z_2$. In addition, we make the convention that z_1 is directly ψ-glued to itself, i.e. $z_1 \overset{\psi}{\simeq} z_1$.

In Definition 5.4.2, the conditions (a)–(c) imply that $\psi'(\sigma_1(1)) = 0$. To see this, assume conversely that $\psi'(\sigma_1(1)) \neq 0$. Let ρ be a local inverse of ψ defined on a neighborhood of $\sigma_1(0)$ and $\rho(\sigma_1(0)) = \sigma_2(0)$. Then by (b) and (c), the continuous map

$$(t, \sigma_1(t)) \mapsto \sigma_2(t),\ t \in [0, 1]$$

naturally induces an analytic continuation $\tilde{\rho}$ of ρ along σ_1, satisfying $\tilde{\rho}(\sigma_1(1)) = \sigma_2(1)$. Since $\sigma_1(1) = \sigma_2(1)$, locally $\tilde{\rho}$ equals the identity map.

By the uniqueness of analytic continuation, ρ is also the identity map, forcing $\sigma_1(0) = \sigma_2(0)$, i.e. $z_1 = z_2$. This is a contradiction. Therefore, $\psi'(\sigma_1(1)) = 0$.

Also, in Definition 5.4.2 one can require that

$$\psi(\sigma_i(t)) \neq \psi(z_1), \ 0 < t < 1,$$

for each i. For this, just replace these paths σ_i with a small permutation of σ_i if $\psi(\sigma_i(t)) \neq \psi(z_1)$ holds for some $t \in (0, 1)$.

We also need the following.

Definition 5.4.3 For a sub-domain Ω of $\mathbb{D}$, if there is a finite sequence $z_0, z_1, \cdots, z_k$ in Ω, such that

$$z_j \overset{\psi}{\simeq} z_{j+1}, \ 0 \leq j \leq k - 1,$$

and all related paths σ_i in Definition 5.4.2 can be chosen in Ω, then z_0 and z_k are called ψ-glued on Ω, denoted by $z_0 \overset{\psi|\Omega}{\sim} z_k$; in short, $z_0 \overset{\Omega}{\sim} z_k$ or $z_0 \sim z_k$. If $w \overset{\Omega}{\sim} z$ for all $z \in \psi^{-1} \circ \psi(w) \cap \Omega$, then Ω is called ψ-glued with respect to w. In particular, if $\mathbb{D}$ is ψ-glued with respect to w, then ψ is called gluable with respect to w.

If ψ is gluable with respect to w for all $w \in E$, then ψ is called *gluable*.

By Definition 5.4.2, each univalent function ψ is gluable. For more examples, write $B(z) = z^n (n \geq 2)$. Then the only critical point of B is 0, and $E = \mathbb{D} - \{0\}$. For two different points $z_1, z_2 \in E$ with $B(z_1) = B(z_2)$, define $\sigma_i(t) = (1-t)z_i$, $i = 1, 2$. It is easy to verify conditions (a), (b) and (c) in Definition 5.4.2 hold, and thus $z_1 \overset{\psi}{\simeq} z_2$. Therefore, B is gluable. In general, we have the following result [GH4].

Proposition 5.4.4 *Each finite Blaschke product B is gluable.*

Proof Given a point z_0 in E, we must show that B is gluable with respect to z_0. Since for each Möbius map φ, the gluable property of $\varphi \circ B$ is equivalent to that of B (with respect to z_0), we may assume that $B(z_0) = 0$.

Write order $B = n + 1$ and

$$B^{-1} \circ B(z_0) = \{z_0, z_1, \cdots, z_n\}.$$

For each $r \in (0, 1)$, consider the components $\mathcal{J}_{i,r}$ of $\{z : |B(z)| < r\}$ containing z_i $(0 \leq i \leq n)$. Clearly, for enough small $r > 0$, those $\mathcal{J}_{i,r}$ are pairwise disjoint and

$$\{z : |B(z)| < r\} = \bigsqcup_{i=0}^{n} \mathcal{J}_{i,r}.$$

Since each $\mathcal{J}_{i,r}$ contains exactly one z_i, it must be directly B-glued. Also, note that by the maximal modulus principle, for $0 < s < 1$, any component $\mathcal{J}_{i,s}$ of $\{z : |B(z)| < s\}$ is simply connected; and the number $N(s)$ of the components $\{\mathcal{J}_{i,s}\}$ decreases in s.

Below, we will enlarge r and make induction on the number of B-glued components of $\{z : |B(z)| < r\}$. By standard analysis, one can show that there exist a number $t(0 < t < 1)$ satisfying the following:

(1) For any r with $0 < r < t$, any two of the components $\mathcal{J}_{i,r}$ are disjoint;
(2) There are at least two distinct i and i' such that $\overline{\mathcal{J}_{i,t}} \cup \overline{\mathcal{J}_{i',t}}$ is connected.

Now we get $m(m < n + 1)$ components of $\cup_{i=0}^{n} \overline{\mathcal{J}_{i,t}}$, and denote by η the minimal distance between any two components among them. Write $\varepsilon = \frac{\eta}{2}$ and there is a constant $\delta > 0$ such that

$$\{z \in \mathbb{D} : |B(z)| < t + \delta\} \subseteq \cup_i O(\mathcal{J}_{i,t}, \varepsilon),$$

where

$$O(\mathcal{J}_{i,t}, \varepsilon) \triangleq \{w \in \mathbb{D} : \text{there is a point } z \in \mathcal{J}_{i,t} \text{ such that } |z - w| < \varepsilon\}.$$

Then there are exactly m components of $\cup_i \mathcal{J}_{i,t+\delta}$, and we deduce that if Ω is a component of $\cup_i \mathcal{J}_{i,t+\delta}$ contains z_j, then Ω is B-glued with respect to z_j.

To see this, note that for any $s(0 < s < 1)$, $\mathcal{J}_{i,s}$ is simply connected, and its boundary $\partial \mathcal{J}_{i,s}$ must be a Jordan curve satisfying

$$\partial \mathcal{J}_{i,s} \subseteq \{z : |B(z)| = s\}.$$

Then by (2) there is at least one point $w \in \overline{\mathcal{J}_{i,t}} \cap \overline{\mathcal{J}_{i',t}}$, see Fig. 5.2. Let Ω denote the component of $\cup_i \mathcal{J}_{i,t+\delta}$ that contains w. Since

$$B(E) = \mathbb{D} - \mathcal{E}_B,$$

by Proposition 5.1.1 $B(E)$ is connected. Thus we can pick a simple curve σ such that $\sigma - \{B(w)\} \subseteq B(E) \cap t\mathbb{D}$, and σ connects $B(z_0)$ with $B(w)$. Since $B^{-1}(\sigma - \{B(w)\})$ consists of disjoint arcs, there exists at least one curve γ_i in $\overline{\mathcal{J}_{i,t}}$ and another $\gamma_{i'}$ in $\overline{\mathcal{J}_{i',t}}$ such that

$$B(\gamma_i) = B(\gamma_{i'}) = \sigma,$$

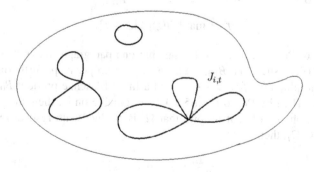

Fig. 5.2 Gluable property of B

where $\gamma_i(1) = \gamma_{i'}(1) = w$ and

$$\gamma_i(0) = z_i \text{ and } \gamma_{i'}(0) = z_{i'}.$$

Therefore, Ω is B-glued with respect to z_i. Similarly, any other component of $\cup_i \mathcal{J}_{i,t+\delta}$ are also B-glued, as desired.

Noting that $m < n + 1$, we have reduced the number of those B-glued components. The next step is to find some t' such that (1) and (2) holds, and the following discussion as the same as the above two paragraphs. After finite procedures, we come to the case of $m = 2$; and by repeating the above procedure, one can prove that there is some $r \in (0, 1)$ such that $\{z : |B(z)| < r\}$ is itself B-glued with respect to z_0; and thus $\mathbb{D}$ is also B-glued. In another word, B is gluable. The proof is complete.   □

Let us make a sketch of the proof of Proposition 5.4.4. For a finite Blaschke product B, write

$$B^{-1} \circ B(z_0) = \{z_0, z_1, \cdots, z_n\}.$$

First we have only small B-glued components of $\{z : |B(z)| < r\}$, and those points z_k in a same component Ω are B-glued on Ω. To glue two points z_j and z_k in different components, we need "longer" paths as in Definition 5.4.2. The idea is to enlarge r and Ω, and then it is possible to glue z_j and z_k. Finally, we get an sufficiently large component that glues all z_k.

We have naturally a corollary and generalization of Proposition 5.4.4.

Corollary 5.4.5 *Each thin Blaschke product is gluable.*

Proof Assume that B is a thin Blaschke product. Recall that

$$E = \{z \in \mathbb{D}; \text{there is no } w \text{ such that } B(w) = B(z) \text{ and } B'(w) = 0\}.$$

Now fix a $z_0 \in E$ and write

$$B^{-1}(B(z_0)) = \{z_n : n = 0, 1, \cdots\}.$$

We will show that $z_0 \sim z_k$ for each given k.

To see this, set $r = \max\{|z_i| : 0 \le i \le k\}$ and put

$$t' = \max\{|B(z)| : |z| \le r\}.$$

In the proof of Proposition 5.3.4, the last but one paragraph shows that there is a number t in $(t', 1)$ such that $B|_{\Omega_t} : \Omega_t \to t\mathbb{D}$ is a proper map; furthermore, there is a biholomorphic map $f : \mathbb{D} \to \Omega_t$ and a finite Blaschke product B_0 such that $B \circ f = tB_0$. Then by Proposition 5.4.4 B_0 is gluable on $\mathbb{D}$, then so is $B \circ f$. By the biholomorphicity of f, it follows that Ω_t is B-glued with respect to z_0. Since $\{z_0, \cdots, z_k\} \subseteq \Omega_t$, then

$$z_0 \overset{\Omega_t}{\sim} z_j, j = 0, \cdots, k,$$

and hence

$$z_0 \overset{\mathbb{D}}{\sim} z_k.$$

By arbitrariness of k, $\mathbb{D}$ is B-glued with respect to z_0. By the arbitrariness of z_0, B is gluable. The proof is complete.   □

In what follows, Bochner's theorem [Wa1, Wa2] will be useful. For a holomorphic function ψ, if $\psi'(w_0) = 0$, then w_0 is called *a critical point* of ψ. For a function ψ over $\mathbb{D}$, all critical points consists of a set, denoted by

$$Z(\psi') = \{z \in \mathbb{D} : \psi'(z) = 0\}.$$

Theorem 5.4.6 (Bochner) *If B is a finite Blaschke product with order n, then the critical set $Z(B')$ of B is contained in $\mathbb{D}$, and B has exactly $n - 1$ critical points, counting multiplicity.*

Using Bochner's theorem, we give the following.

Remark 5.4.7 For a thin Blaschke product B, $Z(B')$ is always an infinite set. To see this, let $\{z_j\}$ denote the zeros of B. By the proof of Proposition 5.3.4, for each n there exists a component Ω_r of $\{z \in \mathbb{D}; |B(z)| < r\}$ satisfying $\overline{\Omega_r} \subseteq \mathbb{D}$ and

$$z_j \in \Omega_r, j = 0, \cdots, n;$$

$B|_{\Omega_r} : \Omega_r \to r\mathbb{D}$ is a k-folds $(k \geq n + 1)$ proper map. In fact, we may regard $B|_{\Omega_r}$ as a finite Blaschke product with order k. Then by Bochner's theorem, $(B|_{\Omega_r})'$ has exactly $k - 1 (k - 1 \geq n)$ zero points, counting multiplicity. By the arbitrariness of n, B' has infinitely many zeros on $\mathbb{D}$. Also, this gives a picture that describes how a thin Blaschke product B is obtained by "gluing" finite Blaschke products.

Below, B denotes a thin or finite Blaschke product. For any component $G[\rho]$ of S_B, if $(z_0, w_0) \in G[\rho]$, then *we also write* $G[(z_0, w_0)]$ *for* $G[\rho]$, *and* $[(z_0, w_0)]$ *for* $[\rho]$.

The following indicates a relation between the gluable property of B and the geometry of components in S_B [GH4].

Lemma 5.4.8 *Let B denote a thin or finite Blaschke product. Given z_0, z_1 and z_2 in E with $B(z_0) = B(z_1) = B(z_2)$, suppose there are three paths σ_i such that*

(1) $\sigma_i(0) = z_i$ for $i = 0, 1, 2$, $\sigma_1(1) = \sigma_2(1)$, and $B'(\sigma_0(1)) \neq 0$;
(2) $B(\sigma_0(t)) = B(\sigma_1(t)) = B(\sigma_2(t)), 0 \leq t \leq 1$;
(3) For $0 \leq t < 1$, $\sigma_i(t) \in E, i = 0, 1, 2$.

Then $G[(z_0, z_1)] = G[(z_0, z_2)]$.

Proof Before giving the proof, we introduce some notations. As mentioned before, a curve means a continuous map from $[0, 1]$ into $\mathbb{C}$. For a curve σ, σ^{-1} denotes its inverse, defined by

$$\sigma^{-1}(t) = \sigma(1 - t), 0 \leq t \leq 1.$$

If γ is another curve satisfying $\gamma(0) = \sigma(1)$, then the product $\sigma \sharp \gamma$ of σ and γ is defined by

$$\sigma \sharp \gamma(t) = \sigma(2t), \ 0 \le t < \frac{1}{2},$$

For simplicity, we write $\sigma \gamma$ for $\sigma \sharp \gamma$, and σ^m for the m-th product of σ.

Suppose the conditions (1)–(3) in Lemma 5.4.8 hold. Without loss of generality, we assume that $\sigma_1(1) = 0 = B(\sigma_1(1))$. From (1) and the comments below Definition 5.4.2, we have $B'(\sigma_1(1)) = 0$. Then by Böttcher's theorem, there is local holomorphic change of coordinate $w = \varphi(z)$ defined on a neighborhood of $\sigma_1(1)$, such that $\varphi \circ B \circ \varphi^{-1}(w) = w^n$. This implies that there are n disjoint paths

$$\gamma_1, \cdots, \gamma_n$$

whose images are contained in a small neighborhood of 0 (see Fig. 5.3), and a loop σ on a neighborhood of $\sigma_0(1)$ such that:

(a) $\gamma_j(1) = \gamma_{j+1}(0), j = 1, \cdots, n - 1$ and $\gamma_n(1) = \gamma_1(0)$;
(b) $B(\sigma(t)) = B(\gamma_j(t)), 0 \le t \le 1, j = 1, \cdots, n$;
(c) For some enough large $t^* \in (0, 1)$, $\sigma(0) = \sigma_0(t^*)$, $\sigma_1(t^*) = \gamma_1(0)$ and $\sigma_2(t^*) = \gamma_{j_0}(1)$ holds for some $j_0(1 \le j_0 \le n)$.

Now let $\widehat{\sigma}_i$ denote the segment of the loop σ_i connecting $\sigma_i(0)$ with $\sigma_i(t^*)$ for $i = 0, 1, 2$. Precisely, put

$$\widehat{\sigma}_i(t) = \sigma_i(t^* t), \ 0 \le t \le 1, \ i = 0, 1, 2.$$

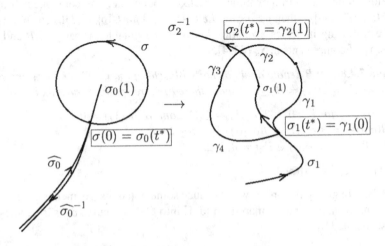

Fig. 5.3 Analytic continuation and gluable property

Note that for any j, there always exists an m such that

$$\sigma^m(t) \mapsto (\gamma_j \cdots \gamma_1)(t)$$

induces a natural analytic continuation along the loop σ^m. Let m_0 be the integer corresponding to $j = j_0$ and consider the loop $\tilde{\sigma} \triangleq \widehat{\sigma_0}^{-1} \sigma^{m_0} \widehat{\sigma_0}$. Also, there is a path

$$\tilde{\gamma} \triangleq \widehat{\sigma_2}^{-1} \gamma_{j_0} \cdots \gamma_2 \gamma_1 \widehat{\sigma_1}$$

which connects z_1 with z_2. Then one can show that $\tilde{\sigma}(t) \mapsto \tilde{\gamma}(t)$ naturally gives an analytic continuation along the loop $\tilde{\sigma}$. Since $G[(\tilde{\sigma}(0), \tilde{\gamma}(0))] = G[(\tilde{\sigma}(1), \tilde{\gamma}(1))]$ and

$$\tilde{\sigma}(0) = \tilde{\sigma}(1) = z_0, \tilde{\gamma}(0) = z_1 \text{ and } \tilde{\gamma}(1) = z_2,$$

we have $G[(z_0, z_1)] = G[(z_0, z_1)]$. The proof of Lemma 5.4.8 is complete.   □

We also need the following lemma [GH4].

Lemma 5.4.9 *Let B be a thin or finite Blaschke product. With the assumptions in Theorem 5.4.1 and F being empty, assume that*

$$B(\widehat{z_0}) = B(\widehat{z_1}) = B(\widehat{z_2}),$$

$$\widehat{z_0} \neq \widehat{z_1}, \widehat{z_0} \neq \widehat{z_2} \text{ and } \widehat{z_1} \overset{B}{\simeq} \widehat{z_2}.$$

Then $G[(\widehat{z_0}, \widehat{z_1})] = G[(\widehat{z_0}, \widehat{z_2})]$.

Proof Inspired by the proofs of Proposition 5.4.4 and Corollary 5.4.5, we first deal with the following special case: there are two different components Ω_1 and Ω_2 of $\{z : |B(z)| < t\}$ such that

$$\widehat{z_i} \in \Omega_i, \quad i = 1, 2;$$

and there is a point $z' \in Z(B')$ such that $z' \in \partial\Omega_1 \cap \partial\Omega_2$, see Fig. 5.4. Clearly, $|B(z')| = t$. Now we can find a simple curve σ connecting $B(\widehat{z_1})$ with $B(z')$, satisfying $\sigma[0, 1) \subseteq E \cap t\mathbb{D}$. Since $B^{-1}(\sigma[0, 1))$ consists of disjoint arcs, among which the endpoints are $\widehat{z_0}, \widehat{z_1}$ and $\widehat{z_2}$, respectively. Thus, there are three curves σ_i with endpoints $\widehat{z_i}(i = 0, 1, 2)$ such that

$$B \circ \sigma_i(t) = \sigma(t), 0 \leq t \leq 1, i = 0, 1, 2. \tag{5.23}$$

Note that $\sigma_1(1) = \sigma_2(1) = z'$, and then $\sigma_0(1) \neq z'$. Otherwise, by the theory of complex analysis $B(\cdot) - B(z')$ would have the order ≥ 3 at the zero z'. Therefore,

$$B'(z') = B''(z') = 0,$$

Fig. 5.4 $B^{-1}(\sigma[0,1])$

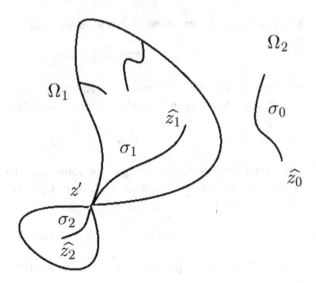

which is a contradiction to the assumption that B' has only simple zeros. Since $B|_{Z(B')}$ is injective, it follows from (5.23) that $B'(\sigma_0(1)) \neq 0$. Then by Lemma 5.4.8, $G[(\widehat{z_0}, \widehat{z_1})] = G[(\widehat{z_0}, \widehat{z_2})]$, as desired.

The general case can be done as above, with a bit modification.   □

Now we are ready to give the proof of Theorem 5.4.1.

Proof of Theorem 5.4.1 The proof is from [GH4]. We will show that S_B has no nontrivial component of finite multiplicity, and then apply Proposition 5.2.8 to finish the proof.   □

In fact, a generalization of Frostman's theorem [Ru4] states that for each nonconstant function $h \in H^{\infty}(\mathbb{D})$, the inner part of $h - h(\lambda)$ is a Blaschke product with distinct zeros, except for λ in a set of capacity zero. Then by Proposition 5.1.4(2), we may assume that B is a thin Blaschke product of simple zeros $\{z_j\}$, without any change of the assumption of Theorem 5.4.1. The proof is divided into two steps because of its length.

Step I. We will prove Theorem 5.4.1 in the case where $F = \emptyset$.

In this case, draw a graph, $\{z_n\}$ being all the vertices. For some preliminaries of the graph theory, one may refer to [Hat, pp. 83–86]. If $z_j \overset{B}{\simeq} z_k$, then by Lemma 5.4.9

$$G[(z_i, z_j)] = G[(z_i, z_k)], \quad \text{if} \quad z_i \neq z_j \quad \text{and} \quad z_i \neq z_k. \tag{5.24}$$

For the pair (z_j, z_k) with $z_j \overset{B}{\simeq} z_k$, draw an abstract edge between z_j and z_k, called an *abstract* ∗-arc. All different ∗-arcs are assumed to be *disjoint*. This can be done because all z_j are in $\mathbb{C}$, which can be regarded as a subspace of $\mathbb{R}^3$; and all ∗-arcs are taken as curves in $\mathbb{R}^3$.

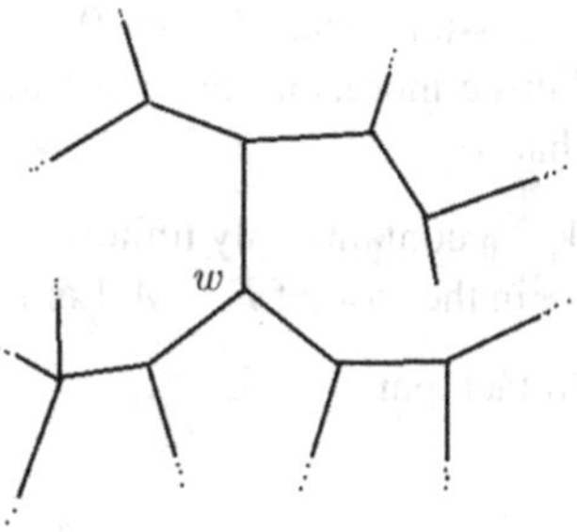

Fig. 5.5 Tree of *-arcs

By the geometry of B (see the details in proofs of Proposition 5.4.4 and Corollary 5.4.5), such *-arcs yields an abstract path (consisting of several adjoining *-arcs) connecting any two given points in $\{z_n\}$, see Fig. 5.5. Therefore, all points z_n and *-arcs consists of a connected graph, S. We may assume that S is simply-connected, since by the graph theory, any connected graph contains a contractible (and thus simply-connected) subgraph, which has all vertices of S. This subgraph is called a maximal tree [Hat, p. 84, Proposition 1A.1]. For each connected subset $\Gamma \subseteq \{z_n\}$ (with some *-arcs), if $z \in \{z_n\} - \Gamma$, then by (5.24) it is not difficult to see that $G[(z, \lambda')] = G[(z, \lambda'')]$ whenever $\lambda', \lambda'' \in \Gamma$. Then one can write $G[(z, \Gamma)]$ to denote $G[(z, \lambda')]$, and we caution the reader that $G[(z, \Gamma)]$ may contain (z, z_n) for some $z_n \notin \Gamma$.

Now pick one $w \in \{z_n\}$. In the graph S, delete the vertex w and all *-arcs beginning at w, and the remaining consists of finitely or infinitely many connected components Γ_i. If each component contains infinitely many points z_j, then by Lemma 5.4.9, except for the identity component $G[(w, w)] \equiv G[z]$, all other components $G[(w, \Gamma_i)]$ of S_B are infinite; that is,

$$\sharp G[(w, \Gamma_i)] = \infty.$$

In this case we are done by Proposition 5.2.8. Otherwise, there is one component Γ^* consisting of only finite points. Since S is a tree and $\Gamma^* \subseteq S$, then Γ^* is also a tree. We distinguish two cases:

Case I $\sharp\Gamma^* \geq 2$. In this case, one can find an endpoint w^* of Γ^*. Replacing w with w^*, from the reasoning of the above paragraph it follows that there are only two components: the identity component and $G[(w^*, \Gamma)]$, which must be infinite. Thus, the only finite component of S_B is the identity component. By Proposition 5.2.8, $V^*(B)$ is trivial, and M_B has no nontrivial reducing subspace.
Case II $\sharp\Gamma^* = 1$. In this case, write $\Gamma^* = \{w^*\}$ and replace w with w^*. By the reasoning as in Case I, one deduces that $V^*(B)$ is trivial, and M_B has no nontrivial reducing subspace. Thus, the case of F being empty is done.

Step II. To complete the proof of Theorem 5.4.1, we must show that $V^*(B)$ is trivial in the case where $F \neq \emptyset$.

Assume that $F \neq \emptyset$. We also have the connected graph S as in Step I. Furthermore, one can construct a connected sub-graph Γ_0 (related with F) of S such that

1. Γ_0 contains only finitely many vertices;
2. in the case of $F \neq \emptyset$, Lemma 5.4.9 still holds provided that $\widehat{z}_0 \notin \Gamma_0$.

In fact, put

$$t' = \max\{|B(z)| : z \in F\}.$$

Following the proof of Proposition 5.3.4, one can take a number t such that

$$t \notin \{|B(z)| : z \in Z(B')\} \quad \text{and} \quad t' < t < 1.$$

Let Ω_t denote the component of $\{z : |B(z)| < t\}$ containing F. Then Ω_t is a simply-connected domain; and there is a biholomorphic map $f : \mathbb{D} \to \Omega_t$, and a finite Blaschke product B_0 such that $B \circ f = t B_0$. Put

$$\Gamma_0 = \{z_j \in \Omega_t : j \in \mathbb{Z}_+\},$$

which is clearly a finite set. Observe that Ω_t is B-glued with respect to z_j for any $z_j \in \Gamma_0$, which implies that Γ_0 is connected. Then one can follow the proof of the claim in Step 1, with a bit modification, to deduce that $G[(\widehat{z}_0, \widehat{z}_1)] = G[(\widehat{z}_0, \widehat{z}_2)]$. To be precise, as one construct the simple curve σ connecting $B(\widehat{z}_1)$ with $B(z')$, one may require that $\sigma[0, 1) \subseteq E \cap \Omega_t$. Since $B^{-1}(\sigma[0, 1))$ consists of disjoint arcs, among which the endpoints are $\widehat{z}_0, \widehat{z}_1$ and $\widehat{z}_2$, respectively; and there are three curves σ_i with endpoints $\widehat{z}_i (i = 0, 1, 2)$ such that

$$B \circ \sigma_i(t) = \sigma(t), 0 \leq t \leq 1, i = 0, 1, 2. \tag{5.25}$$

Since $\sigma_0(0) = \widehat{z}_0 \notin \Omega_t$ and Ω_t is a component of $\{z : |B(z)| < t\}$, $\sigma_0 \cap \Omega_t = \emptyset$. Then we deduce that $\sigma_0(1) \neq z'$ because $z' \in \Omega_t$. Since $\sigma_1(1) = \sigma_2(1) = z'$, by (5.25) and the discussion below Definition 5.4.2,

$$B'(z') = 0.$$

Since $\sigma_0 \cap \Omega_t = \emptyset$,

$$\sigma_0(1) \notin F.$$

Since $B|_{Z(B')-F}$ is injective, it follows from (5.25) and $B'(z') = 0$ that

$$B'(\sigma_0(1)) \neq 0.$$

Then by Lemma 5.4.8, $G[(\widehat{z}_0, \widehat{z}_1)] = G[(\widehat{z}_0, \widehat{z}_2)]$, as desired.

Now regard Γ_0 as one "point" or a whole body. Delete Γ_0 and consider the remaining part $S - \Gamma_0$. If there is some component containing only finitely many points, then by similar discussion as Case I, one can show that the only finite component of S_B is the identity component and thus $\mathcal{V}^*(B)$ is trivial. Otherwise, *each component of $S - \Gamma_0$ has infinitely many points*. In this situation, as follows one can show that $\mathcal{V}^*(B)$ is trivial.

In fact, there are several situations to deal with. Firstly, if $S - \Gamma_0$ has more than two components each with infinitely many points, then let Γ_1 be such a component and pick $w \in \Gamma_1$ (this w plays the same role as the point $w \in \{z_n\}$ in Step I). Regarding Γ_0 as one "point", the conclusion follows from the same discussion as in the case where $F = \emptyset$.

Secondly, if $S - \Gamma_0$ is a connected graph with infinitely many points, then rewrite $\Gamma_1 = S - \Gamma_0$ and pick a point w from Γ_1. If each component of $S - \{w\}$ has infinitely many points, or if $S - \{w\}$ has a finite component which has no intersection with Γ_0, then the conclusion also follows from the same discussion as in the case where $F = \emptyset$. The remaining case is that $S - \{w\}$ has exactly one finite component which necessarily contains Γ_0. Thus, we may assume that for each $\hat{w} \in S - \Gamma_0$, $S - \{\hat{w}\}$ has exactly one finite component which necessarily contains Γ_0; otherwise, it can be done by the above discussions. From this property of Γ_0, it is not difficult to see that *S looks like a half-line, with Γ_0 the endpoint, see Fig. 5.6*, and the proof reduces to this case. Assume conversely that $\mathcal{V}^*(B)$ is nontrivial. Then by Proposition 5.2.8, S_B must contain one nontrivial component of finite multiplicity. Let w be the point next to Γ_0 in S, the only possible components of S_B are as follows: the identity component $G[(w, w)]$, one infinite component and one finite component $G[(w, \Gamma_0)]$ with multiplicity $n = \sharp\Gamma_0$. However, if we denote by w' the point next to w, then the only nontrivial finite component is

$$G[(w', \Gamma_0)] = G[(w', \Gamma_0 \cup \{w\})]$$

whose multiplicity is not less than $\sharp\Gamma_0 + 1$. This derives a contradiction since there is exactly one nontrivial finite component

$$G[(w', \Gamma_0)] = G[(w, \Gamma_0)].$$

Fig. 5.6 Tree like a half line

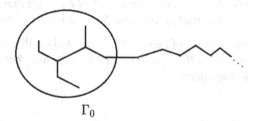

Γ_0

Therefore, $V^*(B)$ is trivial, and hence M_B has no nontrivial reducing subspace. The proof is complete.   □

Let us have a look at the above proof. In the case of a thin Blaschke product, we present a finite subset Γ_0 of S, which is related to a finite set F. However, if B is a finite Blaschke product, it is probable that $\Gamma_0 = S$, which would spoil the proof of Theorem 5.4.1.

In fact, for a finite Blaschke product B, S is a graph of finite vertices. In the case of F being empty, applying the proof of Theorem 5.4.1 shows that there is a point w and a connected subgraph Γ such that S consists of Γ and w. Then S_B has exactly two components: $G[z]$ and $G[(w, \Gamma)]$, which are necessarily of finite multiplicities since B is of finite order. Thus, we get the following result.

Corollary 5.4.10 *For a finite Blaschke product B, if the conditions in Theorem 5.4.1 hold with F being empty, then M_B has exactly two minimal reducing subspaces, and thus dim $V^*(B) = 2$. Therefore, $V^*(B)$ is abelian.*

Very Recently, Douglas et al. [DPW] showed that $V^*(B)$ is abelian and is $*$-isomorphic to q-th direct sum $\mathbb{C} \oplus \cdots \oplus \mathbb{C}$ of $\mathbb{C}$, where q $(2 \leq q \leq n)$ is the number of components of the Riemann surface S_B. Corollary 5.4.10 shows that in most cases $q = 2$, and thus gives some support for Theorem 4.2.1 in the sense of probability. In this case, write $M_0 = \overline{span}\{B'B^n : n = 0, 1, \cdots\}$, and then M_0 and $M_0^\perp$ are the only minimal reducing subspaces for M_B [SZ2, Theorem 15].

In Sect. 5.7 in Chap. 5, we will see more examples of $V^*(B)$ where B dissatisfies the assumption of Corollary 5.4.10, see Proposition 5.8.1 and Example 5.8.2.

5.5   The Construction of an Example

Last section shows that for most thin Blaschke products B, $V^*(B)$ is trivial. This section will give a construction of such a thin Blaschke product B that M_B has no nontrivial reducing subspace.

We need the following lemma, which appeared in [GH4].

Lemma 5.5.1 *Suppose B is a finite Blaschke product with order $B \geq 2$. Then for any $r \in (0, 1)$, there is an $s(0 < s < 1)$ such that for any $\lambda \in \mathbb{D}$ with $|\lambda| > s$, we have the following:*

(1) there is a critical point w^* of $B\varphi_\lambda$ such that $|B(w^*)\varphi_\lambda(w^*)| > r$.
(2) $(B\varphi_\lambda)(w^*)$ is different from the values of $B\varphi_\lambda$ on $Z\big((B\varphi_\lambda)'\big) - \{w^*\}$.

By the proof of Lemma 5.5.1, one will see that with an appropriate choice of s, if the restriction $B|_{Z(B')}$ of B on $Z(B')$ is injective, then the restriction of $B\varphi_\lambda$ on $Z\big((B\varphi_\lambda)'\big)$ is also injective.

Before we give the proof, let us make an observation. For each $\lambda \in \mathbb{D} - \{0\}$ and a compact subset K of $\mathbb{D}$,

$$|\varphi_\lambda(z) - \frac{\lambda}{|\lambda|}| \le |\frac{\lambda - z}{1 - \bar{\lambda}z} - \lambda| + |\lambda - \frac{\lambda}{|\lambda|}|$$

$$= |z\frac{1 - |\lambda|^2}{1 - \bar{\lambda}z}| + 1 - |\lambda|$$

$$\le \frac{|z|(1 - |\lambda|^2) + (1 - |\lambda|)(1 - |z|)}{1 - |z|}$$

$$\le 2\frac{1 - |\lambda|}{1 - |z|} \le \frac{2(1 - |\lambda|)}{1 - r(K)}, \quad z \in K,$$

where $r(K) \triangleq \max\{|z| : z \in K\}$. Write $\psi_\lambda = \frac{|\lambda|}{\lambda}\varphi_\lambda$, and then

$$|\psi_\lambda(z) - 1| \le \frac{2(1 - |\lambda|)}{1 - r(K)}, \quad z \in K. \tag{5.26}$$

Proof of Lemma 5.5.1 The proof comes from [GH4].

Suppose B is a finite Blaschke product with order $B \triangleq n \ge 2$. If (1) is proved, then one will see that the number s in Lemma 5.5.1 can be enlarged to satisfy (2). To see this, given an $r \in (0, 1)$, there is an s such that (1) holds for all $\lambda \in \mathbb{D}$ satisfying $|\lambda| > s$. Pick an $r' \in (r, 1)$ satisfying

$$r' > \max\{|z|; B'(z) = 0\},$$

and choose an $r'' \in (0, 1)$ such that

$$r'' > \max\{|B(z)| : |z| \le r'\} \quad \text{and} \quad r'' > r' > r.$$

For this r'', there is an $s''(0 < s'' < 1)$ such that whenever $|\lambda| > s''$ and $\lambda \in \mathbb{D}$, there is a critical point w^* of $B\varphi_\lambda$ such that

$$|B(w^*)\varphi_\lambda(w^*)| > r''.$$

On another hand, by (5.26) there is an $s' \in (0, 1)$ satisfying the following: for any $\lambda \in \mathbb{D}$ with $|\lambda| > s'$,

$$\|B\psi_\lambda - B\|_{r''\overline{\mathbb{D}},\infty}$$

is enough small so that

$$\|(B\psi_\lambda)' - B'\|_{r'\mathbb{T},\infty} < \min\{|B'(z)| : |z| = r'\}.$$

Then applying Rouche's theorem implies that on $r'\mathbb{D}$, $(B\psi_\lambda)'$ has the same number of zeros as B', counting multiplicity. Since Bochner's theorem (Theorem 5.4.6) says that each Blaschke product with k zeros has exactly $k - 1$ critical points in the unit disk $\mathbb{D}$, counting multiplicity, then $(B\psi_\lambda)'$ has $n - 1$ zeros on $r'\mathbb{D}$. Again by Bochner's theorem, $(B\psi_\lambda)'$ has n zeros in $\mathbb{D}$, and hence there is exactly one zero w_0 outside $r'\mathbb{D}$. That is,

$$\{w_0\} = Z\big((B\psi_\lambda)'\big) - r'\mathbb{D},$$

which gives that

$$r'' > \max\{|B(z)| : |z| \le r'\}$$
$$\ge \max\{|B\psi_\lambda(z)| : |z| \le r'\}$$
$$\ge \max\{|B\psi_\lambda(z)| : z \in Z\big((B\psi_\lambda)'\big) - \{w_0\}\}.$$

As mentioned before, there is a critical point w^* of $B\psi_\lambda$ satisfying

$$|B(w^*)\varphi_\lambda(w^*)| > r''.$$

Since

$$r'' > \max\{|B\psi_\lambda(z)| : z \in Z\big((B\psi_\lambda)'\big) - \{w_0\}\},$$

we get $w^* = w_0$. Thus, for any $\lambda \in \mathbb{D}$ with $\lambda > \max\{s', s''\}$,

$$|B(w_0)\psi_\lambda(w_0)| > r'' > \max\{|B\psi_\lambda(z)| : z \in Z\big((B\psi_\lambda)'\big) - \{w_0\}\}.$$

Also, $|B(w_0)\psi_\lambda(w_0)| > r$ since $r'' > r$. Therefore, write $s = \max\{s', s''\}$ and then both (1) and (2) hold for any $\lambda \in \mathbb{D}$ with $|\lambda| > s$.

It remains to prove (1). Set $t_0 = \max\{|z| : z \in Z(B)\}$ and

$$t_1 = \max\{|B(z)|; z \in t_0\overline{\mathbb{D}}\}.$$

For any $t > t_1$, let Ω_t denote the component of $\{z : |B(z)| < t\}$ containing $t_0\mathbb{D}$. We will see that

$$\Omega_t = \{z : |B(z)| < t\}.$$

The reasoning is as follows. By Lemma 5.1.2, $B|_{\Omega_t} : \Omega_t \to t\mathbb{D}$ is a proper map. By Proposition 2.1.2, $B|_{\Omega_t}$ is a k-folds map for some integer k. Since the restriction $(B - 0)|_{\Omega_t}$ has $n(n = \text{order } B)$ zeros, $k = n$. This implies that there is no component of $\{z : |B(z)| < t\}$ other than Ω_t. Therefore, $\Omega_t = \{z : |B(z)| < t\}$.

Now pick a number t^* in $(0, 1)$ such that

$$t^* > \max\{|z| : z \in Z(B')\} \quad \text{and} \quad t^* > t_1.$$

Without loss of generality, we assume that $r > t^*$. Write $r_0 = r$, and pick real numbers r_1 and r_2 satisfying

$$t^* < r_0 < r_1 < r_2 < 1.$$

As mentioned above, $\Omega_{r_j} = \{z : |B(z)| < r_j\}(j = 0, 1, 2)$. Since

$$\|B\psi_\lambda - B\|_{\Omega_{r_2},\infty} \le \|\psi_\lambda - 1\|_{\Omega_{r_2},\infty},$$

then by (5.26) for enough small $\varepsilon > 0$, there exists an $s \in (r(\overline{\Omega_{r_2}}), 1)$

$$\|B\psi_\lambda - B\|_{\Omega_{r_2},\infty} < \varepsilon, \quad |\lambda| > s. \tag{5.27}$$

Let Ω_1 and Ω_2 denote the components of

$$\{z : |B(z)\psi_\lambda(z)| < r_1\}$$

containing $Z(B)$ and λ, respectively (see Fig. 5.7). The prescribed ε in (5.27) can be enough small so that

$$\Omega_{r_0} \subseteq \Omega_1 \subseteq \Omega_{r_2}.$$

Since $|\lambda| > s > r(\overline{\Omega_{r_2}}) \ge r(\overline{\Omega_1})$, then $\Omega_1 \ne \Omega_2$, and hence $\Omega_1 \cap \Omega_2$ is empty. By Lemma 5.1.2 the restrictions of $B\psi_\lambda$ on Ω_1 and Ω_2 are proper maps

Fig. 5.7 Ω_1 and Ω_2

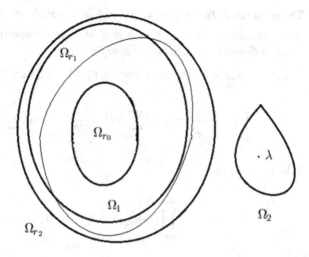

onto $r_1\mathbb{D}$. By (5.27) $B\psi_\lambda|_{\Omega_1}$ has n-folds, and then $B\psi_\lambda|_{\Omega_2}$ is a onefold map, i.e. $B\psi_\lambda|_{\Omega_2}$ is univalent. Combing Bochner's theorem with Theorem 2.1.3 shows that the derivative of $B(z)\psi_\lambda$ has n zeros, with $n-1$ zeros in Ω_1, and that the derivative of $B(z)\psi_\lambda$ never varnishes on Ω_2. Thus, there is exactly one critical point w^* satisfying $w^* \notin \Omega_1 \bigsqcup \Omega_2$. Since

$$\{z : |B(z)\psi_\lambda(z)| < r_1\} = \Omega_1 \bigsqcup \Omega_2,$$

$|B(w^*)\psi_\lambda(w^*)| \geq r_1 > r$, as desired. The proof is complete.   □

Below, we write $\overset{\circ}{D}$ for the interior of a set D. By combining Rouche's theorem with Cauchy's formula for derivatives, we have the following, as mentioned in [GH4].

Lemma 5.5.2 *Suppose B is a finite Blaschke product and*

$$Z(B') \subseteq \overset{\circ}{D_1} \subseteq D_1 \subseteq \overset{\circ}{D_2} \subseteq D_2 \subseteq \mathbb{D},$$

where D_1 and D_2 are two closed disks. Then for any enough small $\varepsilon > 0$, there is a number $\delta > 0$ such that for any holomorphic function h over $\mathbb{D}$ with

$$\|h - B\|_{D_2,\infty} < \delta,$$

we have $Z(h') \cap D_1 \subseteq O(Z(B'), \varepsilon) \subseteq \overset{\circ}{D_1}$.

If in addition, all zeros of B' are simple zeros, then all points in $Z(h') \cap D_1$ are simple zeros of h'; and if $B|_{Z(B')}$ is injective, then the restriction of h on $Z(h') \cap D_1$ is also injective.

By applying Lemmas 5.5.1 and 5.5.2, Guo and Huang are able to give the following [GH4].

Theorem 5.5.3 *For any sequence $\{z_n\}$ in $\mathbb{D}$ satisfying $\lim_{n \to \infty} |z_n| = 1$, there is a Blaschke subsequence $\{z_{n_k}\}$ such that M_B has no nontrivial reducing subspace, where B denotes the Blaschke product for $\{z_{n_k}\}$.*

Let us make an observation. For $x \in (0, \frac{1}{2})$, we have $e^{2x} < 1 + 4x$. Then

$$\prod_{n=0}^{\infty}(1 + \frac{x}{2^n}) = \exp(\sum_{n=0}^{\infty} \ln(1 + \frac{x}{2^n})) < \exp(\sum_{n=0}^{\infty} \frac{x}{2^n}) < 1 + 4x.$$

Take $x = \dfrac{\varepsilon}{2}$, and

$$\prod_{n=1}^{\infty}(1 + \frac{\varepsilon}{2^n}) < 1 + 2\varepsilon, \; 0 < \varepsilon < 1. \tag{5.28}$$

Now let K be a compact subset of $\mathbb{D}$, and pick $\delta \in (0, 1)$. If $\{\lambda_n\}$ is a Blaschke sequence satisfying

$$1 - |\lambda_n| \leq \frac{\delta(1 - r(K))}{2^{n+2}}, \tag{5.29}$$

then by (5.26) we get

$$|\psi_{\lambda_n}(z) - 1| \leq \frac{\delta}{2^{n+1}}, \ z \in K,$$

where

$$\psi_{\lambda_n}(z) = \frac{|\lambda_n|}{\lambda_n} \frac{\lambda_n - z}{1 - \overline{\lambda_n} z}.$$

Write $\varepsilon_n(z) = \psi_{\lambda_n}(z) - 1$. Then by (5.28)

$$\left| \prod_n \psi_{\lambda_n}(z) - 1 \right| = \left| \prod_n (1 + \varepsilon_n(z)) - 1 \right|$$

$$= \left| \sum_{i_1 < i_2 < \cdots < i_k} \varepsilon_{i_1}(z) \varepsilon_{i_2}(z) \cdots \varepsilon_{i_k}(z) \right|$$

$$\leq \sum_{i_1 < i_2 < \cdots < i_k} \frac{\delta}{2^{i_1+1}} \frac{\delta}{2^{i_2+1}} \cdots \frac{\delta}{2^{i_k+1}}$$

$$= \prod_{n=1}^{\infty} \left(1 + \frac{\delta}{2^{n+1}}\right) - 1$$

$$< 1 + \delta - 1 = \delta.$$

Thus, *for any Blaschke product ϕ, if (5.29) holds, then*

$$\left\| \phi \prod_n \psi_{\lambda_n} - \phi \right\|_{K,\infty} \leq \delta. \tag{5.30}$$

Note that (5.30) also holds if $\{\lambda_n\}$ is replaced with any subsequence of $\{\lambda_n\}$.
  The construction of B will be the focus of the discussion as follows.

Proof of Theorem 5.5.3 The proof comes from [GH4].

Assume that $\lim_{n \to \infty} |z_n| = 1$. First we show that $\{z_n\}$ contains a thin Blaschke subsequence. To see this, let d denote the hyperbolic metric over the unit disk, and we have $\lim_{n \to \infty} d(z_k, z_n) = 1$ for each k. Pick two natural numbers m_1 and m_2 such that

$d(z_{m_1}, z_{m_2}) > 1 - \frac{1}{2^2}$, and put $w_j = z_{m_j} (j = 1, 2)$. In general, we will use induction. Assume that we have n points:

$$w_1 = z_{m_1}, \cdots, w_n = z_{m_n},$$

and

$$\prod_{1 \leq j < k} d(w_j, w_k) > 1 - \frac{1}{(k+1)^2}, k = 1, 2, \cdots, n.$$

Since $\lim_{k \to \infty} \prod_{1 \leq j \leq n} d(w_j, z_k) = 1$, there is an integer $m_{n+1} > m_n$ such that

$$\prod_{1 \leq j \leq n} d(w_j, z_{m_{n+1}}) > 1 - \frac{1}{(n+2)^2}.$$

Put $w_{n+1} = z_{m_{n+1}}$, and the induction is done. Therefore, we have a subsequence $\{w_n\}$ satisfying

$$\prod_{1 \leq j < k} d(w_j, w_k) > 1 - \frac{1}{(k+1)^2}, k = 1, 2, \cdots.$$

Since $\prod_{n \geq 2}(1 - \frac{1}{n^2}) > 0$, then

$$\prod_{j; j \neq k} d(w_j, w_k) = \prod_{j<k} d(w_j, w_k) \prod_{j>k} d(w_j, w_k)$$

$$\geq (1 - \frac{1}{(k+1)^2}) \prod_{j>k}(1 - \frac{1}{(j+1)^2}) \to 1, k \to \infty.$$

This immediately shows that $\{w_n\}$ is a thin Blaschke sequence, as desired.

Now we may assume that $\{z_n\}$ is itself a thin Blaschke sequence. We will find a subsequence of $\{z_n\}$ whose Blaschke product B satisfies the conditions (1) and (2) in Theorem 5.4.1 with $F = \emptyset$. For this, we will construct a sequence $\{B_m\}$ of finite Blaschke products with order $B_m = m + 1$ and B equals the limit of B_m.

Write

$$\varphi = \varphi_{z_1} \varphi_{z_2},$$

where

$$\varphi_\lambda(z) = \frac{\lambda - z}{1 - \bar{\lambda} z}.$$

By Theorem 5.4.6, φ has a unique critical point, denoted by w_1. Namely, $\varphi'(w_1) = 0$. Rewrite $B_1 = \varphi$, and one can construct two closed disks $\widetilde{K}_1$, K_1 in $\mathbb{D}$, and a constant $\delta_1 > 0$ satisfying the following:

1. $\widetilde{K}_1$ is a closed disk containing $\{z : |z| < \frac{1}{2}\}$ and w_1.
2. $\widetilde{K}_1 \subseteq \overset{\circ}{K}_1$, δ_1 is enough small such that for any holomorphic function h with $\|h - B_1\|_{K_1,\infty} < \delta_1$, $Z(h') \cap \widetilde{K}_1$ contains exactly one point.
3. Taking $\delta = \delta_1$ and $K = K_1$ in (5.29), there is a subsequence $\{z_{n,1}\}_{n=1}^\infty$ of $\{z_n\}_{n=1}^\infty$ satisfying (5.29).

By induction, suppose we have already constructed a finite Blaschke product B_m with order $B_m = m + 1$ such that:

(i) $B_m|_{Z(B_m')}$ is injective;
(ii) $\widetilde{K}_m$ is a closed disk containing both $\{z : |z| < 1 - \frac{1}{m+1}\}$ and $Z(B_m')$.
(iii) there is a constant $\delta_m > 0$ and two closed disks $\widetilde{K}_m$ and K_m ($\widetilde{K}_m \subseteq \overset{\circ}{K}_m$) in $\mathbb{D}$ such that for any holomorphic function h with $\|h - B_m\|_{K_m,\infty} < \delta_m$, the restriction of h on $Z(h') \cap \widetilde{K}_m$ is injective.
(iv) Taking $\delta = \delta_m$ and $K = K_m$ in (5.29), there is a subsequence $\{z_{n,m}\}_{n=1}^\infty$ of $\{z_{n,m-1}\}_{n=1}^\infty$ satisfying (5.29).

By Lemma 5.5.1, there is a point $\lambda \in \{z_{n,m}\}_{n=1}^\infty$ with $1 - |\lambda|$ enough small such that the restriction of $B_m\varphi_\lambda$ on $Z((B_m\varphi_\lambda)')$ is injective. Then put $B_{m+1} = B_m\psi_\lambda$; that is, B_{m+1} is a constant multiple of $B_m\varphi_\lambda$. Now write

$$r = \max\{\frac{1 + r(K_m)}{2}, 1 - \frac{1}{m+2}\}$$

and $r' = \frac{1+r}{2}$. Set

$$\widetilde{K_{m+1}} = \{z : |z| \leq r\} \quad \text{and} \quad K_{m+1} = \{z : |z| \leq r'\}.$$

Clearly, $\widetilde{K_{m+1}} \subseteq K_{m+1}$. By Lemma 5.5.2 we can find a $\delta_{m+1} > 0$ such that the similar version of (iii) holds for $m + 1$. Taking

$$\delta = \delta_{m+1} \quad \text{and} \quad K = K_{m+1}$$

in (5.29), there is a subsequence $\{z_{n,m+1}\}_{n=1}^\infty$ of $\{z_{n,m}\}_{n=1}^\infty$ satisfying (5.29). Considering that both $z_{1,m+1}$ and λ lie in $\{z_{n,m}\}_{n=1}^\infty$, it is required that $z_{1,m+1}$ is behind λ. Now the induction on m is complete.

By induction, it is clear that B_m converges to a thin Blaschke product B. For each fixed n, there is a Blaschke product $\widetilde{B}_m$ such that $B = B_m\widetilde{B}_m$. Let $\{\lambda_k\}$ be the zero sequence of $\widetilde{B}_m$ (of course, $\{\lambda_k\}$ depends on m). By our construction, $\{\lambda_k\}$ is a subsequence of $\{z_{n,m}\}_{n=1}^\infty$, and then by (iv) (5.29) holds:

$$1 - |\lambda_k| \leq \frac{\delta_m(1 - r(K_m))}{2^{k+2}}.$$

Since $\widetilde{B_m} = \prod_k \psi_{\lambda_k}$, by (5.30) we have

$$\|B_m \widetilde{B_m} - B_m\|_{K_m,\infty} \leq \delta_m.$$

That is,

$$\|B - B_m\|_{K_m,\infty} \leq \delta_m.$$

Thus by (iii), $B|_{Z(B') \cap \widetilde{K_m}}$ is injective. Since $\widetilde{K_m}$ is an increasing sequence of closed domains whose union is $\mathbb{D}$, $B|_{Z(B')}$ is injective. Then applying Lemma 5.5.2 shows that $B'|_{\widetilde{K_m}}$ has only simple zeros for each m, and hence B' has only simple zeros. Thus, B satisfies the assumptions of Theorem 5.4.1, with F being empty. Therefore, the construction of B is complete and the proof is finished.   □

5.6 Another Proof for a Characterization on $\mathcal{V}^*(B)$

This section will provide an alternative proof of a weak version of Theorem 5.3.1. Different ideas will prove to be useful in characterizing the adjoint of a bounded operator in $\mathcal{V}^*(B)$.

In Theorem 5.2.4, a global representation is obtained for unitary operators S in $\mathcal{V}^*(B)$:

$$Sh(z) = \sum_{k=0}^{\infty} c_k h \circ \rho_k(z) \rho_k'(z), \quad h \in L_a^2(\mathbb{D}), z \in \mathbb{D} - L.$$

By Proposition 2.5.1, each operator S in $\mathcal{V}^*(B)$ has the above form, and by Lemma 5.2.6, one may rewrite S as

$$Sh(z) = \sum_{[\rho]} \sum_{\sigma \in [\rho]} c_\rho h \circ \sigma(z) \sigma'(z) \equiv \sum_{[\rho]} c_\rho \mathcal{E}_{[\rho]} h(z), \quad h \in L_a^2(\mathbb{D}), z \in E. \qquad (5.31)$$

The following lemma, due to Guo and Huang [GH4], is our main result in this section.

Lemma 5.6.1 *Given a thin Blaschke product B, suppose S is an operator in $\mathcal{V}^*(B)$ with the form (5.31). If $\overline{Z(B)} \not\supseteq \mathbb{T}$, then*

$$S^*h(z) = \sum_{[\rho]} \overline{c_\rho} \mathcal{E}_{[\rho^-]} h(z), \quad h \in L_a^2(\mathbb{D}), z \in E.$$

Furthermore, for each ρ such that $c_\rho \neq 0$, we have $\sharp[\rho] < \infty$ and $\sharp[\rho^-] < \infty$.

Under a weaker assumption, B is just an interpolating Blaschke product, and in this case, Lemma 5.6.1 also holds, though on a small disk around some zero of B.

To prove Lemma 5.6.1, we need the following claim [GH4].

Claim If $\overline{Z(B)} \not\supseteq \mathbb{T}$, then for any $0 < \delta < 1$, $\overline{B^{-1}(\Delta(0,\delta))} \not\supseteq \mathbb{T}$.

This claim easily follows from a smart observation by the referee of [GH4]: if $\overline{Z(B)} \not\supseteq \mathbb{T}$, then there is an open non-empty arc $\gamma \subseteq \mathbb{T}$ such that B has an analytic continuation across γ and has modulus 1 on γ.

Also, we have an alternative proof for the above claim, as follows. Write $Z(B) = \{z_n\}$. By (5.2), there is an $\varepsilon \in (0,1)$ such that

$$B^{-1}(\Delta(0,\delta)) \subseteq \bigcup_n \Delta(z_n, \varepsilon),$$

Therefore, it suffices to show that $\overline{\bigcup_n \Delta(z_n, \varepsilon)} \not\supseteq \mathbb{T}$. For this, assume conversely that $\overline{\bigcup_n \Delta(z_n, \varepsilon)} \supseteq \mathbb{T}$. Since $\overline{Z(B)} \not\supseteq \mathbb{T}$, there is a $\zeta \in \mathbb{T}$ such that $\zeta \notin \overline{Z(B)}$. Since $\overline{\bigcup_n \Delta(z_n, \varepsilon)} \supseteq \mathbb{T}$, there is a sequence $\{w_k\}$ in $\mathbb{D}$ and a subsequence $\{z_{n_k}\}$ of $\{z_n\}$ such that

$$d(z_{n_k}, w_k) < \varepsilon, \tag{5.32}$$

and

$$\lim_{k\to\infty} w_k = \zeta.$$

Here, $d(z, w) \triangleq \left| \dfrac{z-w}{1-\bar{z}w} \right|$. Since $\zeta \notin \overline{Z(B)}$, we have

$$d(z, \zeta) = 1, z \in \overline{Z(B)}.$$

By the continuity of $d(z, w)$, there is an enough small closed disk K centered at ζ such that

$$d(z, w) \geq \frac{1+\varepsilon}{2}, z \in \overline{Z(B)}, w \in K.$$

For enough large k, $w_k \in K$, and then

$$d(z_{n_k}, w_k) \geq \frac{1+\varepsilon}{2}.$$

This is a contradiction to (5.32). Therefore, $\overline{B^{-1}(\Delta(0,\delta))} \not\supseteq \mathbb{T}$. The proof of the claim is complete. $\square$

By the above claim, for each $z \in E$ there is always an open neighborhood U ($U \subseteq E$) of z that admits a complete local inverse, such that

$$\overline{B^{-1} \circ B(U)} \not\supseteq \mathbb{T}.$$

Then we may stretch U such that $\mathbb{C} - \overline{B^{-1} \circ B(U)}$ is connected. Therefore,

for each $z \in E$, there is an open neighborhood $U(U \subseteq E)$ of z that admits a complete local inverse, and $\mathbb{C} - \overline{B^{-1} \circ B(U)}$ is connected.

With this observation, we can give the proof of Lemma 5.6.1 as below.

Proof of Lemma 5.6.1 The proof is from [GH4].

By Proposition 2.5.1, each operator in a von Neumann algebra is the finite linear span of its unitary operators. Without loss of generality, assume that S is a unitary operator in $\mathcal{V}^*(B)$. Let $\widetilde{M}_B$ denote the multiplication operator by B on $L^2(\mathbb{D}, dA)$ given by

$$\widetilde{M}_B g = Bg, \ g \in L^2(\mathbb{D}, dA).$$

If we can show that $\widetilde{M}_B$ is the minimal normal extension of M_B, then by Conway [Con2, pp. 435–436, Theorem VIII.2.20] and Douglas et al. [DSZ, pp. 552, 553], for each unitary operator $S \in \mathcal{V}^*(B)$, there exists a unitary operator $\tilde{S}$ defined on $L^2(\mathbb{D}, dA)$ such that $\tilde{S}|_{L^2_a(\mathbb{D})} = S$ and $\tilde{S}$ commutes with $\widetilde{M}_B$. This will be demonstrated by the following paragraph in detail.

We will show that $\widetilde{M}_B$ is the minimal normal extension of M_B. Recall that if A is a subnormal operator on $\mathcal{H}$ and $\tilde{A}$ is a normal extension of A to $\mathcal{K}$, where $\mathcal{H}$ and $\mathcal{K}$ are Hilbert spaces, then $\tilde{A}$ is a *minimal normal extension* of A if $\mathcal{K}$ equals the norm-closure of $span\{\tilde{A}^{*k}h : h \in \mathcal{H}, k \geq 0\}$. Now set

$$K = span\{\widetilde{M}_B^{*k}h; \ h \in L^2_a(\mathbb{D}), k \geq 0\}.$$

Define a linear map $\tilde{S}$ on K by requiring that

$$\tilde{S}(\widetilde{M}_B^{*k}h) = \widetilde{M}_B^{*k}Sh, \ h \in L^2_a(\mathbb{D}), k \geq 0.$$

Since S is a unitary operator, it is easy to check that $\tilde{S}$ is well-defined on K and $\tilde{S}$ commutes with $\widetilde{M}_B$. Note that $\tilde{S}$ is an isometric operator whose range equals K, and thus $\tilde{S}$ can be naturally extended to a unitary operator on the closure $\overline{K}$ of K. Since

$$K = span\{\widetilde{M}_B^{*k}\widetilde{M}_B^l h; \ h \in L^2_a(\mathbb{D}), k \geq 0, l \geq 0\},$$

it follows that for any polynomial P, $P(B, \overline{B})h \in K$. Then applying Stone-Weierstrass's Theorem shows that for any $u \in C(\overline{\mathbb{D}})$, $u(B)h \in \overline{K}$. Lusin's Theorem states that for each $v \in L^\infty(\mathbb{D})$, there is a uniformly bounded sequence $\{v_n\}$ in $C(\overline{\mathbb{D}})$ such that $\{v_n\}$ converges to v in measure. Therefore, for any $u \in L^\infty(\mathbb{D})$, $u(B)h \in \overline{K}$. In particular, for any open subset U of $\mathbb{D}$,

$$\chi_{B^{-1} \circ B(U)} h \in \overline{K}.$$

As mentioned above, for each $z \in E$ there is an open neighborhood U ($U \subseteq E$) of z that admits a complete local inverse, such that

$$B^{-1} \circ B(U) = \bigsqcup_j \rho_j(U),$$

and $\mathbb{C} - \overline{B^{-1} \circ B(U)}$ is connected. For each fixed i, $\chi_{\rho_i(U)}$ denotes the characterization function of $\rho_i(U)$. As done in [DSZ], applying the Runge theorem (see Theorem 2.1.18) shows that there is a sequence of polynomials $\{p_k\}$ such that p_k converges to $\chi_{\rho_i(U)}$ uniformly on the closure of $B^{-1} \circ B(U)$. Since $\chi_{B^{-1} \circ B(U)} p_k \in K$, $\chi_{\rho_i(U)} \in \overline{K}$. In particular, $\chi_U \in \overline{K}$. When this U is replaced with any other disk Δ ($\Delta \subseteq U$), we also have $\chi_\Delta \in \overline{K}$. Since all χ_Δ span a dense subspace of $L^2(U)$,

$$L^2(U) \subseteq \overline{K}.$$

Since all such open sets U covers $\mathbb{D}$, we have $L^2(\mathbb{D}) \subseteq \overline{K}$, and hence $L^2(\mathbb{D}) = \overline{K}$. This shows that $\widetilde{M_B}$ is exactly the minimal normal extension of M_B, as desired.

Now assume that the unitary operator $S \in \mathcal{V}^*(B)$ has the form in (5.31):

$$Sh(z) = \sum_{[\rho]} \sum_{\sigma \in [\rho]} c_\rho h \circ \sigma(z) \sigma'(z), \quad h \in L_a^2(\mathbb{D}), \ z \in E.$$

Let U be an open set that admits a complete local inverse, and

$$B^{-1} \circ B(U) = \bigsqcup_j \rho_j(U).$$

Since $\tilde{S}$ commutes with $\widetilde{M_B}$, by the theory of spectral decomposition, $\tilde{S}$ also commutes with M_F, with $F = \chi_{B^{-1} \circ B(\Delta)}$ for any measurable subsets Δ of $\mathbb{D}$ [DSZ, pp. 553–557]. In particular, letting $\Delta = U$, we have

$$\tilde{S} M_{\chi_{B^{-1} \circ B(U)}} = M_{\chi_{B^{-1} \circ B(U)}} \tilde{S}. \tag{5.33}$$

As mentioned in the above paragraph, there is a sequence of polynomials $\{p_k\}$ such that p_k converges to the characteristic function $f \triangleq \chi_{\rho_i(U)}$ uniformly on the closure of $B^{-1} \circ B(U)$. Now by (5.33), we have

$$\tilde{S} \chi_{B^{-1} \circ B(U)} p_k(z) = M_{\chi_{B^{-1} \circ B(U)}} S p_k(z)$$

$$= \sum_{[\rho]} \sum_{\sigma \in [\rho]} \chi_{\sqcup \rho_j(U)}(z) c_\rho p_k \circ \sigma(z) \sigma'(z).$$

Letting $k \to \infty$ gives that

$$\tilde{S}f = \sum_{[\rho]} \sum_{\sigma \in [\rho]} c_\rho f \circ \sigma(z)\sigma'(z). \tag{5.34}$$

By similar reasoning, if $f = \chi_{\rho_i(V)}$ for some measurable subset V of U, (5.34) also holds. Thus, if f is a linear span of $\chi_{\rho_i(V)}(V \subseteq U)$, (5.34) holds. Therefore the identity (5.34) can be generalized to any $f \in C_c(B^{-1} \circ B(U))$. That is, locally, $\tilde{S}$ *has the same representation as S.*

By using the proof of Lemma 5.3.3 and (5.34), we deduce that

$$(\tilde{S})^*f = \sum_{[\rho]} \sum_{\sigma \in [\rho^-]} \overline{c_\rho} f \circ \sigma(z)\sigma'(z), \, f \in C_c(B^{-1} \circ B(U)).$$

Write

$$S^*h(z) = \sum_{[\rho]} \sum_{\sigma \in [\rho^-]} d_\rho h \circ \sigma(z)\sigma'(z), \, h \in L_a^2(\mathbb{D}), \, z \in E,$$

and similarly,

$$\tilde{S}^*f = \sum_{[\rho]} \sum_{\sigma \in [\rho^-]} d_\rho f \circ \sigma(z)\sigma'(z), f \in C_c(B^{-1} \circ B(U)).$$

Since $(\tilde{S})^* = \tilde{S}^*$, by the uniqueness of the coefficients c_ρ, we get $d_\rho = \overline{c_\rho}$ for all ρ. Therefore,

$$S^*h(z) = \sum_{[\rho]} \sum_{\sigma \in [\rho^-]} d_\rho h \circ \sigma(z)\sigma'(z) \equiv \sum_{[\rho]} \overline{c_\rho} \mathcal{E}_{[\rho^-]}h(z), \, h \in L_a^2(\mathbb{D}), \, z \in E.$$

This gives the former part of Lemma 5.6.1. The remaining follows directly from Lemma 5.2.6.   □

Remark 5.6.2 By further effort, one can show that (5.34) holds for any f in $C_c(\mathbb{D})$, a dense subspace of $L^2(\mathbb{D})$.

Since Lemma 5.6.1 can be regarded as a substitute for Proposition 5.3.4 in some sense, the proof of Theorem 5.3.1 can be done as in Sect. 5.5 in Chap. 5, and thus omitted.

The following example, as is well known, is presented to show that there exists a thin Blaschke product B satisfying $\overline{Z(B)} \supseteq \mathbb{T}$. See [GM1] and [CFT, Proposition 4.3(i)], as well as [Ni, Chap. VII].

Example 5.6.3 First let us make an observation from [Hof1, p. 203, 204]: if $\{w_n\}$ is a sequence in $\mathbb{D}$ satisfying

$$\frac{1 - |w_n|}{1 - |w_{n-1}|} \leq c < 1, \tag{5.35}$$

then

$$\prod_{j:j\neq k} d(w_k, w_j) \geq \left(\prod_{j=1}^{\infty} \frac{1-c^j}{1+c^j}\right)^2. \tag{5.36}$$

Note that the right hand side tends to 1 as c tends to 0. The reasoning comes from [Hof1, p. 203, 204]. To show (5.36), note that

$$\left|\frac{a-b}{1-\overline{a}b}\right| \geq \frac{||a|-|b||}{1-|a||b|}, \ a, b \in \mathbb{D}.$$

Then

$$\prod_{j:j\neq k} d(w_k, w_j) \geq \prod_{j:j<k} d(|w_k|, |w_j|) \prod_{j:j>k} d(|w_k|, |w_j|). \tag{5.37}$$

For $j > k$, by (5.35)

$$1 - |w_j| \leq c^{j-k}(1 - |w_k|). \tag{5.38}$$

and thus

$$|w_j| - |w_k| \geq (1 - c^{j-k})(1 - |w_k|).$$

Again by (5.38)

$$1 - |w_j||w_k| = (1 - |w_k|) + (1 - |w_j|)|w_k| \leq (1 + c^{j-k})(1 - |w_k|).$$

Therefore, for $j > k$,

$$d(|w_k|, |w_j|) = \frac{|w_j| - |w_k|}{1 - |w_j||w_k|} \geq \frac{1 - c^{j-k}}{1 + c^{j-k}},$$

and hence

$$\prod_{j:j>k} d(|w_k|, |w_j|) \geq \prod_{j=1}^{\infty} \frac{1-c^j}{1+c^j}.$$

Similarly, we have

$$\prod_{j:j<k} d(|w_k|, |w_j|) \geq \prod_{j=1}^{\infty} \frac{1-c^j}{1+c^j},$$

and then by (5.37) we get (5.36).

Let $\{c_n\}$ be a sequence satisfying $c_n > 0$ and $\lim\limits_{n\to\infty} c_n = 0$. Suppose $\{z_n\}$ is a sequence of points in the open unit disk $\mathbb{D}$ such that

$$\frac{1 - |z_n|}{1 - |z_{n-1}|} = c_n.$$

We will show that $\{z_n\}$ is a thin Blaschke sequence. To see this, given a positive integer m, let $k > m$ and consider the product

$$\prod_{j:j\neq k} d(z_k, z_j) \equiv \prod_{1\leq j\leq m} d(z_k, z_j) \prod_{j>m, j\neq k} d(z_k, z_j). \tag{5.39}$$

Now write

$$d_m = \sup\{c_j : j \geq m\}.$$

Since $\lim\limits_{n\to\infty} c_n = 0$, then $\lim\limits_{m\to\infty} d_m = 0$. For any $k > m$,

$$\frac{1 - |z_k|}{1 - |z_{k-1}|} = c_k \leq d_m,$$

and thus

$$\prod_{j>m, j\neq k} d(z_k, z_j) \geq \left(\prod_{j=1}^{\infty} \frac{1 - d_m^j}{1 + d_m^j}\right)^2, \ k > m,$$

which implies that

$$\prod_{j>m, j\neq k} d(z_k, z_j)$$

tends to 1 as $m \to \infty$. For any $\varepsilon > 0$, there is an m_0 such that

$$\prod_{j>m_0, j\neq k} d(z_k, z_j) \geq 1 - \varepsilon.$$

On the other hand,

$$\lim_{k\to\infty} \prod_{1\leq j\leq m_0} d(z_k, z_j) = 1,$$

which, combined with (5.39), implies that

$$\liminf_{k\to\infty} \prod_{j:j\neq k} d(z_k, z_j) \geq 1 - \varepsilon.$$

Therefore, by the arbitrariness of ε,

$$\lim_{k \to \infty} \prod_{j:j \neq k} d(z_k, z_j) = 1,$$

which implies that $\{z_k\}$ is a thin Blaschke product.

Now choose a dense sequence $\{r_n\}$ in $[0, 1]$. If $\{z_n\}$ is given as above, then $\{|z_n| \exp(2\pi i r_n)\}$ also gives a thin Blaschke sequence. Moreover, we have

$$\overline{\{|z_n| \exp(2\pi i r_n)\}} \supseteq \mathbb{T}.$$

For example, $\{z_n\}$ is a thin Blaschk sequence if we require that $|z_n| = 1 - \frac{1}{n!}$.

5.7 Abelian $\mathcal{V}^*(B)$ for Thin Blaschke Products

It is known that $\mathcal{V}^*(\phi)$ are abelian for finite Blaschke products ϕ, see Theorem 4.2.1. This result can be nontrivially generalized to the class of thin Blaschke products, which is the main result in this section, due to Guo and Huang [GH4].

Theorem 5.7.1 *For any thin Blaschke product B, $\mathcal{V}^*(B)$ is abelian.*

Section 5.6 in this chapter has provided a complete geometric characterization for when $\mathcal{V}^*(B)$ is nontrivial. This section will present the structure of $\mathcal{V}^*(B)$, as follows [GH4]. For the case of finite Blaschke products, see [DSZ, Theorem 7.6].

Theorem 5.7.2 *If B is a thin Blaschke product, then the von Neumann algebra $\mathcal{V}^*(B)$ is generated by all $\mathcal{E}_{[\rho]}$, where $\sharp[\rho] < \infty$.*

Recall that for each local inverse ρ, $G[\rho]$ is the graph of $[\rho]$. Therefore, the dimension of $\mathcal{V}^*(B)$ equals the number of all finite components $G[\rho]$ of S_B. Also, note that Theorem 5.3.1 can be regarded as a direct consequence of Theorem 5.7.2. If B is a finite Blaschke product, each component of S_B is necessarily finite, and thus the dimension of $\mathcal{V}^*(B)$ equals the number of components of S_B [DSZ, DPW].

Before giving the proofs of Theorems 5.7.1 and 5.7.2, we establish the following lemma.

Lemma 5.7.3 *For any thin Blaschke product B, all operators $S_{[\sigma]}$ mutually commutes, where $[\sigma]$ runs over components of S_B satisfying $\sharp[\sigma] < \infty$ and $\sharp[\sigma^-] < \infty$.*

Proof The proof essentially comes from [GH4].

Observe that for any $[\sigma]$ and $[\tau]$, $\mathcal{E}_{[\sigma]}\mathcal{E}_{[\tau]}$ has the form

$$\sum_i \mathcal{E}_{[\sigma_i]},$$

where the sum is finite and some σ_i may lie in the same class. Moreover, Lemma 5.2.6 and Proposition 5.3.4 tells us that each σ_i satisfying $\sharp[\sigma_i] < \infty$ and $\sharp[\sigma_i^-] < \infty$. Based on this observation, we define *the composition*

$$[\tau] \circ [\sigma]$$

to be a formal sum $\sum_i[\sigma_i]$; and its graph $G([\tau] \circ [\sigma])$ is defined to be the *disjoint union* of the graphs $G[\sigma_i]$, denoted by $\sum_i G[\sigma_i]$. For example, if both σ and τ are in Aut($\mathbb{D}$), then $G([\tau] \circ [\sigma]) = G([\tau \circ \sigma])$.

Clearly, $\mathcal{E}_{[\sigma]}\mathcal{E}_{[\tau]} = \mathcal{E}_{[\tau]}\mathcal{E}_{[\sigma]}$ if and only if $[\sigma] \circ [\tau] = [\tau] \circ [\sigma]$; if and only if $G([\sigma] \circ [\tau]) = G([\tau] \circ [\sigma])$. Let us take a close look at $G([\sigma] \circ [\tau])$ and $G([\tau] \circ [\sigma])$. Write $m = \sharp[\sigma]$ and $n = \sharp[\tau]$. Fix a point $z_0 \in E$, there are exactly m points with the form $(z_0, \cdot)$ in $G[\sigma]$:

$$(z_0, \xi_1'), \cdots, (z_0, \xi_m').$$

For each $j = 1, \cdots, m$, there are n points with the form $(\xi_j', \cdot)$ in $G[\tau]$:

$$(\xi_j', \xi_j^k), \ k = 1, \cdots, n.$$

For each j and k, (z_0, ξ_j^k) is in some $G[\sigma_i]$ where $[\sigma_i]$ appears in the sum of $[\tau] \circ [\sigma]$. By omitting the order of (z_0, ξ_j^k), the sequence $\{(z_0, \xi_j^k)\}_{j,k}$ represents $G([\tau] \circ [\sigma])$. Similarly, there are exactly n points with the form $(z_0, \cdot)$ in $G[\tau]$:

$$(z_0, \eta_1'), \cdots, (z_0, \eta_n').$$

For each $j = 1, \cdots, n$, there are m points with the form $(\eta_j', \cdot)$ in $G[\sigma]$:

$$(\eta_j', \eta_j^k), \ k = 1, \cdots, m.$$

Also, the sequence $\{(z_0, \eta_j^k)\}_{j,k}$ represents $G([\sigma] \circ [\tau])$. Then

$$G([\sigma] \circ [\tau]) = G([\tau] \circ [\sigma])$$

if and only if the sequences $\{(z_0, \xi_j^k)\}_{j,k}$ and $\{(z_0, \eta_j^k)\}_{j,k}$ are equal, omitting the order. For example, the sequences

$$(z_0, z_0), (z_0, \rho(z_0)), (z_0, z_0), (z_0, \rho(z_0))$$

and

$$(z_0, z_0), (z_0, z_0), (z_0, \rho(z_0)), (z_0, \rho(z_0))$$

are equal if we omit the order, but they are never equal to the following sequence

$$(z_0, z_0), (z_0, \rho(z_0)).$$

Below, to finish the proof, we need some geometry of thin Blaschke products. As done in the proof of Proposition 5.3.4, let Ω_t be a simply-connected domain whose boundary is an analytic Jordan curve; and Ω_t is B-glued. Also, we may assume that t is enough large such that z_0 and all finitely many points $\xi'_j, \xi^k_j; \eta'_j, \eta^k_j$ lie in Ω_t. Again by the proof of Proposition 5.3.4, there is a biholomorphic map $f : \mathbb{D} \to \Omega_t$ such that $B \circ f = t\phi$ for some finite Blaschke product. The equivalent classes $[\sigma]$ and $[\tau]$ naturally grow on Ω_t, and the compositions $[\tau] \circ [\sigma]$ and $[\sigma] \circ [\tau]$ also make sense on Ω_t. By the identity $B \circ f = t\phi$, the map

$$[\rho] \mapsto [\rho^*] \triangleq [f^{-1} \circ \rho \circ f]$$

maps each component $[\rho] \in \mathcal{E}_{B|\Omega_t}$ to $[\rho^*] \in \mathcal{E}_\phi$. Special attention should be paid to $\rho = \sigma$ or τ. By the proof of Theorem 4.2.1,

$$\mathcal{E}_{[\sigma^*]}\mathcal{E}_{[\tau^*]} = \mathcal{E}_{[\tau^*]}\mathcal{E}_{[\sigma^*]},$$

which implies that

$$G([\sigma^*] \circ [\tau^*]) = G([\tau^*] \circ [\sigma^*]).$$

This immediately gives that $G([\sigma] \circ [\tau]) = G([\tau] \circ [\sigma])$, and hence

$$\mathcal{E}_{[\sigma]}\mathcal{E}_{[\tau]} = \mathcal{E}_{[\tau]}\mathcal{E}_{[\sigma]}.$$

The proof is complete.   □

In some sense, a finite Blaschke product can be regarded as a thin Blaschke product, but with finite zeros. It is of interest that the commutativity of the von Neumann algebras $V^*(B)$ for thin Blaschke products B is connected with that of finite Blaschke products, by the following [DSZ, GH4].

Proposition 5.7.4 *The following are equivalent:*

(1) *For any thin Blaschke product B with $\overline{Z(B)} \not\supseteq \mathbb{T}$, $V^*(B)$ is abelian.*
(2) *For any finite Blaschke product ϕ, $V^*(\phi)$ is abelian.*
(3) *For any finite Blaschke product ϕ, M_ϕ has exactly q minimal reducing subspaces, where q denotes the number of components of S_ϕ.*

Proof First, we show (3) $\Rightarrow$ (2). Given a finite Blaschke product ϕ, let q denote the number of components of S_ϕ. Clearly, $q \leq$ order $\phi < \infty$. By assumption, M_ϕ has exactly q minimal reducing subspaces, which implies that $V^*(\phi)$ has exactly

q minimal projections because dim $V^*(\phi) < \infty$, see Corollary 4.2.4. Then by Proposition 2.6.8, $V^*(\phi)$ is abelian.

To see (2) $\Rightarrow$ (3), assume that $V^*(\phi)$ is abelian, where ϕ is a finite Blaschke product. By [DSZ, Theorem 7.6], dim $V^*(\phi) = q$, where q denotes the number of components of S_ϕ. Then applying Theorem 2.5.11 shows that $V^*(\phi)$ is generated by mutually orthogonal minimal projections, whose number equals q. That is, M_ϕ has exactly q minimal reducing subspaces.

To prove (1) $\Rightarrow$ (2), consider the thin Blaschke product ψ whose zero sequence is $\{1 - \frac{1}{n!}\}$, see Example 5.6.3. For a fixed finite Blaschke product ϕ, write $B \triangleq \psi \circ \phi$. By Example 5.3.5, B is also a thin Blaschke product and it is clear that $V^*(B) \supseteq V^*(\phi)$. Since $V^*(B)$ is abelian, then so is $V^*(\phi)$.

It remains to show that (2) $\Rightarrow$ (1). Suppose B is a thin Blaschke product. By Theorem 5.7.2, $V^*(B)$ is generated by $S_{[\rho]}$, where $G[\rho]$ has finite multiplicity (and hence by Proposition 5.3.4 $G[\rho]$ has bi-finite multiplicities). Therefore, Lemma 5.7.3 shows that $V^*(B)$ is abelian. $\square$

By Theorem 4.2.1, $V^*(\phi)$ is abelian for any finite Blaschke product ϕ. Therefore, for any thin Blaschke product B with $\overline{Z(B)} \not\supseteq \mathbb{T}$, $V^*(B)$ is abelian. In general, one can provide a proof of Theorem 5.7.1 as follows, which essentially comes from [GH4].

Proof of Theorem 5.7.1 Suppose B is a thin Blaschke product. For any $S, T \in V^*(B)$, write

$$S = \sum c_\rho \mathcal{E}_{[\rho]} \quad \text{and} \quad T = \sum d_\rho \mathcal{E}_{[\rho]}. \tag{5.40}$$

Let S_m and T_n denote the corresponding partial sum of the expanding series (5.40) for S and T, respectively. By Lemma 5.7.3, for any two local inverse ρ and σ,

$$\mathcal{E}_{[\rho]}\mathcal{E}_{[\sigma]} = \mathcal{E}_{[\sigma]}\mathcal{E}_{[\rho]}.$$

Then we have

$$T_m S_n h = S_n T_m h, h \in L_a^2(\mathbb{D}).$$

Letting n tend to ∞, both sides converge to holomorphic functions on compact subsets of $\mathbb{D}$. That is,

$$T_m S h = S T_m h, h \in L_a^2(\mathbb{D}). \tag{5.41}$$

As m tends to infinity, the left-hand side of (5.41) converges to TSh uniformly on compact sets. Now consider $h = K_\lambda$ with $\lambda \in E$. We claim that:

$$\lim_{m \to \infty} \|T_m K_\lambda - TK_\lambda\| = 0. \tag{5.42}$$

If (5.42) is proved, then the right-hand side of (5.41) converges in norm (and hence converges uniformly on compact sets) to K_λ. Therefore,

$$TSK_\lambda = STK_\lambda, \quad \lambda \in E.$$

That is, $[S, T]K_\lambda = 0, \quad \lambda \in E$, where

$$[S, T] \triangleq ST - TS.$$

Since $span\{K_\lambda : \lambda \in E\}$ is dense in $L_a^2(\mathbb{D})$, $[S, T] = 0$, completing the proof.

Now it remains to show (5.42). First, observe that for each ρ with $\sharp[\rho] < \infty$,

$$\langle \mathcal{E}_{[\rho]} K_\lambda, f \rangle = \langle K_\lambda, \mathcal{E}_{[\rho]}^* f \rangle, f \in L_a^2(\mathbb{D}),$$

which, combined with Lemma 5.3.3, gives that

$$\mathcal{E}_{[\rho]} K_\lambda = \sum_{\sigma \in [\rho^-]} \overline{\sigma(\lambda)'} K_{\sigma(\lambda)}. \tag{5.43}$$

Now rewrite

$$T = \sum d_\rho' \mathcal{E}_{[\rho^-]} \equiv \sum d_\rho' \mathcal{E}_{[\rho]}^*.$$

On a neighborhood V of λ, all local inverses of B can be written as $\{\rho_k\}$ and rewrite d_k' for d_{ρ_k}'. We have

$$Th(z) = \sum_{k=0}^\infty d_k' h \circ \rho_k(z) \rho_k'(z), z \in V.$$

This, combined with (5.43), yields that

$$TK_\lambda(z) = \sum_{k=0}^\infty d_k' \overline{\rho_k'(z)} K_{\rho_k(\lambda)}(z), z \in V.$$

Below, we will show that $\sum_{k=0}^\infty d_k' \overline{\rho_k'(z)} K_{\rho_k(\lambda)}$ converges in norm, which immediately shows that

$$\lim_{m \to \infty} \|T_m K_\lambda - TK_\lambda\| = 0,$$

as desired, because $\{S_m K_\lambda\}_{m=1}^\infty$ is a subsequence of the partial sums $\{\sum_{k=0}^n d_k' \overline{\rho_k'(z)} K_{\rho_k(\lambda)}\}_{n=1}^\infty$.

In fact, $B \circ \rho_k = B$ holds on V for $k = 0, 1, \cdots$, which gives that

$$|B'(\rho_k(\lambda))||\rho_k'(\lambda)| = |B'(\lambda)|. \tag{5.44}$$

Write $a = B(\lambda)$, by Proposition 5.1.4 $\varphi_a(B)$ is a thin Blaschke product whose zero sequence is $\{\rho_k(\lambda)\}$. Then there is a numerical constant $\delta > 0$ such that

$$\prod_{j,j \neq k} d(\rho_j(\lambda), \rho_k(\lambda)) \geq \delta,$$

and then by computations

$$|(\varphi_a \circ B)'(\rho_k(\lambda))| = \frac{1}{1 - |\rho_k(\lambda)|^2} \prod_{j,j \neq k} d(\rho_j(\lambda), \rho_k(\lambda)) \geq \delta \frac{1}{1 - |\rho_k(\lambda)|^2}.$$

Since $B(\rho_k(\lambda)) = B(\lambda) = a$, the above immediately shows that

$$\frac{1}{1 - |\rho_k(\lambda)|^2} \geq |\varphi_a'(a)||B'(\rho_k(\lambda))| \geq \delta \frac{1}{1 - |\rho_k(\lambda)|^2},$$

and hence

$$\frac{1}{|\varphi_a'(a)|} \frac{1}{1 - |\rho_k(\lambda)|^2} \geq |B'(\rho_k(\lambda))| \geq \frac{\delta}{|\varphi_a'(a)|} \frac{1}{1 - |\rho_k(\lambda)|^2}.$$

Then by (5.44), there is a positive constant C such that

$$\frac{1}{C}(1 - |\rho_k(\lambda)|^2) \leq |\rho_k'(\lambda)| \leq C(1 - |\rho_k(\lambda)|^2). \tag{5.45}$$

On the other hand, $\varphi_a(B)$ is a thin Blaschke product whose zero sequence $\{\rho_k(\lambda)\}$ is uniformly separated. Then by Proposition 2.4.5,

$$A : h \to \{h(\rho_k(\lambda))(1 - |\rho_k(\lambda)|^2)\}$$

is a bounded invertible operator from $(\varphi_a(B)L_a^2(\mathbb{D}))^\perp$ onto l^2. Denote by $\{e_k\}_{k=0}^\infty$ the standard orthogonal basis of l^2, and then

$$A^* e_k = (1 - |\rho_k(\lambda)|^2) K_{\rho_k(\lambda)}, k = 0, 1, \cdots.$$

Then by (5.45),

$$\tilde{A} e_k = \rho_k'(\lambda) K_{\rho_k(\lambda)}$$

naturally induces a (linear) bounded operator $\tilde{A}$. This immediately shows that

$$\sum_{k=0}^{\infty} d_k' \overline{\rho_k'(z)} K_{\rho_k(\lambda)}$$

converges in norm, and the sum does not depend on the order of k. Since $\{T_m K_\lambda\}$ is a subsequence of the partial sum of this series, $\{T_m K_\lambda\}$ converges to TK_λ in norm. $\qquad\square$

Remark 5.7.5 By the proof of Theorem 5.7.1, Lemma 5.6.1 still holds even if the condition $\overline{Z(B)} \not\supseteq \mathbb{T}$ is dropped.

Proof of Theorem 5.7.2 The proof comes essentially from [GH4].

Suppose B is a thin Blaschke product. We must show that the von Neumann algebra $\mathcal{V}^*(B)$ is generated by all $\mathcal{E}_{[\rho]}$, where $\sharp[\rho] < \infty$. By von Neumann bicommutant theorem, it is enough to show that the commutant of *span* $\{\mathcal{E}_{[\rho]}\}$ equals that of $\mathcal{V}^*(B)$. To see this, for a given operator A commuting with all $\mathcal{E}_{[\rho]}$, we must show that A commutes with each operator in $\mathcal{V}^*(B)$.

Let V be a disk centered at 0, and write $\hat{V} = V \cap E$. By the proof of Theorem 5.7.1, for any $T \in \mathcal{V}^*(B)$ there is a sequence $\{T_N\}$ in *span* $\{\mathcal{E}_{[\rho]}; \sharp[\rho] < \infty\}$ such that

$$T_N K_z \to TK_z \quad \text{and} \quad T_N^* K_z \to T^* K_z, \quad z \in \hat{V},$$

where both convergence are in norm. Note that A commutes with T_N, and hence

$$\langle A T_N K_z, K_z \rangle = \langle A K_z, T_N^* K_z \rangle, z \in V.$$

Since $\hat{V}$ is dense in V, then by the continuity of Berezin transformation, we have

$$\langle A T_N K_z, K_z \rangle = \langle A K_z, T_N^* K_z \rangle, z \in V. \qquad (5.46)$$

Take limits in (5.46) and we get

$$\langle A T K_z, K_z \rangle = \langle A K_z, T^* K_z \rangle = \langle T A K_z, K_z \rangle.$$

The property of Berezin transformation yields that $AT = TA$. The proof is complete. $\qquad\square$

The proof of Theorem 5.7.2 can be translated to the case of finite Blaschke products, with some simplification possibly.

5.8   Finite Blaschke Product Revisited

In this section, we will turn back to finite Blaschke products, which will be treated with the techniques developed in this chapter. Below, B denotes a finite Blaschke product, and some examples of $V^*(B)$ will be provided. As we will see, in some cases the dimension $\dim V^*(B)$ of $V^*(B)$ is calculated.

The following shows that even if the assumption of Corollary 5.4.10 fails, $V^*(B)$ may be also two-dimensional. In fact, it is a consequence of the proof of Lemma 5.5.1, see [GH4].

Proposition 5.8.1 *Suppose B is a finite Blaschke product. Then there is an $s(0 < s < 1)$ such that for any $\lambda \in \mathbb{D}$ with $|\lambda| > s$, $\dim V^*(B\varphi_\lambda) = 2$, and hence $V^*(B\varphi_\lambda)$ is abelian.*

Proof By the proof of Lemma 5.5.1, there are two disjoint sub-domains Ω_1 and Ω_2 of $\mathbb{D}$, which are components of $\{z \in \mathbb{D} : |B(z)| < t\}$ for some $t > 0$. For some $w \in \Omega_2 \cap E$, there are n different points in Ω_1: $w_1, \cdots, w_n$, satisfying

$$B(w_j) = B(w), j = 1, \cdots, n.$$

Since Ω_1 is B-glued and $B|_{\Omega_2}$ is univalent, by applying Lemma 5.4.8 one can show that

$$G[(w, w_1)] = G[(w, w_j)], j = 2, \cdots, n.$$

Thus, $G[(w, w_1)]$ and the trivial component $G[(w, w)]$ are the only components of S_B, and hence by a similar version of Theorem 5.7.2, $V^*(B)$ is generated by $S_{[(w,w_1)]}$ and I. Therefore,

$$\dim V^*(B\varphi_\lambda) = 2.$$

By Theorem 2.5.11 a finite dimensional von Neumann algebra is $*$-isomorphic to the direct sum $\bigoplus_{k=1}^{r} M_{n_k}(\mathbb{C})$ of full matrix algebras, and then $V^*(B\varphi_\lambda)$ is $*$-isomorphic to $\mathbb{C} \oplus \mathbb{C}$. Thus, $V^*(B\varphi_\lambda)$ is abelian. $\square$

The following provides more examples of von Neumann algebras $V^*(B)$ [GH4].

Example 5.8.2 Let a_1, a_2, a_3 be three different points in $\mathbb{D}$ and $a_1 a_2 a_3 \neq 0$. Write

$$B = z\varphi_{a_1}^m(z)\varphi_{a_2}^n(z)\varphi_{a_3}^l(z).$$

Pick a point z_0 that is enough close to 0. By Theorem 2.1.13, it is not difficult to see that there are $m + n + l + 1$ points in $B^{-1}(B(z_0))$:

$$z_0; z_1^1, \cdots, z_1^m; z_2^1, \cdots, z_2^n; z_3^1, \cdots, z_3^l,$$

where z_i^j are around a_i ($i = 1, 2, 3$) for all possible j. Then by the proof of Lemma 5.4.8, we have

$$G[(z_0, z_i^1)] = G[(z_0, z_i^j)], \ i = 1, 2, 3, j = 1, 2, \cdots.$$

Thus, there are at most four components of S_B:

$$G[(z_0, z_0)] \equiv G[z], G[(z_0, z_1^1)], G[(z_0, z_2^1)] \text{ and } G[(z_0, z_3^1)].$$

By [DSZ, Theorem 7.6] or the comments below Theorem 5.7.2, $\dim V^*(B) \leq 4$, and hence $V^*(B)$ is abelian. Similarly, if $B = z^n \varphi_{a_1}(z)$, then $\dim V^*(B) = 2$. In this case, M_B has exactly 2 minimal reducing subspaces. In particular, if $n = 3$, this is exactly [SZZ1, Theorem 3.1].

To provide more examples of $V^*(B)$, we need the following technical lemma, whose proof is similar as Lemma 5.4.8.

Lemma 5.8.3 *Given z_0, z_1 and z_2 in E with $B(z_0) = B(z_1) = B(z_2)$, suppose there are three paths σ_i such that*

(i) $\sigma_i(0) = z_i$ for $i = 0, 1, 2$, $\sigma_1(1) = \sigma_2(1)$;
(ii) $B(\sigma_0(t)) = B(\sigma_1(t)) = B(\sigma_2(t)), 0 \leq t \leq 1$;
(iii) For $0 \leq t < 1$, $\sigma_i(t) \in E, i = 0, 1, 2$.
(iv) $\gcd(k, l) = 1$, where k denotes the order of the zero $\sigma_0(1)$ of $B - B(\sigma_0(1))$, and l denotes the order of the zero $\sigma_1(1)$ of $B - B(\sigma_1(1))$.

Then $G[(z_0, z_1)] = G[(z_0, z_2)]$.

All discussions below are based on the fact that $\dim V^*(B)$ equals the number of the components of S_B for a finite Blaschke product B. The following example is provided by Guo and Huang.

Example 5.8.4 Let $B = \varphi_a^m \varphi_b^n$ be a finite Blaschke product, where a and b are different points in $\mathbb{D}$ and $\gcd(m, n) = 1$. Then $\dim V^*(B) = 2$; equivalently, S_B has exactly two components. We only deal with the case where $m > 1$ and $n > 1$.

By Bochner's Theorem, B has $m + n - 1$ critical points, counting multiplicity. Since both a and b are critical points with multiplicity $m - 1$ and $n - 1$, respectively, and $m + n - 1 - (m - 1) - (n - 1) = 1$, then there is exactly one critical point w different from a and b such that

$$B'(w) = 0 \quad \text{and} \quad B''(w) \neq 0.$$

Pick an enough small positive number ε with $\varepsilon < |B(w)|$. By Rouche's theorem and Theorem 2.1.13, there is a small disk around a containing $a_1, \cdots, a_m$, and another disk around b containing $b_1, \cdots, b_n$ (both sequences $\{a_i\}$ and $\{b_j\}$ are in anti-clockwise direction), where a_i and b_j are in $B^{-1}(\varepsilon \mathbb{D})$. Now we can draw a curve L connecting ε and w, such that

$$L \subseteq \left(|B(w)| \mathbb{D} \cap E \right).$$

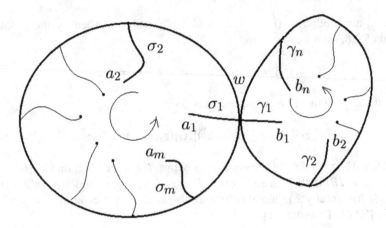

Fig. 5.8 $B^{-1}(L)$

Then $B^{-1}(L)$ consisting of the following curves:

$$\sigma_1, \sigma_2, \cdots, \sigma_m; \gamma_1, \gamma_2, \cdots, \gamma_n,$$

where $a_i \in \sigma_i$ and $b_j \in \gamma_j$ for all i and j. Without loss of generality, assume that $\sigma_1 \cap \gamma_1 = \{w\}$ and $\sigma_i \cap \gamma_j = \emptyset$ for $(i, j) \neq (1, 1)$, see Fig. 5.8.

Take a local inverse ρ such that $\rho(b_2) = b_1$. Clearly, $G[\rho] \neq G[z]$. Since $\sigma_1 \cap \gamma_1 = \{w\}$, $B'(w) = 0$ and $B'(\gamma_2(1)) \neq 0$, then by Lemma 5.4.8

$$G[(b_2, a_1)] = G[(b_2, b_1)].$$

Since $\gcd(m, n) = 1$, then by a careful verification, the proof of Lemma 5.4.8 shows that

$$G[(b_2, a_j)] = G[(b_2, b_1)], \quad j = 1, \cdots, n.$$

Thus $G[\rho]$ contains these points:

$$(*, b_1); (*, a_1), (*, a_2), \cdots, (*, a_m) \quad \text{with} \quad * = b_2.$$

Replace b_2 with b_1 in the above sequence. Along with a rotation from b_2 to b_1 (see the proof of Lemma 5.4.8), applying the techniques of analytic continuation shows that $G[\rho]$ containing the follows:

$$(*, b_n); (*, a_m), (*, a_1) \cdots, (*, a_{m-1}) \quad \text{with} \quad * = b_1,$$

or

$$(*, b_n); (*, a_1), \cdots, (*, a_m) \quad \text{with} \quad * = b_1.$$

Note that the map $(\sigma_1 \sharp \gamma_1)(t) \to (\gamma_1 \sharp \sigma_1)(t)$ naturally induces a local inverse exchanging a_1 with b_1, and thus $G[\rho]$ contains the point sequence:

$$(*, b_n); (*, b_1), (*, a_2) \cdots, (*, a_m) \quad \text{with} \quad * = a_1.$$

Since $\gcd(m, n) = 1$ and $(a_1, b_1) \in G[\rho]$, then by Lemma 5.8.3 all points $(a_1, b_j)(1 \le j \le n)$ lie in $G[\rho]$. Therefore, $G[\rho]$ contains the following $m + n - 1$ points:

$$(*, b_1), \cdots, (*, b_n); (*, a_2) \cdots, (*, a_m) \quad \text{with} \quad * = a_1,$$

which shows that $\sharp[\rho] \ge m + n - 1$. On the other hand,

$$\sum \sharp[\sigma] = \text{order } B = m + n,$$

and then $\sharp[\rho] \le m + n - \sharp[z] = m + n - 1$, forcing $\sharp[\rho] = m + n - 1$. Therefore, $[\rho]$ and $[z]$ are all the components of S_B.

In general, we have the following result due to Guo and Huang.

Proposition 5.8.5 *For two different points $a, b \in \mathbb{D}$, put $B = \varphi_a^m \varphi_b^n$, where $\gcd(m, n) = d$. Then S_B has exactly $d + 1$ components, and thus $\dim V^*(B) = d + 1$.*

Proof By Example 5.8.4, we assume that $\gcd(m, n) = d > 1$, and it suffices to show that S_B has exactly $d + 1$ components. Rewrite

$$m = kd, n = ld \quad \text{with } \gcd(k, l) = 1.$$

Suppose a_i, σ_i and b_j, γ_j $(1 \le i \le m, 1 \le j \le n)$ are the same as in Example 5.8.4, also see Fig. 5.8. For convenience, for any $p, q \in \mathbb{Z}_+$, we write $a_{[p]}$ for $a_{p'}$ with $p = p' \mod m$, and $b_{[q]}$ for $b_{q''}$ with $q = q'' \mod n$.

Now take a local inverse ρ such that $\rho(b_2) = b_1$. Clearly, $[\rho] \ne [z]$. Then by Lemma 5.4.8,

$$G[(b_2, a_1)] = G[(b_2, b_1)].$$

Since $\gcd(m, n) = d$, by a similar version of Lemma 5.8.3

$$G[(b_2, a_1)] = G[(b_2, a_{[1+pd]})], p = 0, 1, \cdots.$$

Therefore, $G[\rho]$ contains the following

$$(b_2, b_1); (b_2, a_{[1+pd]}), p = 0, 1, \cdots,$$

and by the techniques of analytic continuation, $G[\rho]$ also contains

$$(*, b_n); (*, a_{[pd]}), p = 0, 1, \cdots, \quad \text{with} \quad * = b_1,$$

and

$$(*, b_n); (*, a_{[pd]}), p = 0, 1, \cdots, \quad \text{with} \quad * = a_1.$$

Thus, $(a_1, b_n) \in G[\rho]$, then applying the techniques of analytic continuation shows that $G[\rho]$ also contains $(a_1, b_{[pd]})$ for $p = 0, 1, \cdots$. Therefore, $G[\rho]$ contains the following:

$$(*, a_{[pd]}); (*, b_{[pd]}), p = 0, 1, \cdots, \quad \text{with} \quad * = a_1,$$

By a bit more effort, one can show that $G[\rho]$ contains the following:

$$(*, a_{[pd+i-1]}); (*, b_{[pd+i-1]}), p = 0, 1, \cdots, \quad \text{with} \quad * = a_i,$$

and

$$(*, a_{[pd+j-1]}); (*, b_{[pd+j-1]}), p = 0, 1, \cdots, \quad \text{with} \quad * = b_j,$$

for any a_i and b_j. Moreover, this sequence is *complete*; that is, if $(z, w) \in G[\rho]$ with $z = a_i$ (or $z = b_j$), then (z, w) must lie in the above sequence. This immediately gives that

$$\sharp G[\rho] = k + l.$$

Also, by a similar discussion, one can show for $t = 1, \cdots, d-1$, some component $G[\rho_t]$ contains the following points:

$$(*, a_{[pd+i+t]}); (*, b_{[pd+i+t]}), p = 0, 1, \cdots, \quad \text{with} \quad * = a_i,$$

and

$$(*, a_{[pd+j+t]}); (*, b_{[pd+j+t]}), p = 0, 1, \cdots, \quad \text{with} \quad * = b_j.$$

We have $\sharp G[\rho_t] = k + l$, and $G[\rho_t]$ differs for distinct t. In particular, for $t = d - 1$, $G[\rho_{d-1}] = G[\rho]$. There exists one additional component, say $G[\rho_d]$, that contains the following:

$$(*, b_{[pd+1]}); (*, a_{[pd+1]}), p = 0, 1, \cdots, \quad \text{with} \quad * = a_1 \quad \text{excluding}(a_1, a_1),$$

$$(*, b_{[pd+i]}); (*, a_{[pd+i]}), p = 0, 1, \cdots, \quad \text{with} \quad * = a_i \quad \text{excluding}(a_i, a_i),$$

and

$$(*, b_{[pd+j]}); (*, a_{[pd+j]}), p = 0, 1, \cdots, \quad \text{with} \quad * = b_j \quad \text{excluding} (b_j, b_j).$$

Note that $\sharp G[\rho_d] = k + l - 1$, and then

$$\sum_{t=1}^{d} \sharp G[\rho_t] + \sharp G[z] = (k + l)(d - 1) + (k + l - 1) + 1 = (k + l)d = \text{order } B.$$

Therefore, all $G[\rho_t](1 \leq t \leq d)$ and the identity component are all components of S_B. The proof is complete. □

Some words are in order. In the proof of Proposition 5.8.5, we mention the notion of "complete". A finite sequence $\{(z_0, w_j) : j = 0, \cdots, k\}$ is called *complete* if $G[(z_0, w_j)] = G[(z_0, w_0)]$, and whenever $(z_0, w) \in G[(z_0, w_0)]$, w must lie in $\{w_j : j = 0, \cdots, k\}$. For a given sequence $\{(z_0, w_j) : j = 0, \cdots, k\}$, it is complete if and only if $\{w_j : j = 0, \cdots, k\}$ equals the set of all $\tilde{\rho}(z_0)$, where ρ is the local inverse satisfying $\rho(z_0) = w_0$, and $\tilde{\rho}$ is the analytic continuation of ρ along any possible loop γ at z_0. By the theory of algebraic topology, the value $\tilde{\rho}(z_0)$ is invariant under the homotopy equivalence of γ. Thus, the loop γ can be chosen as the product of finitely many simple loops σ_j at z_0, with each winding along exactly one point in $B^{-1} \circ B(Z(B'))$ at most once. This simplification would make our job easier, which has been done in the proof of Proposition 5.8.5. By careful verification, one sees that those mentioned sequences in the proof of Proposition 5.8.5 are complete.

Furthermore, by the same methods one can compute the following example, provided by Guo and Huang. The details are omitted.

Example 5.8.6 For three different points $a_1, a_2, a_3 \in \mathbb{D}$, $B = \varphi_{a_1}^m \varphi_{a_2}^n \varphi_{a_3}^k$ where $\gcd(m, n, k) = 1$. We assume that there is at most one additional critical point of B except for a_1, a_2 and a_3. If $\min(m, n, k) \geq 2$, then S_B has exactly two components. If $m = n$ and $k = 1$, then S_B has exactly three components.

To end this section, we will give another approach for the structures of $\mathcal{V}^*(B)$ where order $B = 3$ and 4.

As done in [GSZZ, DPW], two finite Blaschke products B_1 and B_2 are called similar if there are two members $m_1, m_2 \in \text{Aut}(\mathbb{D})$ such that

$$B_2 = m_1 \circ B_1 \circ m_2.$$

In this case, we write $B_1 \sim B_2$. If so, then there is a unitarily isomorphism between the von Neumann algebras $\mathcal{V}^*(B_1)$ and $\mathcal{V}^*(B_2)$, which is complemented by the unitary operator

$$U_{m_2} : f \to f \circ m_2 m_2', f \in L_a^2(\mathbb{D}).$$

The following was shown in [GSZZ, Theorem 4].

Theorem 5.8.7 (Guo, Sun, Zheng, Zhong) *Suppose B is a finite Blaschke product of order* 3. *If B* $\sim z^3$, *then* dim $V^*(B) = 3$. *Otherwise, there is a nonzero point* $\lambda \in \mathbb{D}$ *such that B* $\sim z^2 \varphi_\lambda$, *and in this case* dim $V^*(B) = 2$.

Proof Suppose B is a finite Blaschke product of order 3. Then by Bochner's theorem (Theorem 5.4.6), B has 2 critical points, counting multiplicity. Assume that z_0 is one of the critical points, and write $b = B(z_0)$. Then it is not difficult to see that $\varphi_b \circ B \circ \varphi_{z_0}$ has a zero at 0, whose order is ≥ 2. Thus, we have either $B \sim z^3$ or $B \sim z^2 \varphi_\lambda$ for some $\lambda \in \mathbb{D} - \{0\}$. Form the discussion below Example 5.8.6, it suffices to consider two cases: $B = z^3$ and $B = z^2 \varphi_\lambda$.

If $B = z^3$, define $\rho_j(z) = e^{\frac{2ji\pi}{3}} z (j = 0, 1, 2)$, and then ρ_0, ρ_1 and ρ_2 are three local inverses of B, belonging to $\text{Aut}(\mathbb{D})$. This shows that S_B has three components

$$\{(z, \rho_j(z)) : z \in \mathbb{D}, z \neq 0\}(0 \leq j \leq 2),$$

and hence dim $V^*(B) = 3$. Otherwise, $B \sim z^2 \varphi_\lambda$ with $\lambda \neq 0$. By Example 5.8.2, S_B has exactly two components, and thus dim $V^*(B) = 2$. The proof is complete. $\quad\square$

It should be pointed out that $B \sim z^3$ if and only if $B = c\varphi_a(\varphi_\lambda^3)$ for some $\lambda, a \in \mathbb{D}$ and $|c| = 1$.

The following context focuses on the case of order $B = 4$. First we need a lemma.

Lemma 5.8.8 *Let B be a finite Blaschke product. Then B* $\sim z^2 \varphi_b^2$ *for some b* $\neq 0$ *if and only if there is some* $\lambda \in \mathbb{D} - \{0\}$ *such that B* $\sim z^2 \varphi_{\lambda^2}(z^2)$.

Proof First assume that B is a finite Blaschke product such that

$$B \sim z^2 \varphi_b^2,$$

where $b \neq 0$. Put $h = z\varphi_b$. Since $z^2 \varphi_b^2 = (z\varphi_b)^2$, we have $B \sim z^2 \circ h$. Since $h \sim z^2$, it is easy to see there is a nonzero $c \in \mathbb{D}$ such that $B \sim \varphi_c^2(z^2)$. Consider $\varphi_{c^2} \circ \varphi_c^2(z^2)$. There is some $\lambda \neq 0$ in $\mathbb{D}$ and some unimodular constant ξ such that $\varphi_{c^2} \circ \varphi_c^2 = \xi z \varphi_{\lambda^2}$, forcing

$$\varphi_{c^2} \circ \varphi_c^2(z^2) = \xi z^2 \varphi_{\lambda^2}(z^2).$$

Then $B \sim z^2 \varphi_{\lambda^2}(z^2)$.

Conversely, assume that there is a constant $\lambda \neq 0$ satisfying

$$B \sim z^2 \varphi_{\lambda^2}(z^2).$$

Write $B_1 = z\varphi_{\lambda^2}$, and $B \sim B_1(z^2)$. Since $B_1 \sim z^2$, there is some constant $a \neq 0$ such that

$$B \sim B_1(z^2) \sim (z^2 \circ \varphi_{a^2})(z^2).$$

Thus,

$$B \sim \varphi_a^2 \varphi_{-a}^2 \sim z^2 \big(\varphi_{-a} \circ \varphi_a \big)^2,$$

which shows that there is some $b \neq 0$ such that $B \sim z^2 \varphi_b^2$. The proof is complete.

□

The following result is from [SZZ1, Theorem 2.1], also restated in [DPW, Sect. 4].

Theorem 5.8.9 (Sun, Zheng, Zhong) *Suppose B is a finite Blaschke product of order 4. If $B \sim z^4$, then* $\dim \mathcal{V}^*(B) = 4$. *If $B \sim z^2 \varphi_\lambda^2$ for some $\lambda \neq 0$, then* $\dim \mathcal{V}^*(B) = 3$. *Otherwise,* $\dim \mathcal{V}^*(B) = 2$.

Proof Suppose B is a finite Blaschke product of order 4. Consider the number q of components of S_B. There are several cases under consideration: $q = 2, 3$ and 4.

If $q = 4 = \operatorname{order} B$, then for each local inverse ρ of B, $\sharp[\rho] = 1$, which implies that ρ is single-valued and holomorphic in $\mathbb{D} - F$ for some finite subset F of $\mathbb{D}$. Since ρ is bounded by 1, it extends analytically to $\mathbb{D}$. Since $B \circ \rho = B$, ρ is a proper map from $\mathbb{D}$ to $\mathbb{D}$. Then it follows from Theorem 2.1.3 that $\rho \in \operatorname{Aut}(\mathbb{D})$. By the proof of Proposition 5.3.6 B has the following form: $B = m(\varphi_a^4)$ for some $m \in \operatorname{Aut}(\mathbb{D})$. Therefore, $B \sim z^4$. Since S_B has four components, then $\dim \mathcal{V}^*(B) = 4$.

To complete the proof, we will show that $\dim \mathcal{V}^*(B) = 3$ if and only if $B \sim z^2 \varphi_\lambda^2 (\lambda \neq 0)$. To see this, we first claim that the following are equivalent.

(i) There is a Blaschke product B_1 of order 2 such that $B \sim B_1 \circ z^2$ and $B \nsim z^4$.
(ii) $B \sim z^2 \varphi_\lambda^2$ for some $\lambda \neq 0$.

(ii)$\Rightarrow$ (i) follows directly from Lemma 5.8.8. To see (i)$\Rightarrow$ (ii), assume that there is some Blaschke product B_1 of order 2 such that $B \sim B_1 \circ z^2$ and $B \nsim z^4$. Note that there is some $h \in \operatorname{Aut}(\mathbb{D})$ such that

$$h \circ B_1 = \xi z \varphi_a(z)$$

for some nonzero $a \in \mathbb{D}$ and $|\xi| = 1$. Then $B \sim z^2 \varphi_a(z^2)$, and by Lemma 5.8.8 $B \sim z^2 \varphi_\lambda^2$ for some $\lambda \neq 0$. Therefore, (i) and (ii) are equivalent.

If $B \sim z^2 \varphi_\lambda^2 (\lambda \neq 0)$, then by Proposition 5.8.5 $q = 3$, forcing $\dim \mathcal{V}^*(B) = 3$.

Conversely, assume that $\dim \mathcal{V}^*(B) = q = 3$. It suffices to show that there is some Blaschke product B_1 of order 2 such that $B \sim B_1(z^2)$ and $B \nsim z^4$. Since $q = 3$, then we may assume that all components of S_B are $G[z], G[\rho_1]$ and $G[\rho_2]$, where $\sharp[\rho_1] = 1$ and $\sharp[\rho_2] = 2$. As discussed above, ρ_1 is a member in $\operatorname{Aut}(\mathbb{D})$ different from id. By a similar version of Proposition 5.3.6 or [DSZ, Lemma 8.1], we get $B \sim B_1(z^2)$ for some Blaschke product B_1. Clearly, $B \nsim z^4$.

The proof is complete. □

By using somewhat different methods based on partition of the local inverses of B, arithmetics of reducing subspace are available when the order of B is not large [DPW]. In particular, [DPW, Theorem 4.2] gives a complete characterization for the numbers of minimal reducing subspaces when order $B = 8$.

5.9   Remarks on Chap. 5

Most materials of this chapter come from Guo and Huang's paper [GH4].

All discussions in this chapter are based on the Bergman space $L_a^2(\mathbb{D})$. We obtain a complete classification for thin Blaschke product B such that $\mathcal{V}^*(B)$ is nontrivial. Also, we construct a first example of a Blaschke product B for which $\mathcal{V}^*(B)$ is trivial; equivalently, M_B is irreducible. This gives a negative answer to the following problem:

For each infinite Blaschke product B, does M_B always have a nontrivial reducing subspace?

This problem is of interest in at least two aspects. Firstly, for a finite product B different from the Möbius map, M_B always has a nontrivial reducing subspace [HSXY]. Secondly, on the Hardy space $H^2(\mathbb{D})$ M_B always has a nontrivial reducing subspaces if B is any Blaschcke product different from the Möbius map.

Section 5.1 of this chapter introduces some necessary preliminaries. Propositions 5.1.1, 5.1.3 and Lemma 5.1.2 are probably known [GH4].

In [GH4], Guo and Huang proved Lemmas 5.2.2, 5.2.3 and Theorems 5.2.4, 5.2.5. Lemmas 5.2.1, 5.2.6, Propositions 5.2.7 and 5.2.8 appeared also in [GH4].

Theorems 5.4.1, Corollaries 5.4.10, 5.4.5, Proposition 5.4.4, Lemma 5.4.8 come from [GH4]. The gluable property is crucial in this chapter, and Definitions 5.4.2, 5.4.3 appeared first in [GH4]. The finite-Blaschke-product version of Proposition 5.3.4 was implied in [DSZ, DPW], though without proof.

Lemma 5.5.1 was proved in [GH4]. Theorems 5.5.3, 5.3.1, Propositions 5.3.2, 5.3.6, Lemma 5.3.3 and Example 5.3.5 come from [GH4]. Also, Lemma 5.6.1 and Example 5.6.3 are proved in [GH4].

Guo and Huang showed Theorems 5.7.1 and 5.7.2. Lemma 5.7.3, Propositions 5.7.4, 5.8.1 and Example 5.8.2 are from [GH4].

Lemma 5.8.3, Examples 5.8.4, 5.8.6 and Proposition 5.8.5 are new.

In Section 5.8 of this chapter we mentioned the work of Douglas et al. [DPW]. They also considered when a finite Blaschke product can be written as the composition of two finite Blaschke products. It is worthy to mention that concerning this question, Cowen wrote an earlier manuscript [Cow5]. He used the techniques of complex analysis and group theory. But the approach and aim in [Cow5] differ a lot from the theme of this chapter. An even earlier paper by Ritt [Ri1] considered the problem of compositional factorization of polynomials and gave the uniqueness of factorization, and this job was later generalized, to some extent, to the case of rational functions by Ritt himself [Ri2].

Concerning reducing subspaces of analytic Toeplitz operators and the question to determine which analytic Toeplitz operators are irreducible, we call the reader's attention to the following references: [A1, Ba, DW, Nor, T1, T3] and [Zhu1, Zhu2]. Here, we have no intention to include a complete literature.

The authors are grateful to Dr. Xie and Dr. Z. Wang for making nice figures in this chapter.

Chapter 6
Covering Maps and von Neumann Algebras

Distinct classes of multiplication operators have been investigated in Chaps. 4 and 5 on the Bergman space, involving their reducing subspaces. Precisely, these multiplication operators arise from finite and thin Blaschke products. The reducing subspaces of a single multiplication operator M_ϕ naturally correspond to those projections, which generate a von Neumann algebra $\mathcal{V}^*(\phi)$. In the above settings, this von Neumann algebra turns out to be abelian, sometimes even trivial, and hence is of type I. However, it is not always the case if the function ϕ varies.

This chapter centers on reducing subspace problem of M_ϕ, where ϕ is a holomorphic covering of any bounded planar domain Ω. One will see how types of such von Neumann algebras $\mathcal{V}^*(\phi)$ are related to topological properties of these planar domains Ω. In most cases, $\mathcal{V}^*(\phi)$ are type II factors. Methods of complex analysis, operator theory and conformal geometry are combined in investigating $\mathcal{V}^*(\phi)$, eventually establishing a fascinating connection to one of the long-standing problems in free group factors.

The context of this chapter involves much of work done by the authors from [GH1] and [GH2]. Briefly, it is divided into two parts: Part I addresses the von Neumann algebras $\mathcal{V}^*(\phi)$ defined on the Bergman space, Part II will extend those results in Part I to weighted Bergman spaces.

Part I: In the Case of the Bergman Space

6.1 Regular Branched Covering Maps and Orbifold Domains

In this section, the definition of orbifold domain is introduced and some examples are provided. Also, the connection between orbifold domains and regular branched covering maps is given, as a classical result in complex dynamics.

© Springer-Verlag Berlin Heidelberg 2015
K. Guo, H. Huang, *Multiplication Operators on the Bergman Space*,
Lecture Notes in Mathematics 2145, DOI 10.1007/978-3-662-46845-6_6

It was mentioned in Sect. 2.1 of Chap. 2 that covering maps share several good properties with regular branched covering maps. For example, given a regular branched covering map $\phi : \mathbb{D} \to \Omega$, it is not difficult to verify that for each $w \in \Omega$, as z varies in $\phi^{-1}(w)$ the multiplicity of the zero point of $\phi - w$ at z only depends on w. Thus, one can naturally define a function v from Ω to $\{1, 2, \cdots\}$, such that $v(w)$ equals the above multiplicity. Also, one sees that v takes the value 1 except on a discrete subset of Ω.

Let Ω be a planar bounded domain, Σ a discrete subset of Ω, and v is a function from Ω to $\{1, 2, \cdots\}$. The triple (Ω, Σ, v) is called *an orbifold domain* if v takes values $\{2, 3, \cdots\}$ on Σ, and 1 on the complementary set $\Omega - \Sigma$. For such a triple (Ω, Σ, v), there exists a regular branched covering map $\phi : \mathbb{D} \to \Omega$ with the following property:

for each $w \in \Omega$, $v(w)$ equals the multiplicity of the zero point of $\phi - w$ at any z in $\phi^{-1}(w)$.

For more details, see [Mi, Theorem E1], [Sc, Theorem 2.3] and [BMP]. Below, in most cases we rewrite (Ω, v) for (Ω, Σ, v). The map $\phi : \mathbb{D} \to (\Omega, v)$ with the desired property as above is unique up to an automorphism of $\mathbb{D}$ [BMP, Theorem 2.6], and ϕ is called *the universal covering* of (Ω, v). *The fundamental group* $\pi_1(\Omega, \Sigma, v)$ of the orbifold (Ω, v) is defined to be the deck transformation group of its universal covering map, that is,

$$\pi_1(\Omega, \Sigma, v) \triangleq G(\phi) = \{\rho \in Aut(\mathbb{D}) : \phi \circ \rho = \phi\},$$

see [Rat, Chap. 13] for an equivalent definition. The function v is called a *ramified function* and Σ is called *the singular locus* of (Ω, v). For an orbifold (Ω, Σ, v), if Σ is empty and $v \equiv 1$, then (Ω, v) is identified with the domain Ω. In this situation, ϕ is the usual universal covering map.

A humble example is the orbifold disk $(\mathbb{D}, \{0\}, v(0) = n)$. Its universal covering map is given by $\phi(z) = z^n : \mathbb{D} \to \mathbb{D}$, and its deck transformation group is

$$\{\rho_k(z) = \exp(\frac{2k\pi i}{n}) z : k = 0, 1, \cdots, n - 1\},$$

as is isomorphic to $\mathbb{Z}_n$.

The following, known as the Seifert-Van Kampen Theorem, enables us to construct more examples. It comes from algebraic topology, see [BMP, Corollary 2.3], [Go] and also [Kap, Theorem 6.8].

Theorem 6.1.1 (Seifert-Van Kampen Theorem) *Let $\mathcal{O}$ be an orbifold and $\mathcal{O}_1, \mathcal{O}_2 \subset \mathcal{O}$ two open suborbifolds such that $\mathcal{O}_1, \mathcal{O}_2$ and $\mathcal{O}_1 \cap \mathcal{O}_2$ are connected. If $\mathcal{O} = \mathcal{O}_1 \cup \mathcal{O}_2$, then $\pi_1\mathcal{O}$ is the amalgamated product,*

$$\pi_1\mathcal{O} \cong \pi_1\mathcal{O}_1 *_\Gamma \pi_1\mathcal{O}_2$$

where $\Gamma = \pi_1(\mathcal{O}_1 \cap \mathcal{O}_2)$.

In what follows, it is supposed that all orbifolds and suborbifolds are open, and Γ reduces to a trivial group, in which case $\pi_1 \mathcal{O}$ is isomorphic to $\pi_1 \mathcal{O}_1 * \pi_1 \mathcal{O}_2$, the free product of $\pi_1 \mathcal{O}_1$ and $\pi_1 \mathcal{O}_2$.

Free groups prove to play an important role in studying the structure of von Neumann algebras and its commutant arising from multiplication operators on $L_a^2(\mathbb{D})$ defined by bounded holomorphic covering maps. Here, we refer to [Ja, Sect. 2.9] for a relatively easy definition of free product of groups G_α, $\alpha \in \mathcal{I}$. Due to van der Waerden, *the free product* $*_\alpha G_a$ of $\{G_\alpha : \alpha \in \mathcal{I}\}$ can be regarded as the set of *reduced words*, by which we mean either the identity e or the words $x_1 x_2 \cdots x_n$ which satisfies that x_i and $x_{i+1}(1 \leq i \leq n-1)$ do not belong to a same G_α and all $x_i \neq e$. A *free group* F_k on k generators can always be regarded as the free product of k groups, with each isomorphic to $\mathbb{Z}$. Roughly speaking, each free group G arises in this way: there is a set E such that any element of G can be written in one and only one way as a product of finitely many elements of E and their inverses (disregarding trivial variations such as $st^{-1} = su^{-1}ut^{-1}$). For its definition, one can also refer to [Ro] or [Hat, Chap. 1]. Later in Sect. 6.5, we will see that the theory of free groups plays an important role in the study of $\mathcal{V}^*(\phi)$ and $\mathcal{W}^*(\phi)$, where ϕ is a regular branched covering map.

More examples of orbifold domains emerges as follows.

Example 6.1.2 Take distinct points in $\mathbb{D}$, $\Sigma = \{w_1, w_2, \cdots, w_n\}$ $(1 \leq n < \infty)$, and give positive integers $m_1, m_2, \cdots, m_n$. Define $\upsilon(w_k) = m_k$ for $k = 1, \cdots, n$. Applying Seifert-Van Kampen Theorem shows that

$$\pi_1(\mathbb{D}, \Sigma, \upsilon) \cong \mathbb{Z}_{m_1} * \mathbb{Z}_{m_2} \cdots * \mathbb{Z}_{m_n}.$$

More generally, put $n = \infty$, and then by the idea in [Mi, p. 258, Problem E-4],

$$\pi_1(\mathbb{D}, \Sigma, \upsilon) \cong *_k \mathbb{Z}_{m_k}.$$

From Example 6.1.2, it follows that for an orbifold domain $(\mathbb{D}, \Sigma, \upsilon)$, $\pi_1(\mathbb{D}, \Sigma, \upsilon)$ is finite if and only if Σ contains at most one point. Since each holomorphic regular branched covering from $\mathbb{D}$ onto $\mathbb{D}$ can be realized by a universal covering of some $(\mathbb{D}, \Sigma, \upsilon)$. Moreover, it is easy to see that if a finite Blaschke product B is a regular branched covering, then $B(z) = c\varphi_\alpha^k(z)$ for some orbifold $\alpha \in \mathbb{D}$ and $|c| = 1$, where

$$\varphi_\alpha(z) = \frac{\alpha - z}{1 - \overline{\alpha}z}$$

is the Möbius map. This can be regarded as a generalization of Lemma 4.6.3.

6.2   Representations of Operators in $\mathcal{V}^*(\phi)$

This section aims to formulate those operators in $\mathcal{V}^*(\phi)$ in a function-theoretic form, where ϕ is a holomorphic regular branched covering map from the unit disk $\mathbb{D}$ onto a bounded planar domain Ω. Also, some lemmas will be established.

As done in Sect. 2.1 of Chap. 2, $G(\phi)$ denotes the deck transformation group of ϕ; that is, $G(\phi)$ consists of those automorphisms ρ of $\mathbb{D}$ satisfying $\phi \circ \rho = \phi$. In this case, for any points z_1, z_2 in $\mathbb{D}$, $\phi(z_1) = \phi(z_2)$ if and only if there is a member ρ in $G(\phi)$ such that $z_2 = \rho(z_1)$. Since $G(\phi)$ is countable, one can write $G(\phi) = \{\rho_k\}$. Let $\mathcal{E}_\phi$ denote the critical value set of ϕ; that is,

$$\mathcal{E}_\phi = \{\phi(z) : \text{there is a } z \in \mathbb{D} \text{ such that } \phi'(z) = 0\}.$$

The following lemma comes from [Cow1].

Lemma 6.2.1 *Suppose $\phi : \mathbb{D} \to \Omega$ is a holomorphic regular branched covering map and $G(\phi)$ is infinite. Then for each $w_0 \in \mathbb{D} - \phi^{-1}(\mathcal{E}_\phi)$, $\{\rho_k(w_0)\}$ is an interpolating sequence for $H^\infty(\mathbb{D})$.*

Proof The proof is from that of [Cow1, Theorem 6].

Since $\phi - \phi(w_0)$ is a bounded holomorphic function with zeros $\{\rho_k(w_0)\}_{k=0}^{\infty}$, $\{\rho_k(w_0)\}_{k=0}^{\infty}$ is a Blaschke sequence, which gives a Blaschke product B, see [Hof1]. For each j, $B \circ \rho_j$ is also a Blaschke product, whose zero sequence equals $\{\rho_j^{-1}(\rho_k(w_0))\}_{k=0}^{\infty} = \{\rho_k(w_0)\}_{k=0}^{\infty}$. Therefore, there is some unimodular constant ξ such that $B \circ \rho_j = \xi B$, which immediately gives

$$B'(\rho_j(w_0))\rho_j'(w_0) = \xi B'(w_0).$$

For simplicity, rewrite w_j for $\rho_j(w_0)$, and then $B'(w_j)\rho_j'(w_0) = \xi B'(w_0)$. On another hand, since $\rho_j \in \text{Aut}(\mathbb{D})$, by direct computations

$$|\rho_j'(w_0)| = \frac{1 - |w_j|^2}{1 - |w_0|^2},$$

which gives that

$$|B'(w_j)|(1 - |w_j|^2) = |B'(w_0)|(1 - |w_0|^2).$$

Therefore,

$$\inf_j |B'(w_j)|(1 - |w_j|^2) = |B'(w_0)|(1 - |w_0|^2) > 0.$$

This is equivalent to say that B is an interpolating Blaschke product, completing the proof. $\square$

In very restricted cases, the deck transformation group $G(\phi)$ is finite, as will be demonstrated in Theorem 6.6.2.

In general, for each $\rho_k \in G(\phi)$, define U_{ρ_k} on $L_a^2(\mathbb{D})$ by setting

$$U_{\rho_k} h = h \circ \rho_k \, \rho_k', \, h \in L_a^2(\mathbb{D}).$$

Clearly, U_{ρ_k} is a unitary operator satisfying

$$M_\phi U_{\rho_k} = U_{\rho_k} M_\phi,$$

and hence all U_{ρ_k} are in $\mathcal{V}^*(\phi)$. For each $w \in \mathbb{D}$, put

$$\pi_w : h \mapsto \{U_{\rho_k} h(w)\}_k, \, h \in L_a^2(\mathbb{D}).$$

Soon we will see that each π_w defines a bounded operator from the Bergman space to l^2 or $\mathbb{C}^n$. The following theorem, established by Guo and Huang [GH2], gives the representation of unitary operators in $\mathcal{V}^*(\phi)$.

Theorem 6.2.2 *Suppose $\phi : \mathbb{D} \to \Omega$ is a holomorphic regular branched covering map, and S is a unitary operator on the Bergman space which commutes with M_ϕ. If $G(\phi)$ is infinite (or finite), then there is a unique operator $W : l^2 \to l^2$ (or $W : \mathbb{C}^n \to \mathbb{C}^n$) such that*

$$W\pi_w(h) = \pi_w(Sh), \, h \in L_a^2(\mathbb{D}), \, w \in \mathbb{D}. \tag{6.1}$$

This W is necessarily a unitary operator. Moreover, there is a unique square-summable sequence $\{c_k\}$ satisfying

$$Sh(z) = \sum_k c_k U_{\rho_k} h(z), \, h \in L_a^2(\mathbb{D}), \, z \in \mathbb{D}, \tag{6.2}$$

where ρ_k run over $G(\phi)$.

If ϕ is just a branched covering map and its critical value set $\mathcal{E}_\phi$ is discrete, then for any $w_0 \in \mathbb{D} - \phi^{-1}(\mathcal{E}_\phi)$ there is a neighborhood of w_0 on which (6.1) holds. However in this general setting, all ρ_k, coming from branches of $\phi^{-1} \circ \phi$, are just locally holomorphic and it remains unknown whether W is unique. In particular, for the situation when ϕ is a finite Blaschke product, a similar version of (6.2) is obtained by Sun [Sun1] and Douglas et al. [DSZ], with a different proof; also see Lemma 4.2.2.

Put $A_r = \{z \in \mathbb{C} : r < |z| < 1\}$ and $\phi(z) = z^n$. A similar version of Theorem 6.2.2 also holds on $L_a^2(A_r)$. In this case, M_ϕ defined on $L_a^2(A_r)$ has exactly n minimal reducing subspaces, as shown in [DK].

If ϕ is a bounded holomorphic function and there is an open set V such that $\phi^{-1} \circ \phi(V) = V$, then the only U_{ρ_k} appearing in (6.2) must be the identity I. To put it in another way, M_ϕ has no nontrivial reducing subspace, which was first shown by Zhu [Zhu2, Theorem 1]. See Appendix B for a Hardy-space analogue of this fact.

Proof of Theorem 6.2.2 The proof is from [GH2].

Suppose $\phi : \mathbb{D} \to \Omega$ is a holomorphic regular branched covering map. Without loss of generality, we assume that $G(\phi)$ is infinite. As one will see, the discussion for the case of $G(\phi)$ being finite is similar and easier. For the reader's convenience, the proof is divided into several steps because of its length, and the difficulty lies in establishing (6.1).

Step 1 First we make a claim as follows.

Claim For each $z_0 \in \mathbb{D}$, there exists a disk $\Delta \subseteq \mathbb{D}$ containing z_0 such that $\{U_{\rho_k} h\}$ is a square-summable sequence of $L_a^2(\Delta)$ for any $h \in L_a^2(\mathbb{D})$. As a consequence, for any $\{c_i\} \in l^2$,

$$\sum_{i=0}^{\infty} c_i U_{\rho_i} h(z), z \in \mathbb{D}$$

converges uniformly on each compact set in $\mathbb{D}$, and hence is holomorphic in $\mathbb{D}$.

To see this, for each $z_0 \in \mathbb{D}$, there exists a connected neighborhood U_0 of $\phi(z_0)$ such that each connected component $V_i (i \geq 0)$ of $\phi^{-1}(U_0)$ maps onto U_0 by a proper map, namely $\phi|_{V_i}$. Without loss of generality, assume $z_0 \in V_0$. The following statement needs to be highlighted:

for each V_j, there are m distinct $\rho_k \in G(\phi)$ satisfying $\rho_k(V_0) = V_j$; moreover, this m does not depend on j, and m is finite.

The reasoning is as follows. Assume that for each V_j, there are exactly m_j distinct members ρ_k of $G(\phi)$ satisfying $\rho_k(V_0) = V_j$. Note that $m_0 \geq 1$ because the identity is in $G(\phi)$. Since $\phi|_{V_0} : V_0 \to U_0$ is a proper map, for each point $w \in U_0 - \mathcal{E}_\phi$, $(\phi|_{V_0})^{-1}(w)$ is a compact set, and hence is finite. Observe that for a point $z \in (\phi|_{V_0})^{-1}(w)$,

$$(\phi|_{V_0})^{-1}(w) \supseteq \{\rho_k(z); \, \rho_k \in G(\phi) \text{ and } \rho_k(V_0) = V_0\},$$

which implies that

$$m_0 \leq \sharp(\phi|_{V_0})^{-1}(w) < \infty.$$

Thus, m_0 is finite. It remains to show that $m_j = m_0$ for each j. To see this, note that $\phi|_{V_j} : V_j \to U_0$ is surjective, and hence there exists at least one point z_j in V_j satisfying $\phi(z_j) = \phi(z_0) \in V_0$. Since ϕ is regular, there is a member ρ in $G(\phi)$ such that

$$\rho(z_j) = z_0.$$

Also noting that all V_i are pairwisely disjoint, $\rho(V_j) = V_0$. Therefore, for each $\rho_k \in G(\phi)$ satisfying $\rho_k(V_0) = V_j$, we have

$$\rho \circ \rho_k(V_0) = V_0.$$

This leads to $m_j \leq m_0$, and similarly $m_0 \leq m_j$, forcing $m_j = m_0$. Therefore, m_j does not depend on j.

For m_0 distinct members ρ_k satisfying $\rho_k(V_0) = V_j$, we have

$$\int_{V_0} |U_{\rho_k}h(z)|^2 dA(z) = \int_{V_0} |h \circ \rho_k(z) \, \rho_k'(z)|^2 dA(z) = \int_{V_j} |h(z)|^2 dA(z),$$

which implies that

$$\sum_{k=0}^{\infty} \int_{V_0} |U_{\rho_k}h(z)|^2 dA(z) = m_0 \int_{\bigsqcup_{j\geq 0} V_j} |h(z)|^2 dA(z) \leq m_0 \int_{\mathbb{D}} |h(z)|^2 dA(z) < \infty.$$

That is, $U_{\rho_k}h$ is a square-summable sequence of $L_a^2(V_0)$. Now take an open disk $\Delta \subseteq V_0$ containing z_0, and clearly, $U_{\rho_k}h$ is a square-summable sequence of $L_a^2(\Delta)$.

Note that for each $\{c_i\} \in l^2$ and $h \in L_a^2(\mathbb{D})$, $\sum_{i=0}^{\infty} c_i U_{\rho_i}h$ converges to some function in the norm of $L_a^2(\Delta)$, and hence it converges uniformly locally in Δ. By the arbitrariness of z_0, it follows that for each $\{c_i\} \in l^2$,

$$\sum_{i=0}^{\infty} c_i U_{\rho_i}h(z)$$

converges uniformly locally in $\mathbb{D}$; equivalently, it converges uniformly on each compact subset of $\mathbb{D}$. Thus it is holomorphic in $\mathbb{D}$.

Step 2 Next we will show that for each $z_0 \in \mathbb{D} - \phi^{-1}(\mathcal{E}_\phi)$, there is a small disk containing z_0, on which (6.1) holds and W is unique.

By Proposition 2.1.16 $\phi^{-1}(\mathcal{E}_\phi)$ is discrete. For each $z_0 \in \mathbb{D} - \phi^{-1}(\mathcal{E}_\phi)$, there is a disk Δ containing z_0 such that $\Delta \cap \phi^{-1}(\mathcal{E}_\phi)$ is empty. By regularity of ϕ, we have

$$\phi^{-1}(\phi(\Delta)) = \bigcup_k \rho_k(\Delta).$$

Furthermore, we can require that for any $j \neq k$,

$$\rho_j(\Delta) \cap \rho_k(\Delta) = \emptyset. \tag{6.3}$$

Then one can write

$$\phi^{-1}(\phi(\Delta)) = \bigsqcup_k \rho_k(\Delta).$$

Now it will be demonstrated why (6.3) can be satisfied. In fact, (6.3) is equivalent to the following condition:

$$\Delta \cap \rho_k(\Delta) = \emptyset, \ k \geq 1.$$

Assume conversely that there is some $k_0 \geq 1$ satisfying $\Delta \cap \rho_{k_0}(\Delta) \neq \emptyset$. Let U denote the component of $\phi^{-1}(\phi(\Delta))$ containing Δ and $\rho_{k_0}(\Delta)$, and then by definition of branched covering map, $\phi|_U : U \rightarrow \phi(\Delta)$ is a proper map. By Proposition 2.1.2, $\phi|_U$ is a finite-folds map, which implies that $\rho_k(\Delta) \subseteq U$ holds only for finitely many k. Then stretching Δ we have

$$\Delta \cap \rho_k(\Delta) = \emptyset, \ k \geq 1.$$

Therefore, (6.3) holds.

Write $H = L_a^2(\Delta)$ and $\Lambda = L_a^2(\mathbb{D})$. For each $h \in \Lambda$, set

$$e_h^k(z) = U_{\rho_k} h(z), \quad f_h^k(z) = U_{\rho_k}(Sh)(z), z \in \Delta, \ k = 0, 1, \cdots.$$

Clearly, e_h^k and f_h^k are in H. By Step 1, for any $h \in L_a^2(\mathbb{D})$, $\{e_h^k\}$ is a square-summable sequence of H. Following the proof of (5.5) we shall get a unitary operator $W : l^2 \rightarrow l^2$ such that

$$W\{e_g^k(w)\} = \{f_g^k(w)\}, \ g \in \Lambda, w \in \Delta.$$

This is exactly (6.1), $W\pi_w(h) = \pi_w(Sh)$, as desired. In addition, the condition (6.1) uniquely determines the operator W. The reasoning is the same as the end of the proof of (5.5).

Step 3 Now we are ready to show (6.1) and (6.2).

To prove (6.1) is equivalent to show that for any $h \in L_a^2(\mathbb{D})$,

$$\langle W\pi_w h, e_i \rangle = \langle \pi_w(Sh), e_i \rangle, \ w \in \mathbb{D} \ \text{and} \ i = 1, 2, \cdots$$

where $\{e_i; i \in \mathbb{Z}_+\}$ denotes the usual orthonormal basis of l^2. By Step 2, the above identity holds on an open subset Δ of $\mathbb{D}$, and hence on $\mathbb{D}$, because both sides of the above are holomorphic functions in $\mathbb{D}$. Therefore (6.1) holds.

Notice that $\rho_0(z) = z$, and $U_{\rho_0} = I$. Denote the first row of the matrix W by $\{c_k\}$, and then expanding (6.1) yields

$$Sh(z) = \sum_{k=0}^{\infty} c_k U_{\rho_k} h(z), \ h \in L_a^2(\mathbb{D}), \ z \in \mathbb{D}.$$

For the uniqueness of the coefficients c_k, just note that for any fixed $w_0 \in \mathbb{D} - \phi^{-1}(\mathcal{E}_\phi)$, $\{\rho_k(w_0)\}$ is an interpolating sequence. Then following the last paragraph of the proof of (5.5), c_k are unique. The proof is complete.   □

By Proposition 2.5.1 any von Neumann algebra is the finite linear span of its unitary operators, and hence each operator in $V^*(\phi)$ has the representation (6.2). Then it is immediate to obtain the following, due to Guo and Huang [GH2].

Corollary 6.2.3 *Suppose $\phi : \mathbb{D} \to \Omega$ is a holomorphic regular branched covering map, and $S \in V^*(\phi)$. Then there is a unique square-summable sequence $\{c_k\}$ satisfying*

$$Sh(z) = \sum_k c_k U_{\rho_k} h(z), \ h \in L_a^2(\mathbb{D}), \ z \in \mathbb{D},$$

where ρ_k run over $G(\phi)$.

The next lemma, due to Guo and Huang [GH2], gives elementary analysis for operators in $V^*(\phi)$.

Lemma 6.2.4 *Suppose $\phi : \mathbb{D} \to \Omega$ is a holomorphic regular branched covering map, and $G(\phi)$ is infinite. For two operators S, T in $V^*(\phi)$ we write*

$$Sh(z) = \sum_{k=0}^{\infty} c_k U_{\rho_k} h(z) \quad \text{and} \quad Th(z) = \sum_{k=0}^{\infty} c'_k U_{\rho_k} h(z), \ h \in L_a^2(\mathbb{D}), \ z \in \mathbb{D}.$$

where $\{c_k\}$ and $\{c'_k\}$ belong to l^2. Then we have

(1)

$$S^* h(z) = \sum_{k=0}^{\infty} \overline{c_k} U_{\rho_k}^* h(z), \ h \in L_a^2(\mathbb{D}), \ z \in \mathbb{D}; \tag{6.4}$$

Note that $U_{\rho_k}^* = U_{\rho_k^{-1}}$.

(2) *Define $S_N = \sum_{k=0}^{N} c_k U_{\rho_k}$, then we have*

$$\lim_{N \to \infty} \|S_N^* K_z - S^* K_z\| = \lim_{N \to \infty} \|S_N K_z - S K_z\| = 0, \ a.e. \ z \in \mathbb{D};$$

(3)

$$STh(z) = \sum_{i=0}^{\infty} d_i U_{\rho_i} h(z), h \in L_a^2(\mathbb{D}), \ z \in \mathbb{D}.$$

where

$$d_i = \sum_{\rho_j \circ \rho_k = \rho_i} c_k c_j'.$$

When $G(\phi)$ is finite, similar versions of (1), (2) and (3) are trivial.

Proof The proof comes from [GH2].

(1) Before proving (6.4), let us make an observation. For an operator $V \in \mathcal{V}^*(\phi)$, Corollary 6.2.3 shows that there is a unique vector $\{d_k\} \in l^2$ such that

$$Vh(z) = \sum_{k=0}^{\infty} d_k U_{\rho_k}^* h(z), \ h \in L_a^2(\mathbb{D}), \ z \in \mathbb{D}.$$

For each $\lambda \in \mathbb{D} - \phi^{-1}(\mathcal{E}_\phi)$, we have $VK_\lambda(w) = \sum_{k=0}^{\infty} d_k U_{\rho_k}^* K_\lambda(w)$ and

$$VK_\lambda(w) = \sum_{k=0}^{\infty} d_k \overline{\rho_k'(\lambda)} K_{\rho_k(\lambda)}(w), w \in \mathbb{D}.$$

Below one will see that

$$VK_\lambda = \sum_{k=0}^{\infty} d_k \overline{\rho_k'(\lambda)} K_{\rho_k(\lambda)}, \tag{6.5}$$

where the right hand side converges in $L_a^2(\mathbb{D})$-norm and the coefficients of $K_{\rho_k(\lambda)}$ are unique.

To see this, for each $\lambda \in \mathbb{D} - \phi^{-1}(\mathcal{E}_\phi)$, let B denote the Blaschke product for $\{\rho_k(\lambda)\}$. By Lemma 6.2.1, $\{\rho_k(\lambda)\}$ is an interpolating sequence, that is, a uniformly separated sequence. Then Proposition 2.4.5 shows that

$$h \mapsto \{(1 - |\rho_k(\lambda)|^2) h(\rho_k(\lambda))\}$$

induces a bounded invertible linear map A from $L_a^2(\mathbb{D}) \ominus BL_a^2(\mathbb{D})$ onto l^2. Denote by $\{e_k\}_{k=0}^{\infty}$ the standard orthogonal basis of l^2, and

$$A^* e_k = (1 - |\rho_k(\lambda)|^2) K_{\rho_k(\lambda)}, \ k = 0, 1, \cdots.$$

Then it is clear that for any $\{d_k'\} \in l^2$, the series

$$\sum_{k=0}^{\infty} d_k' (1 - |\rho_k(\lambda)|^2) K_{\rho_k(\lambda)}$$

converges in $L_a^2(\mathbb{D})$-norm. This, along with a computational result

$$|\rho_k'(\lambda)| = \frac{1 - |\rho_k(\lambda)|^2}{1 - |\lambda|^2},$$

shows that the right hand side of (6.5) converges in $L_a^2(\mathbb{D})$-norm. The uniqueness of
the coefficients follows from Lemma 6.2.1.

With the above discussion, we deduce that: *for a fixed $\lambda \in \mathbb{D} - \phi^{-1}(\mathcal{E}_\phi)$, the
representation (6.5) of VK_λ determines the coefficients d_k in the representation of V.*

Now put $V = S^*$ in (6.5), and write

$$S^*h(z) = \sum_{k=0}^{\infty} d_k U_{\rho_k}^* h(z), \; h \in L_a^2(\mathbb{D}), \; z \in \mathbb{D},$$

where $\{d_k\} \in l^2$. To prove (6.4), it suffices to show that $d_i = \overline{c_i}$ for all i. Rewrite
(6.5) as

$$S^* K_\lambda = \sum_{k=0}^{\infty} d_k \overline{\rho_k'(\lambda)} K_{\rho_k(\lambda)}. \tag{6.6}$$

Since by Lemma 6.2.1 $\{\rho_k(\lambda)\}$ is an interpolating sequence for H^∞, for each i there
is a bounded holomorphic function h such that

$$h(\rho_k(\lambda)) = \delta_i^k.$$

Noting that

$$\langle S^* K_\lambda, h \rangle = \langle K_\lambda, Sh \rangle,$$

by (6.6) and the representation of Sh we conclude that $d_i = \overline{c_i}$. Thus S^* has the
representation (6.4), as desired.

(2) Set $S_N = \sum_{k=0}^{N} c_k U_{\rho_k}$. In fact, the discussion in (1) has shown that for each
$\lambda \in \mathbb{D} - \phi^{-1}(\mathcal{E}_\phi)$,

$$\lim_{N \to \infty} \|S_N^* K_\lambda - S^* K_\lambda\| = 0. \tag{6.7}$$

Furthermore, by (6.4) we can also define $(S^*)_N \overset{\triangle}{=} \sum_{k=0}^{N} \overline{c_k} U_{\rho_k}^*$. Note that
$(S^*)_N = S_N^*$ and replace S^* with S in (6.7). Then we have

$$\lim_{N \to \infty} \|S_N K_\lambda - S K_\lambda\| = 0, \lambda \in \mathbb{D} - \phi^{-1}(\mathcal{E}_\phi).$$

The conclusion thus follows from the discreteness of $\phi^{-1}(\mathcal{E}_\phi)$ (it is worthy to point out that if ϕ is a holomorphic covering map, then $\phi^{-1}(\mathcal{E}_\phi)$ is empty).

(3) Noting that ST is in $\mathcal{V}^*(\phi)$, we can assume

$$STh(z) = \sum_{i=0}^{\infty} d_i' U_{\rho_i} h(z), \ h \in L_a^2(\mathbb{D}), \ z \in \mathbb{D}.$$

It remains to determine the coefficients d_i'.

Put $T_N = \sum_{k=0}^{N} c_k' U_{\rho_k}$. It is easy to check that

$$ST_N h(z) = \sum_{i=0}^{\infty} d_i^{(N)} U_{\rho_i} h(z), \ h \in L_a^2(\mathbb{D}), \ z \in \mathbb{D},$$

where

$$d_i^{(N)} = \sum_{j \leq N, \rho_j \circ \rho_k = \rho_i} c_k c_j'.$$

Note that the right hand side is a finite sum.

Fix a point $\lambda \in \mathbb{D} - \phi^{-1}(\mathcal{E}_\phi)$. By Lemma 6.2.1, $\{\rho_k^{-1}(\lambda)\}$ is an interpolating sequence, and hence for each i there is a bounded holomorphic function h satisfying

$$h(\rho_k^{-1}(\lambda)) = \delta_k^i. \tag{6.8}$$

By the discussion in (2),

$$\lim_{N \to \infty} \|T_N K_\lambda - T K_\lambda\| = 0.$$

Thus $\lim_{N \to \infty} \|ST_N K_\lambda - STK_\lambda\| = 0$, forcing

$$\lim_{N \to \infty} \langle ST_N K_\lambda, h \rangle = \langle STK_\lambda, h \rangle. \tag{6.9}$$

In the proof of (1), put $V = ST_N$ and by (6.5) we have

$$ST_N K_\lambda = \sum_k d_k^{(N)} \overline{(\rho_k^{-1})'(\lambda)} K_{\rho_k^{-1}(\lambda)},$$

and similarly

$$STK_\lambda = \sum_k d_k' \overline{(\rho_k^{-1})'(\lambda)} K_{\rho_k^{-1}(\lambda)}.$$

Combining (6.8) and (6.9) with the above two identities, we have

$$\lim_{N\to\infty} d_i^{(N)}\overline{(\rho_i^{-1})'(\lambda)} = d_i'\overline{(\rho_i^{-1})'(\lambda)}.$$

Since $\rho_i \in \mathrm{Aut}(\mathbb{D})$, $(\rho_i^{-1})'(\lambda) \neq 0$, forcing

$$d_i' = \lim_{N\to\infty} d_i^{(N)} = \sum_{\rho_j\circ\rho_k=\rho_i} c_k c_j'.$$

The proof of the lemma is complete. □

Observe that for each $\rho \in G(\phi)$, U_ρ belongs to $\mathcal{V}^*(\phi)$. Guo and Huang attained the following [GH2].

Corollary 6.2.5 *The von Neumann algebra $\mathcal{V}^*(\phi)$ equals the SOT-closure (and WOT-closure) of span $\{U_\rho : \rho \in G(\phi)\}$.*

Proof The proof is from [GH2].

It suffices to prove Corollary 6.2.5 in the case of $G(\phi)$ being infinite. To see this, write $G(\phi) = \{\rho_k\}$. By von Neumann bicommutant theorem, it is enough to show that the commutant of *span* $\{U_{\rho_k}\}$ equals $\mathcal{W}^*(\phi)$, the commutant of $\mathcal{V}^*(\phi)$. For this, for a given operator A commuting with all U_{ρ_k}, we must show that A commutes with each operator in $\mathcal{V}^*(\phi)$.

For any $S \in \mathcal{V}^*(\phi)$, assume S has the form in Lemma 6.2.4:

$$Sh(z) = \sum_{k=0}^{\infty} c_k U_{\rho_k} h(z), \quad h \in L_a^2(\mathbb{D}), z \in \mathbb{D},$$

and define

$$S_N = \sum_{k=0}^{N} c_k U_{\rho_k}.$$

Since A commutes with all U_{ρ_k}, A commutes with S_N, and hence

$$\langle AS_N K_z, K_z \rangle = \langle AK_z, S_N^* K_z \rangle, \ z \in \mathbb{D}. \tag{6.10}$$

By Lemma 6.2.4, we have

$$\lim_{N\to\infty} \|S_N^* K_z - S^* K_z\| = \lim_{N\to\infty} \|S_N K_z - SK_z\| = 0, a.e.\, z \in \mathbb{D}.$$

Taking limits in (6.10) gives

$$\langle ASK_z, K_z \rangle = \langle AK_z, S^* K_z \rangle = \langle SAK_z, K_z \rangle, z \in \mathbb{D}.$$

By the continuity of both sides in variable z, the above holds for every $z \in \mathbb{D}$. By Theorem A.1 (see Appendix A), the property of Berezin transformation yields that $AS = SA$. The proof is complete.   □

Some words are in order. In Theorem 5.7.2 we get a similar result which states that for any thin Blaschke product B, the von Neumann algebra $V^*(B)$ is generated by $\mathcal{E}_{[\rho]}$, where ρ run over all local inverses of B for which $G[\rho]$ have finite multiplicity. Those operators $\mathcal{E}_{[\rho]}$ are not necessarily unitary, and here they are substituted with unitary operators U_ρ with each ρ belonging to Aut($\mathbb{D}$). Later, we will see that $V^*(\phi)$ possesses a far richer structure than $V^*(B)$.

The rest of this section elaborates on the existence of a ultraweakly continuous faithful trace on $V^*(\phi)$; that is, there exists a ultraweakly continuous linear map $Tr : V^*(\phi) \to \mathbb{C}$ satisfying the following:

(1) $Tr(I) = 1$;
(2) $Tr(S^*S) \geq 0$, and $Tr(S^*S) = 0$ if and only if $S = 0$;
(3) $Tr(ST) = Tr(TS)$, $S, T \in V^*(\phi)$.

The construction of this map Tr comes from [GH2]. Recall that ρ_0 always denotes the identity function on $\mathbb{D}$, i.e. $\rho_0(z) = z, z \in \mathbb{D}$. By Lemma 6.2.1 $\{\rho_i(0)\}$ is an interpolating sequence, and hence there is a bounded holomorphic function h_0 satisfying

$$h_0(\rho_i(0)) = \delta_0^i.$$

Now put

$$Tr(S) = \langle Sh_0, 1 \rangle, \ S \in V^*(\phi),$$

and it is easy to see that the trace Tr is a ultraweakly continuous linear functional on $V^*(\phi)$.

To verify the above properties (1)–(3), let us make an observation. To begin with, for each operator $S \in V^*(\phi)$, there is a unique square-summable sequence $\{c_k\}$ such that

$$Sh(z) = \sum_k c_k U_{\rho_k} h(z), h \in L_a^2(\mathbb{D}), \ z \in \mathbb{D},$$

where ρ_k run over $G(\phi)$. It is easy to check that $Tr(S) = c_0$, and hence $Tr(I) = 1$. Besides, for any two operators $S, T \in V^*(\phi)$, by Lemma 6.2.4(3) it is not difficult to verify that

$$Tr(ST) = Tr(TS).$$

Moreover, by Lemma 6.2.4 we have

$$S^*Sh(z) = \sum_i d_i U_{\rho_i} h(z), h \in L_a^2(\mathbb{D}), \ z \in \mathbb{D},$$

where

$$d_i = \sum_{\rho_j \circ \rho_k^{-1} = \rho_i} \overline{c_k} c_j.$$

In particular,

$$Tr(S^*S) = d_0 = \sum_k |c_k|^2 \geq 0. \tag{6.11}$$

This shows that Tr is a trace on $\mathcal{V}^*(\phi)$. Furthermore, (6.11) immediately gives that if $Tr(S^*S) = 0$, then $S^*S = 0$, forcing $S = 0$. Therefore, Tr is a faithful trace. Then applying Theorem 2.5.3 gives the finiteness of $\mathcal{V}^*(\phi)$. In conclusion, we have the following.

Proposition 6.2.6 *The von Neumann algebra $\mathcal{V}^*(\phi)$ is finite.*

6.3 Abelian $\mathcal{V}^*(\phi)$

Based on the formulation of those operators in $\mathcal{V}^*(\phi)$, this section will present a complete characterization of when the von Neumann algebra $\mathcal{V}^*(\phi)$ is abelian.

To begin with, let us recall a fact about holomorphic covering maps. Throughout this section, Ω always denotes a bounded planar domain.

Theorem 6.3.1 (The Koebe Uniformization Theorem) *Given $z_0 \in \mathbb{D}$ and $w_0 \in \Omega$, there always exists a unique holomorphic covering map ϕ of $\mathbb{D}$ onto Ω with $\phi(z_0) = w_0$ and $\phi'(z_0) > 0$, see [Gol, V].*

The above theorem shows that if ψ is another holomorphic covering map from $\mathbb{D}$ onto Ω, then there is a $\varphi \in \text{Aut}(\mathbb{D})$ such that $\psi = \phi \circ \varphi$. Therefore, it is easy to construct a natural spatial isomorphism (and thus a C^* isomorphism) between $\mathcal{V}^*(\phi)$ and $\mathcal{V}^*(\psi)$.

The following theorem, due to Guo and Huang, shows that the commutativity of $\mathcal{V}^*(\phi)$ only depends on the geometric property of Ω [GH2].

Theorem 6.3.2 *If $\phi : \mathbb{D} \to \Omega$ is a holomorphic covering map, then the following are equivalent:*

(1) the von Neumann algebra $\mathcal{V}^*(\phi)$ is abelian;
(2) the fundamental group $\pi_1(\Omega)$ of Ω is abelian;
(3) Ω is conformally isomorphic to one of the disk, annuli or the punctured disk.

Proof By a hyperbolic surface, we mean a connected Riemann surface S whose universal covering is conformally isomorphic to the unit disk; that is, there is a holomorphic covering map from $\mathbb{D}$ onto S [Mi, p. 15]. In particular, all bounded planar domains are hyperbolic surfaces. Up to a conformal isomorphism, disk, annuli and the punctured disk are the only hyperbolic surfaces with abelian fundamental group [Mi]. So (2) ⇔ (3) is immediate.

Now it remains to show (1) ⇔ (2). Recall that $G(\phi)$ is the group consisting of those automorphisms ρ of $\mathbb{D}$ which satisfy $\phi \circ \rho = \phi$. For any ρ and γ in $G(\phi)$, we have

$$U_\rho U_\gamma = U_{\gamma \circ \rho}.$$

Thus all U_ρ compose a group, denoted by $\mathcal{U}_{G(\phi)}$; and the above identity shows that $\mathcal{U}_{G(\phi)}$ is abelian if and only if $G(\phi)$ is abelian; if and only if $\pi_1(\Omega)$ is abelian since $\pi_1(\Omega)$ is isomorphic to $G(\phi)$ as groups [GHa, Theorem 5.8].

By Corollary 6.2.5, the von Neumann algebra $\mathcal{V}^*(\phi)$ equals the SOT-closure of span $\{U_\rho : \rho \in G(\phi)\}$. Thus, $\mathcal{V}^*(\phi)$ is abelian if and only if $\mathcal{U}_{G(\phi)}$ is abelian, if and only if $\pi_1(\Omega)$ is abelian. The proof is complete.   □

The following theorem tells us when $\mathcal{V}^*(\phi)$ is nontrivial, or when M_ϕ has a nontrivial reducing subspace, see [GH2].

Theorem 6.3.3 Let $\phi : \mathbb{D} \to \Omega$ be a holomorphic covering map. Then the following are equivalent:

(1) $\mathcal{V}^*(\phi)$ is nontrivial;
(2) M_ϕ has a nontrivial reducing subspace;
(3) ϕ is not univalent;
(4) Ω is not simply connected.

Proof Clearly, (1)⇔ (2). We will show (3) ⇒ (4) ⇒ (1) ⇒ (3) to complete the proof.

(3) ⇒ (4). It is well-known that a domain is simply connected if and only if its covering map is a conformal isomorphism. Therefore, if $\phi : \mathbb{D} \to \Omega$ is not univalent, then ϕ is not a conformal isomorphism, and hence Ω is not simply connected.

(4) ⇒ (1). Now assume that Ω is not simply connected. Equivalently, the fundamental group $\pi_1(\Omega)$ is nontrivial. Since $G(\phi)$ is isomorphic to $\pi_1(\Omega)$ as groups, $G(\phi)$ contains a nontrivial element, say ρ, and hence U_ρ is a nontrivial unitary operator in $\mathcal{V}^*(\phi)$.

(1) ⇒ (3). Assume conversely that ϕ is univalent. Since ϕ is also a covering map, this implies that ϕ is a conformal isomorphism. Thereby, both $\pi_1(\Omega)$ and $G(\phi)$ are trivial. By Corollary 6.2.5, $\mathcal{V}^*(\phi)$ is trivial. This is a contradiction to (1), forcing (1) ⇒ (3) as desired.   □

Theorem 6.3.3 also holds in the Hardy space $H^2(\mathbb{D})$, which was first noted by Abrahamse [A1], Abrahamse and Ball [AB].

Remark 6.3.4 Under a mild assumption, Theorem 6.3.3 can be generalized to multi-variable case. Precisely, let Ω_0 and Ω be two domains in $\mathbb{C}^d$ and Ω_0 is simply connected. If $\phi : \Omega_0 \to \Omega$ is a holomorphic covering map, then the conclusions in Theorem 6.3.3 hold.

However, in multi-variable case things become more complicated, since there is no similar result as the Koebe uniformization theorem [Mi, MT] for Riemann surfaces. As mentioned in [Kran, p. 10], in the sense of category most domains in $\mathbb{C}^d$ that are close to the ball (say, in the C^∞-topology) are not biholomorphic to the ball. Furthermore, uncountable is the set of biholomorphic equivalent classes arising from domains close to the ball in any reasonable sense; for this, one can refer to [BSW] and [GK1, GK2]. As for the universal covering map, there is a conjecture which says the universal covering of some special manifold should be the ball [MS]. Here, we have no intention to give a complete list of literature on this line. Instead, we would like to mention two related results for instance. One result proved by Wong [Wo] is that a bounded domain in $\mathbb{C}^d$ with smooth strongly pseudoconvex boundary must be the ball provided that its automorphism group is noncompact. Besides, as pointed in [MS], Wong's method also shows that a bounded domain in $\mathbb{C}^d$ with smooth boundary which covers a compact manifold must be the ball. In multi-variable case, see [Guo5] and Chap. 7 for some consideration on the commutant algebra $\mathcal{V}^*(\Phi)$ and $\mathcal{W}^*(\Phi)$ in a more general setting. Except for Sect. 6.7, this chapter will not go further in multi-variable case, and below we turn back to the complex plane. Some examples will be given then.

In the following two examples, $\mathcal{V}^*(\phi)$ defined on $L_a^2(\mathbb{D})$ are nontrivial and abelian.

Example 6.3.5 Let Ω be the punctured disk $\mathbb{D} - \{0\}$, and write

$$\phi(z) = e^{-\frac{1+z}{1-z}}.$$

Observe that ϕ is the composition of exponent map and the linear fraction map $\kappa(z) = -\frac{1+z}{1-z}$ from the unit disk onto the left half plane, and thus ϕ is a covering map. Applying Theorems 6.3.2 and 6.3.3 shows that $\mathcal{V}^*(\phi)$ is nontrivial and abelian. Furthermore, by a simple computation, the deck transformation group of ϕ is

$$\{\kappa^{-1} \circ h_n \circ \kappa : n \in \mathbb{Z}\} = \{\frac{n\pi i - (n\pi i + 1)z}{n\pi i - 1 - n\pi i z} : n \in \mathbb{Z}\},$$

where $h_n(z) = z + 2n\pi i$ are automorphisms of the left half plane.

For $\lambda \in \mathbb{D}$, write $g_\lambda(z) = \frac{\lambda - \phi(z)}{1 - \bar{\lambda}\phi(z)}$. By Frostman's theorem (Theorem 2.1.4), except for those λ in a subset of capacity zero of $\mathbb{D}$, $g_\lambda(z)$ are infinite Blaschke products. For such Blaschke products, $\mathcal{V}^*(g_\lambda) = \mathcal{V}^*(\phi)$, both being abelian.

Note that ϕ is an inner function. Therefore, on the Hardy space of the disk $\mathcal{V}^*(\phi)$ is not abelian. Consequently, Theorem 6.3.2 fails in the case of $H^2(\mathbb{D})$.

Example 6.3.6 For $0 < r < 1$, write $\Omega_r = \{z \in \mathbb{C} : r < |z| < 1\}$, and let ϕ_r denote the covering map from $\mathbb{D}$ onto Ω_r. As done in [A1] and [Sa],

$$\phi_r(z) = \exp\left(\ln r \,(\frac{i}{\pi} \ln \frac{1+z}{1-z} + \frac{1}{2})\right).$$

By Theorems 6.3.2 and 6.3.3, $\mathcal{V}^*(\phi_r)$ is nontrivial and abelian.

On the Hardy space Cowen considered the von Neumann algebra $\mathcal{V}^*(\phi_r)$, where ϕ_r is defined in Example 6.3.6. He gave the following, see [Cow2, Theorem 4] and its corollary.

Theorem 6.3.7 (Cowen) *Let ϕ be a covering map from $\mathbb{D}$ onto*

$$\Omega_r \triangleq \{z \in \mathbb{C} : r < |z| < 1\},$$

and ρ denotes the generator of $G(\phi)$. Set

$$U_\rho f(z) = f \circ \rho(z) \sqrt{\rho'(z)}, f \in H^2(\mathbb{D}), z \in \mathbb{D}.$$

Then on $H^2(\mathbb{D})$ the von Neumann algebra $\mathcal{V}^*(\phi)$ is generated by U_ρ, and hence $\mathcal{V}^*(\phi)$ is abelian. Precisely, each member in $\mathcal{V}^*(\phi)$ exactly has the following form

$$\sum_{n=-\infty}^{\infty} a_n U_\rho^n \ (in \ the \ sence \ of \ functional \ calculus),$$

where $\sum_{n=-\infty}^{\infty} a_n e^{i\theta}$ is the Fourier series of some L^∞-function.

Note that by a simple calculation $\rho'(z)$ is the square of a holomorphic function, by which $\sqrt{\rho'(z)}$ denotes.

The following is an immediate consequence of Theorem 6.3.7, which is of interest by contrast with Theorem 6.3.2.

Corollary 6.3.8 *If $\phi : \mathbb{D} \to \Omega$ is a holomorphic covering map, then the following are equivalent:*

(1) the von Neumann algebra $\mathcal{V}^*(\phi)$ defined on $H^2(\mathbb{D})$ is abelian;
(2) Ω is conformally isomorphic to the disk or annuli.

The next example is given by Guo and Huang [GH2]. It shows that there do exist interpolating Blaschke products B such that the von Neumann algebras $\mathcal{V}^*(B)$ are not abelian.

Example 6.3.9 Let E be a relatively closed subset of $\mathbb{D} - \{0\}$ with capacity zero. A discrete subset of $\mathbb{D} - \{0\}$ is a case in point. If E contains at least two points, then the fundamental group of $\mathbb{D} - E$ is not abelian. In this case, let ϕ be a holomorphic covering map from $\mathbb{D}$ onto $\mathbb{D} - E$. By Theorem 6.3.2, $\mathcal{V}^*(\phi)$ is not abelian.

Below it will be illustrated that ϕ is an interpolating Blaschke product. In fact, by Proposition 2.1.17 and Lemma 6.2.1, if ψ is a holomorphic covering map from $\mathbb{D}$ onto a bounded domain Ω, then for each $\lambda \in \mathbb{D}$, the inner part of $\psi - \psi(\lambda)$ is an interpolating Blaschke product, also see [Cow1, Theorem] and Stout [St]. Note that $0 \in \phi(\mathbb{D})$, and then the inner part of ϕ is an interpolating Blaschke product. Furthermore, by Proposition 2.2.4 or the proof of [GM2, Theorem 1.1], ϕ is itself an inner function, and hence an interpolating Blaschke product.

In addition, one has the following proposition. Its former part is deduced by Collingwood and Lohwater [CL], Garnett [Ga] and its latter part is a consequence of Theorem 6.3.2.

Proposition 6.3.10 *If $\phi : \mathbb{D} \to \Omega$ is a holomorphic covering map, then ϕ is an inner function if and only if $\Omega = \mathbb{D} - E$, where E is a relatively closed subset of $\mathbb{D}$ with capacity zero. In this case, $V^*(\phi)$ is abelian if and only if E contains no more than one point.*

Proof The former part is Proposition 2.2.4.

For the latter part, note that by Theorem 6.3.2, $V^*(\phi)$ is abelian if and only if the fundamental group $\pi_1(\Omega)$ is abelian. Therefore, it suffices to show that if E is a relatively closed subset of $\mathbb{D}$ with capacity zero, and E contains at least two points, then $\pi_1(\mathbb{D} - E)$ is not abelian.

Below, we present a fundamental proof, which comes from [GH2]. Note that a set of capacity zero is of linear measure zero, see Sect. 2.2 in Chap. 2. A subset F of $\mathbb{C}$ has liner measure zero if and only if for any $\varepsilon > 0$, there is a sequence of disks $O(z_k, r_k)$ which is a cover of F satisfying $\sum_k r_k < \varepsilon$. Since E has linear measure zero, $\{|z| : z \in E\}$ is a subset of $(-1, 1)$ with linear measure zero. Therefore, there is an $r_0(0 < r_0 < 1)$ such that $r_0 \mathbb{D} \cap E$ contains at least two points, and $r_0 \mathbb{T} \cap E$ is empty. Since $\{Re\, z : z \in E\}$ has linear measure zero, by omitting a rotation, we can assume that there is an $r_1 \in (-1, 1)$ and an $\varepsilon_0 > 0$ satisfying the following conditions:

(1) $[r_1 - \varepsilon_0, r_1 + \varepsilon_0] \subseteq (-1, 1)$;
(2) $E \cap \{z \in \mathbb{D} : Re\, z \in [r_1 - \varepsilon_0, r_1 + \varepsilon_0]\}$ is empty;
(3) Neither $\{z \in r_0 \mathbb{D} \cap E : Re\, z < r_1 - \varepsilon_0\}$ nor $\{z \in r_0 \mathbb{D} \cap E : Re\, z > r_1 + \varepsilon_0\}$ is empty.
(4) $r_0 \mathbb{T} \cap E = \emptyset$.

Now take

$$V_1 = \{z \in \mathbb{D} - E : Re\, z < r_1 + \varepsilon_0\}$$

and

$$V_2 = \{z \in \mathbb{D} - E : Re\, z > r_1 - \varepsilon_0\}.$$

See Fig. 6.1, which is provided by Dr. Xie.

As follows we will apply Seifert-Van Kampen's Theorem [GHa] to show that $\pi_1(\mathbb{D}-E)$ is not abelian. In fact, by (3) and (4) the complement of V_1 in the Riemann sphere $\hat{\mathbb{C}}$ has at least two components: one is the unbounded component, and another is contained in

$$\{z \in r_0 \mathbb{D} : Re\, z < r_1 - \varepsilon_0\}.$$

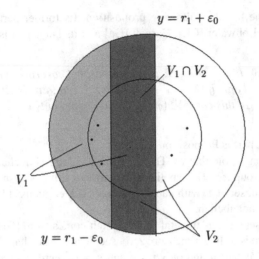

$$y = r_1 + \varepsilon_0$$

$V_1 \cap V_2$

V_1

$$y = r_1 - \varepsilon_0$$

V_2

Fig. 6.1 $V_1 \cap V_2$

As is well-known, an open subset X of $\mathbb{C}$ is simply connected if and only if both X and its complement in $\hat{\mathbb{C}}$ are connected. Therefore, V_1 is not simply connected; that is, $\pi_1(V_1)$ is nontrivial. Similarly, $\pi_1(V_2)$ is nontrivial.

Since $V_1 \cap V_2$ is simply connected and $\mathbb{D} - E = V_1 \cup V_2$, by Seifert-Van Kampen's Theorem [GHa], $\pi_1(\mathbb{D} - E)$ is isomorphic to the free product of $\pi_1(V_1)$ and $\pi_1(V_2)$. Therefore, $\pi_1(\mathbb{D} - E)$ is not abelian, completing the proof.   □

Theorems 6.3.2 and 6.3.3 can be naturally generalized to the case of holomorphic regular branched covering map [GH2].

Proposition 6.3.11 *Suppose $\phi : \mathbb{D} \to \Omega$ is a holomorphic regular branched covering map. Then M_ϕ has a nontrivial reducing subspace if and only if ϕ is not univalent, and $\mathcal{V}^*(\phi)$ is abelian if and only if $G(\phi)$ is abelian.*

In the case of $\mathcal{V}^*(\phi)$ being abelian, the structure of $\mathcal{V}^*(\phi)$ is characterized by the following proposition, due to Guo and Huang [GH2]. Later in Sect. 6.5, we will present another treatise for this.

Proposition 6.3.12 *Suppose $\phi : \mathbb{D} \to \Omega$ is a holomorphic covering map, and Ω is conformally isomorphic to one of the punctured disk or annuli. Then $\mathcal{V}^*(\phi)$ is $*$-isomorphic to $L^\infty[0, 1]$.*

Proof Suppose $\phi : \mathbb{D} \to \Omega$ is a holomorphic covering map, and Ω is conformally isomorphic to either the punctured disk or an annulus. Note that the deck transformation group $G(\phi)$ of ϕ is an infinite cyclic group, whose generator is denoted by ρ. By Corollary 6.2.5, $\mathcal{V}^*(\phi)$ is generated by the unitary operator U_ρ.

Before continuing, we mention a fact on multiplier. For a separable σ-finite measurable space (X, μ) and a function f in $L^\infty(X, \mu)$, M_f defines a multiplier on $L^2(X, \mu)$. It is known that the von Neumann algebra $\mathcal{W}^*(M_f)$ generated by M_f

has no minimal projection if and only if $\mathcal{W}^*(M_f)$ is $*$-isomorphic to $L^\infty[0,1]$, if and only if M_f has no nonzero eigenvector. Since a normal operator on a separable space is always unitarily equivalent to some M_f on a (necessarily separable) σ-finite measurable space (X, μ) [Ar1], the proof reduces to showing that U_ρ has no nonzero eigenvector.

To see this, suppose conversely that h is a nonzero eigenvector of U_ρ. Since U_ρ is a unitary operator, there is a unimodular constant ξ such that $U_\rho h = \xi h$. Write $G(\phi) = \{\rho_k\}$. Since $G(\phi)$ is cyclic, for each k there is a unimodular constant ξ_k such that

$$U_{\rho_k} h = \xi_k h. \tag{6.12}$$

As done in the proof of Theorem 6.2.2, we can pick a disk Δ such that all $\rho_k(\Delta)$ are pairwise disjoint and $\bigsqcup_k \rho_k(\Delta) \subseteq \mathbb{D}$. Then

$$\int_{\bigsqcup_k \rho_k(\Delta)} |h(z)|^2 dA(z) = \sum_{k=0}^{\infty} \int_{\rho_k(\Delta)} |h(w)|^2 dA(w)$$

$$= \sum_{k=0}^{\infty} \int_{\Delta} |h \circ \rho_k(z)|^2 |\rho_k'(z)|^2 dA(z)$$

$$= \sum_{k=0}^{\infty} \int_{\Delta} |h(z)|^2 dA(z).$$

The last identity follows from (6.12). Therefore h is constantly zero, which is a contradiction. Thus, $\mathcal{V}^*(\phi)$ is $*$-isomorphic to $L^\infty[0,1]$, completing the proof. $\square$

Note that Theorem 6.3.12 lives on $L_a^2(\mathbb{D})$. An analogue of Theorem 6.3.12 on $H^2(\mathbb{D})$ makes difference. In more detail, Theorem 6.3.7 indicates that if ϕ is a holomorphic covering map from $\mathbb{D}$ onto an annulus, then $\mathcal{V}^*(\phi)$ defined on $H^2(\mathbb{D})$ is $*$-isomorphic to $L^\infty[0,1]$. However, if ψ is a holomorphic covering map from $\mathbb{D}$ onto the punctured disk, then ψ is an inner function; in this case, $\mathcal{V}^*(\psi)$ defined on $H^2(\mathbb{D})$ is $*$-isomorphic to $B(l^2)$, see Example 2.6.6 for details.

6.4 Type II Factors Arising from Planar Domains

Last section shows that if $\phi : \mathbb{D} \to \Omega$ is a bounded holomorphic covering map, then the following are equivalent:

1. $\mathcal{V}^*(\phi)$ is abelian;
2. the fundamental group $\pi_1(\Omega)$ of Ω is abelian;
3. Ω is conformally isomorphic to one of the disk, annuli or the punctured disk.

This section mainly deals with the remaining case where $\pi_1(\Omega)$ is nonabelian. The following is the main result in this section, proved by Guo and Huang [GH2].

Theorem 6.4.1 *Suppose* $\phi : \mathbb{D} \to \Omega$ *is a holomorphic covering map and* $\pi_1(\Omega)$ *is not abelian, then* $\mathcal{V}^*(\phi)$ *is a type* II_1 *factor, and* $\mathcal{W}^*(\phi)$ *is a type* II_∞ *factor.*

To establish Theorem 6.4.1, we need a statement from algebraic topology, which is implied in [AS] (also see [Gr, Theorem 3.2]).

Lemma 6.4.2 *Suppose* Ω *is a bounded domain in* $\mathbb{C}$ *such that* $\pi_1(\Omega)$ *is nontrivial. Then* $\pi_1(\Omega)$ *is a free group on finite or countably many generators.*

The non-triviality of $\pi_1(\Omega)$ means exactly the non simply-connectedness of Ω. If $\mathbb{C} - \Omega$ has n bounded components, then $\pi_1(\Omega)$ is a free group on n generators. If $\mathbb{C} - \Omega$ has infinitely many (sometimes, uncountably many) bounded components, then $\pi_1(\Omega)$ is a free group on countably many generators.

By Theorem 6.3.2 and Lemma 6.4.2, we find the following are equivalent.

1. Ω is not conformally isomorphic to one of the disk, annuli or the punctured disk;
2. $\pi_1(\Omega)$ is not abelian;
3. $\pi_1(\Omega)$ is a free group on n generators ($n \geq 2$, allowed as ∞).

Before continuing, let us see an example.

Example 6.4.3 For $n \geq 2$, set $\Omega = \mathbb{D} - \{\frac{1}{k+1} : 1 \leq k \leq n\}$ and let $\phi : \mathbb{D} \to \Omega$ be a holomorphic covering map. As shown in Example 6.3.9, ϕ is an interpolating Blaschke product. Applying Theorem 6.4.1 shows that $\mathcal{V}^*(\phi)$ is a type II_1 factor, and $\mathcal{W}^*(\phi)$ is a type II_∞ factor.

Note that Example 6.4.3 is related to the Invariant Subspace Problem. As mentioned in the introduction, the Invariant Subspace Problem is equivalent to the problem whether there exists some single operator in the $\mathbb{A}_{\aleph_0}$-class that is saturated [BFP]. That is, for an $\mathbb{A}_{\aleph_0}$-operator T, and two invariant subspaces M and N of T satisfying $N \subseteq M$ and $\dim M/N = \infty$, is there another invariant subspace L satisfying $N \subsetneqq L \subsetneqq M$? It is well-known that the Bergman shift M_z is an $\mathbb{A}_{\aleph_0}$-operator. Also, for each ϕ in Example 6.4.3, M_ϕ is in $\mathbb{A}_{\aleph_0}$. The reasoning is as follows. A contraction T in a Hilbert space is called to belong to the class C_{00} if both $\{T^n\}$ and $\{T^{*n}\}$ converge to zero in the strong operator topology. It is easy to verify that $M_\phi \in C_{00}$ and that the spectrum of M_ϕ equals $\overline{\mathbb{D}}$. Recall that if $T \in C_{00}$ and $\sigma(T) = \overline{\mathbb{D}}$, then $T \in \mathbb{A}_{\aleph_0}$, see [BFP, Corollary 6.9]. Thus, $M_\phi \in \mathbb{A}_{\aleph_0}$.

The Bergman shift M_z has trivial reducing subspace lattice. However, for such an ϕ as in Example 6.4.3, $\mathcal{V}^*(\phi)$ is a type II_1 factor. By [Jon, Corollary 6.1.17], there is an order isomorphism between the interval $[0, 1]$ and the totally ordered set of equivalent classes of projections on the type II_1 factor $\mathcal{V}^*(\phi)$. Therefore, if M and N are two reducing subspaces of M_ϕ such that $N \subsetneqq M$, then there always exist uncountably many reducing subspaces L lying strictly between M and N. This would bring some helpful information in studying the lattice of invariant subspace for M_ϕ, a concrete $\mathbb{A}_{\aleph_0}$-operator.

Remark 6.4.4 Examples 6.3.5 and 6.4.3 illustrates that there exist an infinite Blaschke product B such that $V^*(B)$ is abelian or a type II_1 factor. This shows that for an infinite Blaschke product B, the structure of $V^*(B)$ can be considerably complicated. Also, one can refer to Chap. 5, where $V^*(B)$ are investigated via a rather geometric approach in the case of B being thin Blaschke products. In that case, $V^*(B)$ is usually trivial.

Now it remains to prove Theorem 6.4.1. Several preliminary results are in order.

Suppose G is a group. Two elements a and b of G are called conjugate if there exists an element g in G such that $gag^{-1} = b$. Clearly, conjugacy is an equivalence relation on G. The equivalence class containing the element a in G is $[a] = \{gag^{-1} : g \in G\}$, called *the conjugacy class* of a. If every nontrivial conjugacy class of the group G is infinite, then G is called an *i.c.c. group*. For example, let F_n denote the free group on n generators. Then all $F_n (n \geq 2)$ are i.c.c. groups.

The next result essentially comes from [GH2, Lemma 4.4].

Lemma 6.4.5 *Suppose $\phi : \mathbb{D} \to \Omega$ is a holomorphic covering map and $\pi_1(\Omega)$ is an i.c.c. group, then $V^*(\phi)$ is a factor.*

Proof The proof is essentially from [GH2].

Assume that $\phi : \mathbb{D} \to \Omega$ is a holomorphic covering map and $\pi_1(\Omega)$ is an i.c.c. group. To prove that $V^*(\phi)$ is a factor, it suffices to show that each member in the center $Z(V^*(\phi))$ of $V^*(\phi)$ must be a constant multiple of the identity I. For this, assume that T is a unitary operator in $Z(V^*(\phi))$. Applying Theorem 6.2.2 shows that there is a vector $\{c_i\} \in l^2$ such that

$$Th(z) = \sum_{i=0}^{\infty} c_i U_{\rho_i} h(z), h \in L_a^2(\mathbb{D}), z \in \mathbb{D},$$

where ρ_i run over $G(\phi)$. As done before, ρ_0 denotes the identity e. Since $U_{\rho_0} = I$, it suffices to show that $c_i = 0$ for $i \geq 1$.

For each $k \in \mathbb{Z}_+$, set

$$\Lambda_k' = \{i \in \mathbb{Z}_+ : c_i = c_k \text{ and } c_i \neq 0\}.$$

Since $\{c_i\} \in l^2$, each Λ_k' is either empty or finite. Furthermore, either $\Lambda_i' = \Lambda_j'$ or $\Lambda_i' \cap \Lambda_j' = \emptyset$. Therefore, one can find a subsequence $\{\Lambda_k\}$ of $\{\Lambda_k'\}$ such that each $\Lambda_k \neq \emptyset$, any two of Λ_k are disjoint, and the union of Λ_k equals

$$\{i \in \mathbb{Z}_+ : c_i \neq 0\}.$$

Since $T \in Z(V^*(\phi))$, T commutes with U_η for every $\eta \in G(\phi)$, and thus

$$\sum_{i=0}^{\infty} c_i U_{\rho_i} U_\eta h(z) = \sum_{i=0}^{\infty} c_i U_\eta U_{\rho_i} h(z), h \in L_a^2(\mathbb{D}), z \in \mathbb{D}.$$

That is,

$$\sum_{i=0}^{\infty} c_i U_{\eta \circ \rho_i} h(z) = \sum_{i=0}^{\infty} c_i U_{\rho_i \circ \eta} h(z), h \in L_a^2(\mathbb{D}), \ z \in \mathbb{D}.$$

By the uniqueness of the coefficients c_i, we get

$$\{\eta \circ \rho_i : i \in \Lambda_k\} = \{\rho_i \circ \eta : i \in \Lambda_k\},$$

and hence

$$\{\eta^{-1} \circ \rho_i \circ \eta : i \in \Lambda_k\} = \{\rho_i : i \in \Lambda_k\}.$$

This implies that the set $E_k \triangleq \{\rho_i : i \in \Lambda_k\}$ satisfies $\eta^{-1} E_k \eta = E_k$ for each $\eta \in G(\phi)$.

To finish the proof, it suffices to show that for each k, either $E_k = \emptyset$ or $E_k = \{e\}$. To see this, assume that $E_k \neq \emptyset$. Note that $\{c_i\} \in l^2$, which implies that each E_k is a finite set. Since $\eta^{-1} E_k \eta = E_k$ for each $\eta \in G(\phi)$, for each $\rho_j \in E_k$,

$$[\rho_j] \subseteq E_k.$$

Since $G(\phi) \cong \pi_1(\Omega)$ is an i.c.c. group, $[\rho_j] = \{e\}$, forcing $\rho_j = e$. Therefore, $E_k = \{e\}$. It then follows that T is a constant multiple of the identity I, and hence $V^*(\phi)$ is a factor. The proof is complete.   □

From the proof of Lemma 6.4.5, we immediately get a stronger result [GH2].

Corollary 6.4.6 *Suppose* $\phi : \mathbb{D} \to \Omega$ *is a regular branched covering map. Then the following are equivalent:*

(1) $G(\phi)$ *is an i.c.c. group;*
(2) $V^*(\phi)$ *is a factor.*

Remark 6.4.7 If $G(\phi)$ is not an i.c.c. group, then one can pick $\sigma \in G(\phi)$ such that $[\sigma]$ is a finite conjugacy class. In this case, $\sum_{\rho \in [\sigma]} U_\rho$ defines an operator in the center of $V^*(\phi)$.

Now we are ready to prove Theorem 6.4.1.

Proof of Theorem 6.4.1 The proof comes from [GH2].

Combing Lemma 6.4.2 with Lemma 6.4.5 shows that $V^*(\phi)$ is a factor. Proposition 6.2.6 states that $V^*(\phi)$ is finite. Then by Proposition 2.5.4, $V^*(\phi)$ is either type I_n or type II_1. If it were type I_n, it would be $*$-isomorphism to $M_n(\mathbb{C})$ for some n, which is a contradiction to the fact that $\dim V^*(\phi) = \infty$. Thus $V^*(\phi)$ is a type II_1 factor.

It remains to prove that $W^*(\phi)$ is a type II_∞ factor.

First we show that $W^*(\phi)$, as a von Neumann algebra, is infinite. To this aim, it suffices to construct a non-unitary isometric operator S in $W^*(\phi)$. We may assume that $\phi(0) = 0$ because otherwise one can replace ϕ with $\phi - \phi(0)$, leaving $W^*(\phi)$ unchanged. This immediately gives

$$\overline{\text{Range}\, M_\phi} = \overline{\phi\, L_a^2(\mathbb{D})} \neq L_a^2(\mathbb{D}).$$

Let $M_\phi = U|M_\phi|$ be the polar decomposition of M_ϕ. Then by Proposition 2.5.1 both U and $|M_\phi|$ are in $W^*(\phi)$. It is easy to check that the initial space of U is $L_a^2(\mathbb{D})$, and the final space is $\overline{\phi\, L_a^2(\mathbb{D})}$. Hence U is a non-unitary isometry, as desired. This shows that $W^*(\phi)$ is infinite.

Furthermore, by Proposition 2.5.4 the commutant of a type II_1 factor is either a type II_1 factor or a type II_∞ factor. This fact, along the above discussion, shows that $W^*(\phi)$ is a type II_∞ factor, finishing the proof.   □

Remark 6.4.8 The above proof indeed shows that for a bounded holomorphic function ϕ on $\mathbb{D}$, whenever $V^*(\phi)$ is a type II_1 factor, $W^*(\phi)$ is always a type II_∞ factor. Also note that when $V^*(\phi)$ is a type II_1 factor, M_ϕ is a completely reducible operator. That is, for each nonzero reducing subspace M of M_ϕ, $M_\phi|_M$ has a nontrivial reducing subspace [Ros]. This is a direct consequence of [Jon, Corollary 6.1.14], which states that if $\mathcal{A}$ is a type II_1 factor on H and $p \in \mathcal{A}$ is a nontrivial projection, then $p\mathcal{A}p$ is a type II_1 factor on pH.

Combing the proof of Theorem 6.4.1 with Corollary 6.4.6 gives the following proposition, due to Guo and Huang [GH2].

Proposition 6.4.9 *If $\phi : \mathbb{D} \to \Omega$ is a holomorphic regular branched covering map, then $V^*(\phi)$ is a type II_1 factor if and only if $G(\phi)$ is an i.c.c. group.*

Let $\phi_i : \mathbb{D} \to \Omega (i = 0, 2)$ be two holomorphic covering maps, and then there exists a $\rho \in \text{Aut}(\mathbb{D})$ such that $\phi_1 \circ \rho = \phi_2$. This implies that

$$U_\rho^* M_{\phi_2} U_\rho = M_{\phi_2 \circ \rho^{-1}} = M_{\phi_1},$$

and hence

$$U_\rho^* W^*(\phi_2) U_\rho = W^*(\phi_1), \qquad U_\rho^* V^*(\phi_2) U_\rho = V^*(\phi_1).$$

Therefore, by omitting a unitary isomorphism, we can assign two von Neumann algebras $W^*(\Omega)$, $V^*(\Omega)$ for each Ω; both arise from the domain Ω, independent from the choice of covering maps.

Given two planar domains Ω_1 and Ω_2, let G be a biholomorphic map from Ω_1 to Ω_2. If $\phi : \mathbb{D} \to \Omega_1$ is a holomorphic covering map, then $G \circ \phi : \mathbb{D} \to \Omega_2$ is a holomorphic covering map. Then it is easy to verify that

$$W^*(G \circ \phi) = W^*(\phi) \quad \text{and} \quad V^*(G \circ \phi) = V^*(\phi).$$

This shows that if Ω_1 and Ω_2 is conformally isomorphic, then $\mathcal{W}^*(\Omega_1)$ and $\mathcal{W}^*(\Omega_2)$ are unitarily isomorphic, and so is it with $\mathcal{V}^*(\Omega_1)$ and $\mathcal{V}^*(\Omega_2)$.

Given a holomorphic covering map $\phi : \mathbb{D} \rightarrow \Omega$, one has $G(\phi) \cong \pi_1(\Omega)$. Therefore, if Ω is not conformally isomorphic to one of the disk, annuli or the punctured disk, then $G(\phi)$ is a free group on n generators $(2 \leq n \leq \infty)$; that is, $G(\phi) = F_n$. As one knows, $\mathcal{L}(F_n)$ is the von Neumann algebra generated by left regular representation of F_n on $l^2(F_n)$, and by Corollary 6.2.5, $\mathcal{V}^*(\phi)$ is the von Neumann algebra generated by representation U_ρ of $G(\phi)$ on $L^2_a(\mathbb{D})$. However, $\mathcal{L}(F_n)$ is not unitarily isomorphic to $\mathcal{V}^*(\phi)$ because $\mathcal{L}'(F_n) = \mathcal{R}(F_n)$, a type II_1 factor, but $\left(\mathcal{V}^*(\phi)\right)' = \mathcal{W}^*(\phi)$, a type II_∞ factor. This observation naturally gives rise to a problem: is $\mathcal{V}^*(\phi)$ $*$-isomorphic to $\mathcal{L}(F_n)$? As we will see, the answer is affirmative. This will be the main focus of the next section.

6.5 $\mathcal{V}^*(\phi)$ and Free Group Factors

This section will further the study of the structure of $\mathcal{V}^*(\phi)$ where $\phi : \mathbb{D} \rightarrow \Omega$ is a holomorphic covering map. It will be shown that in this case $\mathcal{V}^*(\phi)$ is $*$-isomorphic to the group von Neumann algebra $\mathcal{L}(\pi_1(\Omega))$. Furthermore, for two bounded domains Ω_1 and Ω_2, if $\pi_1(\Omega_1) \cong \pi_1(\Omega_2)$ and $\pi_1(\Omega_1)$ is not abelian, then $\mathcal{V}^*(\Omega_1)$ and $\mathcal{V}^*(\Omega_2)$, $\mathcal{W}^*(\Omega_1)$ and $\mathcal{W}^*(\Omega_2)$ are unitarily isomorphic, respectively.

We adopt the notations in [Con1, Sect. 43]. For a group G, let $\mathcal{R}(G)$ be the WOT-closure of the span of all $R_a : l^2(G) \rightarrow l^2(G)(a \in G)$ defined by

$$R_a f(x) = f(xa), x \in G, f \in l^2(G);$$

similarly, denote by $\mathcal{L}(G)$ the WOT-closure of the span of all $L_a : l^2(G) \rightarrow l^2(G)$ defined by

$$L_a f(x) = f(a^{-1}x), x \in G, f \in l^2(G).$$

It is well-known that $\mathcal{R}(G)$ is unitarily isomorphic to $\mathcal{L}(G)$. If G is the free group on $n(n \geq 2)$ generators, then it is an i.c.c. group, and hence by Conway [Con1, Theorem 53.1] $\mathcal{R}(G)$ and $\mathcal{L}(G)$ are type II_1 factors.

Let us recall some preliminaries of group von Neumann algebras. As done in [Con1, p. 248, 250], for each $b \in G$, let ϵ_b denote the characteristic function of $\{b\}$, and put

$$f^*(x) = \overline{f(x^{-1})}, \; x \in G.$$

For $g, h \in l^2(G)$, the involution $g * h$ of g and h is defined as follows:

$$g * h(a) = \sum_{x \in G} g(x)h(x^{-1}a), a \in G.$$

For each f in $l^2(G)$, we can define a densely-defined operator L_f via

$$L_f g = f * g,$$

where $g \in l^2(G)$ satisfies $f * g \in l^2(G)$. As is known, L_f defines a bounded operator on $l^2(G)$ if and only if $f * g \in l^2(G)$ for every $g \in l^2(G)$; $\mathcal{L}(G)$ consists of all bounded operators with the form L_f [Con1, p. 250]. Similar results hold for $\mathcal{R}(G)$. For any two operators in $\mathcal{L}(G)$, say L_g and L_h, we have

$$L_g^* = L_{g^*} \text{ and } L_g L_h = L_{g*h}. \tag{6.13}$$

The following lemma [GH2] is of independent interest. It will be applied to prove our main result in the sequel.

Lemma 6.5.1 *Suppose G is a countable group and f is in $l^2(G)$. Then $\langle f, L_a f \rangle = 0$ for all $a \neq e$ if and only if $\langle f, R_a f \rangle = 0$ for all $a \neq e$, where e is the identity of G.*

Proof The proof is from [GH2].

Without loss of generality, assume that f is nonzero and $\|f\| = 1$. By a simple computation, Lemma 6.5.1 reduces to the following statement: $f^* * f = \epsilon_e$ if and only if $f * f^* = \epsilon_e$. By symmetry, it suffices to prove that if $f^* * f = \epsilon_e$, then $f * f^* = \epsilon_e$. To see this, we first make the following claim.

Claim If $f^* * f = \epsilon_e$, then L_f is a bounded operator on $l^2(G)$.

As before-mentioned, to each function h in $l^2(G)$, we can assign a densely-defined operator L_h. In particular, both L_f and L_{f^*} are well defined on the linear span of $\{\epsilon_b : b \in G\}$, denoted by $\mathcal{P}$.

To prove the above claim, we first show that if $f^* * f = \epsilon_e$, then

$$\langle L_f p, L_f p \rangle = \langle p, p \rangle, p \in \mathcal{P}.$$

Since $f^* * f = \epsilon_e$, by a simple computation

$$\langle L_f \epsilon_b, L_f \epsilon_a \rangle = \langle L_{f^*} L_f \epsilon_b, \epsilon_a \rangle = \langle \epsilon_b, \epsilon_a \rangle, a, b \in G.$$

Thus for each $p \in \mathcal{P}$, we get

$$\langle L_f \epsilon_b, L_f p \rangle = \langle \epsilon_b, p \rangle, b \in G,$$

forcing $\langle L_f p, L_f p \rangle = \langle p, p \rangle, p \in \mathcal{P}$, as desired.

Next we are to prove that for any $g \in l^2(G), f * g \in l^2(G)$. Once this is done, we can apply the closed graph theorem to deduce that L_f is a bounded operator. To see $f * g \in l^2(G)$, pick a sequence $\{p_n\}$ in $\mathcal{P}$ such that $\|p_n - g\|_2 \to 0$ as $n \to \infty$. This implies that $f * p_n$ converges to $f * g$ pointwise, and hence for each finite set F in G, we have

$$\sum_{a \in F} |f * g(a)|^2 \le \sup_n \sum_{a \in F} |f * p_n(a)|^2 \le \sup_n \|f * p_n\|_2^2$$

$$= \sup_n \|L_f p_n\|_2^2 = \sup_n \|p_n\|_2^2 < \infty.$$

By the arbitrariness of F, we have $f * g \in l^2(G)$, as desired. The proof of the claim is complete.

Now L_f is a bounded operator, and by (6.13) $L_f^* = L_{f^*}$. Besides, the proof of [Con1, Theorem 53.1] shows that for a countable group G, both $\mathcal{R}(G)$ and $\mathcal{L}(G)$ are finite von Neumann algebras. Therefore, for any operator S in $\mathcal{L}(G)$ satisfying $S^*S = I$, we must have $SS^* = I$. Since $f^* * f = \epsilon_e$, by a straightforward computation, $L_f^* L_f = L_{f^* * f} = I$, and hence $L_f L_f^* = I$. This guarantees $L_{ff^*} = \epsilon_e$, forcing $f * f^* = \epsilon_e$. The proof is complete. $\square$

By the proof of Lemma 6.5.1, one gets the following proposition [GH2].

Proposition 6.5.2 *Suppose G is a countable group and f is in $l^2(G)$. Then $f * f^* = \epsilon_e$ if and only if $f^* * f = \epsilon_e$. In this situation, L_f is a unitary operator.*

Some observations are in order. Given a holomorphic covering map $\phi : \mathbb{D} \to \Omega$, write $G(\phi) = \{\rho_k\}$. In the proof of Theorem 6.2.2, the claim in Step 1 shows that for any $f \in l^2(G(\phi))$, the formal sum

$$\sum_k f(\rho_k) U_{\rho_k}^* : h(z) \mapsto \sum_k f(\rho_k)(U_{\rho_k}^* h)(z), \ h \in L_a^2(\mathbb{D})$$

defines a linear map from $L_a^2(\mathbb{D})$ to holomorphic functions over $\mathbb{D}$. Note that for each $k \in \mathbb{Z}_+$, $U_{\rho_k}^* = U_{\rho_k^{-1}}$, and the above map can be rewritten as

$$\sum_k f(\rho_k) U_{\rho_k}^* : h(z) \mapsto \sum_k f(\rho_k) h \circ \rho_k^{-1}(z)(\rho_k^{-1})'(z), \ h \in L_a^2(\mathbb{D}).$$

For simplicity, write

$$\Theta(f) \triangleq \sum_k f(\rho_k) U_{\rho_k}^*.$$

In addition, based on the fact that $L_{\epsilon_x} = L_x, x \in G(\phi)$, for each $f \in l^2(G(\phi))$ we identify the formal sum $\sum_k f(\rho_k) L_{\rho_k}$ with the densely-defined operator L_f in $l^2(G(\phi))$.

One sees from Corollary 6.2.3 that for each $S \in V^*(\phi)$ there exists a unique $f \in l^2(G(\phi))$ satisfying $S = \Theta(f)$. Now define a linear map Λ on $V^*(\phi)$ as follows:

$$\Lambda : \Theta(f) \mapsto L_f;$$

that is,

$$\Lambda : \sum_k f(\rho_k) U_{\rho_k}^* \mapsto \sum_k f(\rho_k) L_{\rho_k}.$$

Namely, Λ maps each member $\Theta(f)$ in $V^*(\phi)$ to a densely-defined operator L_f in $l^2(G(\phi))$. In fact, these L_f are bounded and defined on the whole space $L_a^2(\mathbb{D})$, as will be shown right away. The following is due to Guo and Huang [GH2].

Proposition 6.5.3 *The linear map Λ is an injective $*$-homomorphism from $V^*(\phi)$ to $\mathcal{L}(G(\phi))$.*

Proof The proof comes from [GH2].

We first claim that Λ maps each member in $V^*(\phi)$ to $\mathcal{L}(G(\phi))$. For this, note that Proposition 2.5.1 states that any von Neumann algebra is the finite linear span of its unitary operators, and it suffices to show that if $\Theta(f)$ is a unitary operator, then L_f is a bounded operator in $\mathcal{L}(G(\phi))$. In fact, by Lemma 6.2.4 it is easy to check that for any two operators $\Theta(f_1)$ and $\Theta(f_2)$ in $V^*(\phi)$,

$$\Theta(f_1)^* = \Theta(f_1^*),\tag{6.14}$$

and

$$\Theta(f_1)\Theta(f_2) = \Theta(f_1 * f_2).\tag{6.15}$$

Since $\Theta(f)^*\Theta(f) = I$, we have

$$\Theta(f^* * f) = \Theta(f^*)\Theta(f) = \Theta(f)^*\Theta(f) = I,$$

forcing $f^* * f = \epsilon_{\rho_0}$, where ρ_0 denotes the identity of $G(\phi)$. By Proposition 6.5.2, L_f is a bounded operator in $\mathcal{L}(G(\phi))$.

Furthermore, combing (6.13) with (6.14) shows that for any operators $\Theta(f_1), \Theta(f_2)$ in $V^*(\phi)$,

$$\Lambda(\Theta(f_1)^*) = \Lambda(\Theta(f_1^*)) = L_{f_1^*} = L_{f_1}^* = \left(\Lambda(\Theta(f_1))\right)^*,$$

and by (6.13) and (6.15),

$$\begin{aligned}\Lambda(\Theta(f_1)\Theta(f_2)) &= \Lambda(\Theta(f_1 * f_2))\\&= L_{f_1 * f_2} = L_{f_1} L_{f_2}\\&= \Lambda(\Theta(f_1))\,\Lambda(\Theta(f_2)).\end{aligned}$$

Therefore Λ is $*$-homomorphism from $V^*(\phi)$ to $\mathcal{L}(G(\phi))$, and its injectivity is trivial. The proof is complete. $\qquad\square$

The above map Λ turns out to be a $*$-isomorphism, which was proved by Guo and Huang [GH2].

Theorem 6.5.4 *The map Λ is a $*$-isomorphism from $V^*(\phi)$ onto $\mathcal{L}(G(\phi))$.*

Proof The proof comes from [GH2].

Given two von Neumann algebras $\mathcal{A}$ and $\mathcal{B}$, a positive map $\gamma : \mathcal{A} \to \mathcal{B}$ is called *normal* if for any increasing net $\{A_\alpha\}$ in $\mathcal{A}$ converging strongly to $A \in \mathcal{A}$, $\gamma(A_\alpha)$ converges strongly to $\gamma(A)$. We will first show that Λ is normal.

To see this, note that by Proposition 6.5.3 Λ is a $*$-homomorphism, and hence Λ is positive. Now assume that $\{\Theta(f_\alpha)\}$ is an increasing net in $V^*(\phi)$ which converges strongly to $\Theta(f)$. Since Λ is positive, $\{\Lambda(\Theta(f_\alpha))\}$ is an increasing net satisfying

$$\Lambda(\Theta(f_\alpha)) \le \Lambda(\Theta(f));$$

that is, $\{L_{f_\alpha}\}$ is an increasing net with an upper bound L_f. Therefore $\{L_{f_\alpha}\}$ converges strongly to some operator L_g in $\mathcal{L}(G(\phi))$. In particular,

$$\lim_\alpha \langle L_{f_\alpha}\epsilon_e, L_{\rho_k} \rangle = \langle L_g\epsilon_e, L_{\rho_k} \rangle,$$

which immediately gives

$$\lim_\alpha f_\alpha(\rho_k) = g(\rho_k), \ k = 0, 1, \cdots. \tag{6.16}$$

Besides, by Lemma 6.2.1 $\{\rho_i^{-1}(0)\}$ is an interpolating sequence. Hence, for each k there is a bounded holomorphic function h satisfying

$$h(\rho_i^{-1}(0)) = \delta_k^i. \tag{6.17}$$

Let K_0 be the reproducing kernel in $L_a^2(\mathbb{D})$ at 0 (actually, $K_0 \equiv 1$). Since $\{\Theta(f_\alpha)\}$ converges strongly to $\Theta(f)$,

$$\lim_\alpha \ \langle \Theta(f_\alpha)h, K_0 \rangle = \langle \Theta(f)h, K_0 \rangle.$$

which, combined with (6.17), gives that

$$\lim_\alpha f_\alpha(\rho_k) = f(\rho_k), \ k = 0, 1, \cdots. \tag{6.18}$$

Then by (6.16) and (6.18), $f = g$, and hence $\Lambda(\Theta(f_\alpha))$ converges strongly to $\Lambda(\Theta(f))$. Therefore Λ is normal.

Since Λ is normal, by Conway [Con1, Theorem 46.8] $Range(\Lambda)$ is ultraweakly closed, and thus $Range(\Lambda)$ is a von Neumann algebra. Then $Range(\Lambda) = \mathcal{L}(G(\phi))$, and by Proposition 6.5.3, Λ is a $*$-isomorphism from $V^*(\phi)$ to $\mathcal{L}(G(\phi))$. The proof is complete. $\square$

For a holomorphic covering map $\phi : \mathbb{D} \rightarrow \Omega$, one has $G(\phi) \cong \pi_1(\Omega)$. Therefore, Theorem 6.5.4 shows that $V^*(\Omega)$ is $*$-isomorphic to $\mathcal{L}(\pi_1(\Omega))$. Then by Lemma 6.4.2, $V^*(\Omega)$ is $*$-isomorphic to $\mathcal{L}(F_n)$, where n is the cardinality of generators of $\pi_1(\Omega)$. In particular, if $\mathbb{C} - \Omega$ has $q(q < \infty)$ bounded components, then $n = q$. The special case of $n = 1$ has been demonstrated by Proposition 6.3.12.

As is well known, $\mathcal{L}(G)$ is a type II_1 factor if and only if G is an i.c.c. group. Therefore, Proposition 6.4.9 is a direct consequence of Theorem 6.5.4, whose proof is independent of the discussions in Sect. 6.3 of this chapter.

The following example indicates that the structure of $V^*(\Omega)$ is very complicated.

Example 6.5.5 Fix an $r \in (0, 1)$, and set $\Omega_r = \{z \in \mathbb{C} : r < |z| < 1\}$. As done in Example 6.3.6, let ϕ_r denote the covering map from $\mathbb{D}$ onto Ω_r:

$$\phi_r(z) = \exp\left(\ln r \left(\frac{i}{\pi} \ln \frac{1+z}{1-z} + \frac{1}{2}\right)\right).$$

Take a point $w_0 \in \Omega_r$ and put $\Omega' = \mathbb{D} - \phi^{-1}(w_0)$. Now let φ denote the covering map from $\mathbb{D}$ onto Ω'. By our convention,

$$V^*(\varphi) = V^*(\Omega') \text{ and } V^*(\phi_r \circ \varphi) = V^*(\Omega_r - \{w_0\}).$$

Therefore, $V^*(\Omega')$ is a SOT-closed $*$-subalgebra of $V^*(\Omega_r - \{w_0\})$. But $\pi_1(\Omega') \cong F_\infty$ and $\pi_1(\Omega_r - \{w_0\}) \cong F_2$, and hence

$$V^*(\Omega') \stackrel{*}{\cong} \mathcal{L}(F_\infty), \quad V^*(\Omega_r - \{w_0\}) \stackrel{*}{\cong} \mathcal{L}(F_2).$$

Thus, $\mathcal{L}(F_2)$ contains a $*$-subalgebra that is $*$-isomorphic to $\mathcal{L}(F_\infty)$.

Next we will give the following statement, due to Guo and Huang [GH2]. Because of its length, the proof is deferred to the end of this section.

Theorem 6.5.6 *Suppose both Ω_1 and Ω_2 are not conformally isomorphic to one of the disk, annuli and the punctured disk. Then the following are equivalent:*

1. *$V^*(\Omega_1) \stackrel{*}{\cong} V^*(\Omega_2)$;*
2. *$V^*(\Omega_1)$ is unitarily isomorphic to $V^*(\Omega_2)$;*
3. *$W^*(\Omega_1)$ is unitarily isomorphic to $W^*(\Omega_2)$.*

Consequently, by Theorem 6.5.4, if $\pi_1(\Omega_1) \cong \pi_1(\Omega_2)$, then $V^*(\Omega_1)$ is unitarily isomorphic to $V^*(\Omega_2)$, and $W^*(\Omega_1)$ is unitarily isomorphic to $W^*(\Omega_2)$.

Remark 6.5.7 If Ω_1 is homotopy equivalent to Ω_2, then $\pi_1(\Omega_1) \cong \pi_1(\Omega_2)$ [Le, Theorem 7.24]. Therefore, both $W^*(\Omega)$ and $V^*(\Omega)$ are completely determined by the homotopy class of the domain Ω.

We need some preliminary results on type II_1 factors. For a type II_1 factor $\mathcal{M}$, there is always a unique ultraweakly continuous normalized trace tr, which is necessarily faithful. This trace gives rise to a GNS construction, and the corresponding Hilbert space is denoted by $L^2(\mathcal{M})$, which contains a dense linear subspace, $\mathcal{M}$. By [Jon, Corollary 7.1.8], $\mathcal{M}$ acts on $L^2(\mathcal{M})$ as a von Neumann algebra. In this situation, we say $\mathcal{M}$ is in standard form. For example, if G is an i.c.c. group, then both $\mathcal{L}(G)$ and $\mathcal{R}(G)$ are in standard forms.

The following fact about standard form is likely well-known.

Lemma 6.5.8 *For two type* II_1 *factors* $\mathcal{A}$ *and* $\mathcal{B}$ *in standard forms, if* $\mathcal{A} \stackrel{*}{\cong} \mathcal{B}$, *then* $\mathcal{A}$ *is unitarily isomorphic to* $\mathcal{B}$.

Proof The proof is from [GH2].

Let $\mathcal{A}$ and $\mathcal{B}$ be two II_1 factors in standard forms, and assume that $\theta : \mathcal{A} \to \mathcal{B}$ is a $*$-isomorphism. Denote by tr_B the unique ultraweakly continuous normalized trace over $\mathcal{B}$, and then $tr_B \circ \theta$ is the unique ultraweakly continuous normalized trace over $\mathcal{A}$ (for details, see [Bla, Jon]). This observation immediately gives a unitary operator u from $L^2(\mathcal{A})$ onto $L^2(\mathcal{B})$, which maps each a in $\mathcal{A}(\mathcal{A} \subseteq L^2(\mathcal{A}))$ to $\theta(a)$. Then one can verify directly that

$$ua = \theta(a)u, \ a \in \mathcal{A}.$$

Therefore, as standard forms, $\mathcal{A}$ is unitarily isomorphic to $\mathcal{B}$. $\qquad\qquad \square$

The next result is also needed, which follows from the proof of [Jon, Theorem 10.1.1].

Lemma 6.5.9 ([Jon]) *Suppose* $\mathcal{M}$ *is a type* II_1 *factor acting on a separable Hilbert space* $\mathcal{H}$, *and* $\mathcal{M}'$ *is infinite. Then there exists a unitary operator* $U : \mathcal{H} \to L^2(\mathcal{M}) \otimes l^2$ *satisfying*

$$UA = (A \otimes I)U, \ A \in \mathcal{M}.$$

Therefore, $\mathcal{M}$ *is unitarily isomorphic to* $\mathcal{M}|_{L^2(\mathcal{M})} \otimes I$.

Now we are ready to prove Theorem 6.5.6.

Proof of Theorem 6.5.6 The proof is from [GH2].

Both $(2) \Leftrightarrow (3)$ and $(2) \Rightarrow (1)$ are trivial. We will show $(1) \Rightarrow (2)$ to complete the proof.

To see this, assume that Ω_1 and Ω_2 are not conformally isomorphic to one of the disk and $\mathcal{V}^*(\Omega_1) \stackrel{*}{\cong} \mathcal{V}^*(\Omega_2)$. Since $\mathcal{V}^*(\Omega_1)$ is a type II_1 factor and its commutant $\mathcal{W}^*(\Omega_1)$ is infinite, by Lemma 6.5.9

$$\mathcal{V}^*(\Omega_1) \stackrel{\substack{unitarily\ isomorphic}}{\cong} \mathcal{M}_1 \otimes I, \qquad\qquad (6.19)$$

where $\mathcal{M}_1$ denotes the standard form of $\mathcal{V}^*(\Omega_1)$, i.e. $\mathcal{M}_1 = \mathcal{V}^*(\Omega_1)|_{L^2(\mathcal{V}^*(\Omega_1))}$. Similarly,

$$\mathcal{V}^*(\Omega_2) \stackrel{\substack{unitarily\ isomorphic}}{\cong} \mathcal{M}_2 \otimes I. \qquad\qquad (6.20)$$

where $\mathcal{M}_2$ denotes the standard form of $V^*(\Omega_2)$. Recall that $V^*(\Omega_1) \overset{*}{\cong} V^*(\Omega_2)$. Combining (6.19) with (6.20) gives

$$\mathcal{M}_1 \otimes I \overset{*}{\cong} \mathcal{M}_2 \otimes I,$$

forcing

$$\mathcal{M}_1 \overset{*}{\cong} \mathcal{M}_2.$$

Then by Lemma 6.5.8, $\mathcal{M}_1$ is unitarily isomorphic to $\mathcal{M}_2$. In view of (6.19) and (6.20), $V^*(\Omega_1)$ is unitarily isomorphic to $V^*(\Omega_2)$. The proof of (1) $\Rightarrow$ (2) is complete.

The remaining part follows immediately from an observation: for a holomorphic covering map $\phi : \mathbb{D} \rightarrow \Omega$, $\pi_1(\Omega) \cong G(\phi)$, and by Theorem 6.5.4, $V^*(\phi) \overset{*}{\cong} \mathcal{L}(\pi_1(\Omega))$. $\qquad\square$

Theorems 6.5.4 and 6.5.6 are closely related to an unsolved problem in von Neumann algebras: whether

$$\mathcal{L}(F_n) \overset{*}{\cong} \mathcal{L}(F_m)$$

hold for $n \neq m$ and $n, m \geq 2$. This problem is thus equivalent to the following:

Problem 6.5.10 If $\pi_1(\Omega_1) \not\cong \pi_1(\Omega_2)$, then is $V^*(\Omega_1)$ unitarily isomorphic to $V^*(\Omega_2)$; or equivalently, is $W^*(\Omega_1)$ unitarily isomorphic to $W^*(\Omega_2)$?

To close this section, we review some results from Abrahamse and Douglas [AD]. Actually in [AD], some von Neumann algebra closely related to $V^*(\phi)$ was constructed from M_ϕ which acts on a vector-valued Hardy space. However, the techniques developed in [AD] depend more on the method of vector bundle, and less on the structure of the Bergman space $L_a^2(D)$. It is worthwhile to point out that the ideas in [AD] are completely different from those in this chapter. As done in [AD], let $\mathcal{U}$ be the group of all unitary operators on the Bergman space, G be a group, and $\alpha : G \rightarrow \mathcal{U}$ be some $*$-homomorphism. Now write $G = G(\phi)$, where ϕ is a holomorphic covering map from $\mathbb{D}$ onto Ω, a bounded planar domain. Let $\mathcal{H}$ denote $H^2(\mathbb{D}) \otimes L_a^2(\mathbb{D})$, the space of $L_a^2(\mathbb{D})$-valued Hardy space over the unit disk, and $\mathcal{H}_\alpha$ denotes its subspace consisting of all functions f in $\mathcal{H}$ satisfying

$$f \circ \rho(z) = \alpha(\rho)(f(z)), \quad z \in \mathbb{D}, \rho \in G.$$

Observe that the vector-valued Hardy subspace $\mathcal{H}_\alpha$ is invariant for M_ϕ, and this subspace relies on $G = G(\phi)$, and essentially on ϕ. Now set $T_\alpha = M_\phi|_{\mathcal{H}_\alpha}$, and denote by $W^*(T_\alpha)$ the von Neumann algebra generated by T_α. By [AD, Theorem 8],

$$W^*(T_\alpha) \overset{\underset{\text{unitarily isomorphic}}{}}{\cong} W^*(\alpha) \otimes B(l^2),$$

where $\mathcal{W}^*(\alpha)$ is the von Neumann algebra generated by $\{\alpha(\rho) : \rho \in G\}$. In particular, if $\alpha(\rho) = U_\rho^*(\rho \in G)$, then $\mathcal{W}^*(\alpha) = \mathcal{V}^*(\phi)$.

6.6   Type II Factors and Orbifold Domains

In last several sections we focus on the von Neumann algebras which essentially arise from holomorphic covering maps. It turns out that the structure of them has close link with the fundamental group of the image of those maps. This section will give a slight generalization to regular branched covering maps, which has close connection with orbifold domains. In particular, this section present a generalization of Theorem 6.4.1 by using Proposition 6.4.9. Also, some applications are given to special inner functions.

Before we present the main theorem, a lemma from [GH2] will be established. It concerns with free product of groups.

Lemma 6.6.1 *The following hold:*

(1) *The free product G of group family $\{G_\alpha : \alpha \in \mathcal{I}\}$ is an i.c.c. group if at least 3 groups G_α are nontrivial;*
(2) *For two integers $m, n \geq 2$, if $\max(m, n) \geq 3$, then $\mathbb{Z}_m * \mathbb{Z}_n$ is an i.c.c. group. Also, $\mathbb{Z} * \mathbb{Z}_n (n \geq 2)$ is an i.c.c. group;*
(3) *$\mathbb{Z}_2 * \mathbb{Z}_2$ is not an i.c.c. group. In fact, it has infinitely many nontrivial finite conjugacy classes.*

The group $\mathbb{Z}_2 * \mathbb{Z}_2$ is known as the *infinite dihedral group*.

Proof The following discussion is based on reduced words [GH2].

(1) Without loss of generality, let G be the free product of three nontrivial groups G_1, G_2 and G_3. For each $u \in G - \{e\}$, let $u = x_1 x_2 \cdots x_k (k \geq 1)$ be the *reduced form* of u; that is, for each $i(1 \leq i \leq k-1)$, x_i and x_{i+1} do not lie in one of the same groups: G_1, G_2 or G_3. There must be a group, say G_3, such that $x_1 \notin G_3$ and $x_k \notin G_3$. Pick $\rho \in G_3$ and $\tau \in G_1$ satisfying $\rho \neq e$ and $\tau \neq e$. Then it is easy to check that

$$\rho u \rho^{-1}, \quad (\tau\rho)u(\tau\rho)^{-1}, \quad (\rho\tau\rho)u(\rho\tau\rho)^{-1}, \quad (\tau\rho\tau\rho)u(\tau\rho\tau\rho)^{-1}, \quad \cdots$$

is an infinite sequence of words, no two of which are equal. Since these words are in the conjugacy class of u, the arbitrariness of u shows that G is an i.c.c. group.

(2) First we prove the former statement. Assume $G_1 \cong \mathbb{Z}_m$ and $G_2 \cong \mathbb{Z}_n$ with $m \geq 3$ and $n \geq 2$. Denote by ρ and τ the generators of G_1 and G_2 respectively, and then

$$\rho^m = e \text{ and } \tau^n = e.$$

Now assume $u = x_1 x_2 \cdots x_k (u \neq e)$ is a reduced word. It is enough to show that the conjugacy class of u is infinity. There are several cases under consideration.

(i) Both x_1 and x_k are in G_1 or G_2. First assume that x_1 and x_k are in G_1. Then

$$\tau u \tau^{-1}, \quad (\rho\tau)u(\rho\tau)^{-1}, \quad (\tau\rho\tau)u(\tau\rho\tau)^{-1}, \quad \cdots$$

is an infinite sequence whose members are in the conjugacy of u. The case that both x_1 and x_k are in G_2 can be handled in a similar way.

(ii) There is an i such that $x_1 = \rho^i (1 \leq i \leq m-1)$ and $x_k \in G_2$. Then there is an $i' (1 \leq i' \leq m-1)$ such that $\rho^{i'} \rho^i \neq e$. Therefore $u' = \rho^{i'} u \rho^{m-i'}$ is a word in the conjugacy class of u and it suffices to consider the conjugacy class of u'. This is case (i) since both $\rho^{i'} \rho^i$ and $\rho^{m-i'}$ are nontrivial element in G_1. Thus the conjugacy class of u is infinite, as desired.

(iii) There is an i such that $x_k = \rho^i (1 \leq i \leq m-1)$ and $x_1 \in G_2$. This case can be done in a similar way as case (ii).

With a similar argument, one can show that $\mathbb{Z} * \mathbb{Z}_n (n \geq 2)$ is an i.c.c. group.

(3) As done in (2), write $G_1 \cong \mathbb{Z}_2$ and $G_2 \cong \mathbb{Z}_2$, and assume ρ and τ are the generators of G_1 and G_2, respectively. We have

$$\rho^2 = e \text{ and } \tau^2 = e.$$

Then it is easy to verify that for each positive integer n, $\{(\rho\tau)^n, (\tau\rho)^n\}$ is a conjugacy class in $G_1 * G_2$, and no two of which are equal. The proof is complete. $\square$

Now we are ready to establish the following theorem, due to Guo and Huang.

Theorem 6.6.2 *Let (Ω, v) be an orbifold domain with the singular locus Σ, and suppose $\phi : \mathbb{D} \to (\Omega, v)$ is a holomorphic universal covering. Then $\mathcal{V}^*(\phi)$ is a type II_1 factor with the following exceptions:*

1. *$\Sigma = \emptyset$, and Ω is conformally isomorphic to one of the disk, annuli or the punctured disk. In this case, ϕ is a covering map.*
2. *$\Sigma = \{w_0\}$ is a singleton and Ω is conformally isomorphic to the unit disk $\mathbb{D}$. In this case, $\phi = \psi(\rho^n)$, where ψ is a conformal map from $\mathbb{D}$ onto Ω, and $\rho \in Aut(\mathbb{D})$, $n = v(w_0)$.*
3. *$\Sigma = \{w_0, w_1\}$ with $v(w_0) = v(w_1) = 2$ and Ω is conformally isomorphic to the unit disk $\mathbb{D}$. In this case, $\mathcal{V}^*(\phi)$ is a finite von Neumann algebra whose center is infinite dimensional.*

Whenever $\mathcal{V}^*(\phi)$ is a type II_1 factor, $\mathcal{W}^*(\phi)$ always is a type II_∞ factor.

It is clear that $\mathcal{V}^*(\phi)$ is abelian only in situations (1) and (2). This is a clearer version of Proposition 6.3.11.

Proof The proof comes from [GH2].

For simplicity, we write $\pi_1(\Omega, \nu)$ for $\pi_1(\Omega, \Sigma, \nu)$. We shall first discuss when the fundamental group $\pi_1(\Omega, \nu)$ is an i.c.c. group. Recall that $\pi_1(\Omega, \nu)$ is defined to be the deck transformation group of ϕ.

For this, it is better to get some instructive ideas of the existence for ϕ from [Mi, p. 258, Problem E-4] and [Sc, p. 423, 424], where the arguments and ideas give a simple fact: let w be a point in the singular locus Σ, $\Sigma' = \Sigma - \{w\}$, and define ν' on Ω such that

$$\nu'|_{\Sigma'} = \nu \text{ and } \nu|_{\Omega - \Sigma'} = 1.$$

Then (Ω, ν') is also an orbifold domain and

$$\pi_1(\Omega, \nu) \cong \mathbb{Z}_{\nu(w)} * \pi_1(\Omega, \nu').$$

Note that if Σ' is empty, then (Ω, ν') is just a usual domain and $\pi_1(\Omega, \nu')$ is nothing but $\pi_1(\Omega)$.

If there are at least k points in Σ, say $w_1, \cdots, w_k$, then by iterative use of the above fact there is a ramified function ν'' such that

$$\pi_1(\Omega, \nu) \cong \left(\mathbb{Z}_{\nu(w_1)} * \cdots * \mathbb{Z}_{\nu(w_k)} \right) * \pi_1(\Omega, \nu''). \tag{6.21}$$

Therefore, if the number of points in Σ satisfies $\sharp\Sigma \geq 3$, then $\pi_1(\Omega, \nu)$ is the free product of at least three nontrivial groups, and by Lemma 6.6.1 $\pi_1(\Omega, \nu)$ is an i.c.c. group. Thus it remains to deal with the following three cases: $\sharp\Sigma = 0, 1$ and 2.

First assume $\sharp\Sigma = 0$. That is, $\Sigma = \emptyset$ and ϕ is a holomorphic covering map. This situation has been done by Theorems 6.4.1 and 6.3.2, and the only exception is stated as in (1).

Next we shall deal with the case of $\sharp\Sigma = 1$. Suppose $\Sigma = \{w_1\}$ and $n = \nu(w_1)$. Then (6.21) is reduced to

$$\pi_1(\Omega, \nu) \cong \mathbb{Z}_n * \pi_1(\Omega), \quad n \geq 2.$$

In this situation, if $\pi_1(\Omega)$ is trivial, then as shown in Example 6.1.2, we reach exception as stated as in (2). Otherwise, $\pi_1(\Omega)$ is a free group with m generators $(1 \leq m \leq \infty)$. Therefore $\pi_1(\Omega)$ is isomorphic to the free product of m integer groups $\mathbb{Z}$ [Ja, p. 87, 88]. Applying Lemma 6.6.1(1) and (2) shows that $\pi_1(\Omega, \nu)$ is an i.c.c. group.

It remains to handle the case of $\sharp\Sigma = 2$. In this situation, either $\pi_1(\Omega)$ is trivial or not. If $\pi_1(\Omega)$ is nontrivial, then combining Lemma 6.6.1(1) with (6.21) shows that $\pi_1(\Omega, \nu)$ is an i.c.c. group. If $\pi_1(\Omega)$ is trivial, then by Lemma 6.6.1(3), $\pi_1(\Omega, \nu)$ is not an i.c.c. group only if $\pi_1(\Omega, \nu) = \mathbb{Z}_2 * \mathbb{Z}_2$. This happens only in the situation (3). Furthermore, by Remark 6.4.7 $\mathcal{V}^*(\phi)$ is a von Neumann algebra whose center is infinite dimensional. Then applying Proposition 6.2.6 shows that $\mathcal{V}^*(\phi)$ is finite.

In summary, except for (1)–(3) in Theorem 6.6.2, $\pi_1(\Omega, v)$ is an i.c.c. group. Since the holomorphic universal covering ϕ for (Ω, v) is always a regular branched covering map, the former part of Theorem 6.6.2 follows from Proposition 6.4.9. From the latter part of the proof of Theorem 6.4.1, it follows that whenever $V^*(\phi)$ is a type II_1 factor, $W^*(\phi)$ always is a type II_∞ factor. The proof is complete. $\square$

Remark 6.6.3 By Propositions 2.7 and 2.8 in [BMP], the group $\pi_1(\Omega, \Sigma, v)$ is isomorphic to

$$*_{w \in \Sigma} \mathbb{Z}_{v(w)} * \pi_1(\Omega).$$

Note that by Lemma 6.4.2, $\pi_1(\Omega)$ is either trivial or isomorphic to the free product of m integer groups $\mathbb{Z}$ with $1 \leq m \leq \infty$.

The situation (3) in Theorem 6.6.2 can happen, see the following suggested by Professor Qiu.

Example 6.6.4 For a fixed $r \in (0, 1)$, set

$$W_r = \{x + iy : \frac{x^2}{(r + \frac{1}{r})^2} + \frac{y^2}{(r - \frac{1}{r})^2} < 1, \ x, y \in \mathbb{R}\},$$

where i is the imaginary unit. Put $\Sigma = \{-1, 1\}$, and $v(-1) = v(1) = 2$.
Consider the Zhukovski function:

$$f(z) = \frac{1}{2}(z + \frac{1}{z}), \quad z \in V_r \triangleq \{z : r < |z| < \frac{1}{r}\}.$$

It is easy to know that f maps V_r onto W_r. Let ϕ_r be the covering map from $\mathbb{D}$ onto $\{r < |z| < 1\}$ defined by

$$\phi_r(z) = \exp\left(\ln r \ (\frac{i}{\pi} \ln \frac{1+z}{1-z} + \frac{1}{2})\right), \ z \in \mathbb{D}.$$

Write $\phi(z) = f(\frac{1}{r}\phi_{r^2}(z)), z \in \mathbb{D}$. Since W_r has exactly two ramified points of f; -1 and 1, at which local orders of f are 2, it is then not difficult to verify that $\phi : \mathbb{D} \to (W_r, \Sigma, v)$ is a holomorphic universal covering map.

Now we turn to special domains $\Omega = \mathbb{D} - E$, where E are relatively closed subsets of $\mathbb{D}$ with capacity zero. We have the following corollary of Theorem 6.6.2, due to Guo and Huang [GH2].

Corollary 6.6.5 *Suppose $\Omega = \mathbb{D} - E$, where E is a relatively closed subset of $\mathbb{D}$ with capacity zero and (Ω, v) is an orbifold domain with the singular locus Σ. Then there is a universal covering ϕ of (Ω, v), which is necessarily an inner function. Moreover, $V^*(\phi)$ is a type II_1 factor with the following exceptions:*

1. *Σ is empty and E is a singlet or empty. In this situation, Ω is the disk or the punctured disk, and ϕ is a covering map.*

2. $\Sigma = \{w_0\}$ *is a singlet and* E *is empty. In this situation,*

$$\phi(z) = c\varphi_{w_0}^n(z) \equiv c\Big(\frac{w_0 - z}{1 - \overline{w_0}z}\Big)^n,$$

where $|c| = 1$ *and* $n = \nu(w_0)$*;*
3. $\Sigma = \{w_0, w_1\}$ *with* $\nu(w_0) = \nu(w_1) = 2$*, and* $E = \emptyset$*.*

In Corollary 6.6.5, if $0 \notin E$, ϕ is always an interpolating Blaschke product, which is also a covering map. Remind that in Chap. 5 we have investigated another type of interpolating Blaschke products, thin Blascke products ϕ, for which $\mathcal{V}^*(\phi)$ are always abelian; and in most cases, $\mathcal{V}^*(\phi) = \mathbb{C}I$.

Remark 6.6.6 By using the same method in Sect. 6.5 of this chapter, one can prove that for any universal covering map $\phi : \mathbb{D} \to (\Omega, \Sigma, \nu)$, $\mathcal{V}^*(\phi)$ is $*$-isomorphic to the group von Neumann algebra $\mathcal{L}(\pi_1(\Omega, \Sigma, \nu))$.

6.7   Applications to Multi-variable Case

In the preceding sections, we deal essentially with covering maps. As we will see, these methods developed there can be naturally imported to multi-variable case under a mild setting.

Section 2.1 of Chap. 2 presents the definition of covering map, which can be naturally generalized in multi-variable case. Suppose Ω_0 and Ω are domains in $\mathbb{C}^d (d \geq 1)$. If $\Phi : \Omega_0 \to \Omega$ is a holomorphic map and every point of Ω has a connected open neighborhood U in Ω such that Φ maps each component of $\Phi^{-1}(U)$ bi-holomorphically onto U, then Φ is called a *holomorphic covering map*. If in addition Ω_0 is simply-connected, then Φ is called *universal*. A holomorphic covering map $\Phi : \Omega_0 \to \Omega$ is called *regular* if for any points $z, w \in \Omega_0$, $\Phi(z) = \Phi(w)$ implies there is a member ρ in $\mathrm{Aut}(\Omega_0)$ satisfying $\Phi \circ \rho = \Phi$ and $\rho(z) = w$. It is well-known that all universal covering maps are regular, see [GHa] and [Hat, Proposition 1.39]. In particular, when Ω is a bounded planar domain, and $\phi : \mathbb{D} \to \Omega$ is a holomorphic covering map, then ϕ is universal, and hence ϕ is regular.

Throughout this section, we make the following assumption:

$\Phi = (\phi_1, \cdots, \phi_d) : \Omega_0 \to \Omega$ is a holomorphic regular covering map, where Ω_0 and Ω are domains in $\mathbb{C}^d$.

In particular, if $\Phi : \Omega_0 \to \Omega$ is a holomorphic covering map and Ω_0 is simply-connected, then it is necessarily regular; and in this case, the fundamental group $\pi_1(\Omega)$ is isomorphic to the deck transformation group $G(\Phi)$ of Φ, which consists of all holomorphic automorphisms ρ satisfying $\Phi \circ \rho = \Phi$, see [GHa]. For each $\rho \in G(\phi)$, let $J\rho$ denote the determination for the Jacobian of ρ. Then

$$U_\rho h = h \circ \rho \, J\rho, \; h \in L_a^2(\Omega_0)$$

defines a unitary operator on the Bergman space $L_a^2(\Omega_0)$. Denote by $V^*(G(\Phi))$ the von Neumann algebra generated by $\{U_\rho; \rho \in G(\phi)\}$, and put

$$V^*(\Phi) = \{M_{\phi_1}, \cdots, M_{\phi_d}; M_{\phi_1}^*, \cdots, M_{\phi_d}^*\}'.$$

By using the techniques in Sect. 6.2 of this chapter, one can give the representation of operators in $V^*(\Phi)$. Precisely, for each operator S in $V^*(\Phi)$, there is a sequence $\{c_k\}$ in l^2 such that

$$Sh(z) = \sum_{k=0}^{\infty} c_k U_{\rho_k} h(z), \ h \in L_a^2(\Omega_0), \ z \in \Omega_0. \tag{6.22}$$

However, it is not yet known whether the coefficients c_k are unique.

Inspired by Theorem 6.2.2 and Lemma 6.2.4, we make some assumptions for operators in $V^*(\Phi)$.

I. For each operator S in $V^*(\Phi)$, the coefficients c_k in the formulation (6.22) are unique;
II. If an operator S in $V^*(\Phi)$ has the form (6.22), then

$$S^*h(z) = \sum_{k=0}^{\infty} \overline{c_k} U_{\rho_k}^* h(z), \ h \in L_a^2(\Omega_0), \ z \in \Omega_0,$$

where $U_{\rho_k}^* = U_{\rho_k^{-1}}$;
III. For any two operators S, T in $V^*(\phi)$ with

$$Sh(z) = \sum_{k=0}^{\infty} c_k U_{\rho_k} h(z) \quad \text{and} \quad Th(z) = \sum_{k=0}^{\infty} c_k' U_{\rho_k} h(z), \ h \in L_a^2(\Omega_0), \ z \in \Omega_0.$$

where both $\{c_k\}$ and $\{c_k'\}$ belonging to l^2, then we have

$$STh(z) = \sum_{i=0}^{\infty} d_i U_{\rho_i} h(z), h \in L_a^2(\Omega_0), z \in \Omega_0,$$

where

$$d_i = \sum_{\rho_j \circ \rho_k = \rho_i} c_k c_j'.$$

Roughly speaking, the conditions II and III means that the adjoint and composition of operators in $V^*(\Phi)$ equals their formal calculations.

In some special cases, conditions I–III hold. For example, if $d = 1$, $\Omega_0 = \mathbb{D}$ and Φ is a holomorphic covering map, then the conditions I–III are met. It seems that these conditions are also satisfied in a more general setting. For example, as pointed out by Professor Zheng, if there is a point $a \in \Omega_0$ such that $\{\rho_k(a)\}$ is a Blaschke sequence, then condition I is valid. The reasoning is similar as that in Sect. 6.2 of this chapter. There exists some possible connection between the conditions II, III and condition I.

We have the following lemma.

Lemma 6.7.1 *Suppose* $\Phi : \Omega_0 \to \Omega$ *is a holomorphic regular covering map and the conditions I–III hold. Then* $V^*(G(\Phi)) = V^*(\Phi)$. *That is,* $V^*(\Phi)$ *is generated by* U_ρ *with* $\rho \in G(\Phi)$. *In this case,* $V^*(\Phi)$ *is $*$-isomorphic to the group von Neumann algebra* $\mathcal{L}(G(\Phi))$.

The proof of Lemma 6.7.1 will be deferred a bit.

Let $L^2(\Omega_0)$ be the Hilbert space consisting of all Lebesgue-measurable functions over Ω_0, which are square integrable with respect to the Lebesgue measure dV. Set

$$u_{\rho_k} f = f \circ \rho_k \, J\rho_k, f \in L^2(\Omega_0),$$

The set $E = \{u_{\rho_k} : k = 0, 1 \cdots \}$ generates a von Neumann algebra, say $\mathcal{A}$. Let p denote the projection onto $L_a^2(\Omega_0)$, and then $p \in \mathcal{A}'$. Also, $\mathcal{A}p$ is a von Neumann algebra, that is generated by $\{U_{\rho_k} : k = 0, 1, \cdots \}$, where

$$U_{\rho_k} = p \, u_{\rho_k}|_{pL^2(\Omega)}, \, k = 0, 1, \cdots .$$

Below we will construct a unitary isomorphism which maps u_ρ^* to $L_\rho \otimes I$, complemented by a unitary operator W.

The idea is from [GHJ, Chap. 3]. As pointed out in [GHJ, Chap. 3], there always exists a subdomain Δ of Ω_0 such that

(a) $\Omega_0 - \cup_k \rho_k(\Delta)$ has null measure;
(b) $\rho_j(\Delta) \cap \rho_k(\Delta)$ has null measure for $j \neq k$.

Let dm be the restriction of the Lebesgue measure dV on Δ. There is a unitary operator W from $L^2(\Omega_0, dV)$ onto $l^2(G(\phi)) \otimes L^2(\Delta, dm)$, mapping each function f in $L^2(\Omega_0, dV)$ to $\sum_k \epsilon_{\rho_k} \otimes f_{\rho_k}$, where ϵ_{ρ_k} is the characteristic function of $\{\rho_k\}$ and

$$f_{\rho_k}(z) = (u_{\rho_k} f)(z), z \in \Delta.$$

Then for each $f \in L^2(\Omega_0, dA)$ and $\rho \in G(\phi)$, we have

$$Wu_\rho f = W(f \circ \rho \, J\rho)$$

$$= \sum_{k=0}^{\infty} \epsilon_{\rho_k} \otimes f_{\rho \circ \rho_k}$$

$$= \sum_{k=0}^{\infty} \epsilon_{\rho^{-1} \circ \rho_k} \otimes f_{\rho_k}$$

$$= (L_{\rho^{-1}} \otimes I)Wf.$$

Here $L_{\rho^{-1}}$ is the left regular representation as defined in Sect. 6.5 of this chapter. Replacing ρ with ρ^{-1} immediately gives

$$Wu_\rho^* = (L_\rho \otimes I)W. \tag{6.23}$$

Since both $\rho \mapsto u_{\rho*}$ and $\rho \mapsto L_\rho \otimes I$ are unitary representations, by (6.23) we deduce that the von Neumann algebra generated by u_{ρ_k} is unitarily isomorphic to the von Neumann algebra generated by $L_\rho \otimes I(\rho \in G(\phi))$; that is, $\mathcal{A}$ is unitarily isomorphic to $\mathcal{L}(G(\phi)) \otimes I$. Note that each operator L_f in $\mathcal{L}(G(\phi))$ can also be written as $L_f = \sum_{k=0}^\infty c_k L_{\rho_k}$, where

$$f = \sum_{k=0}^\infty c_k \epsilon_{\rho_k}.$$

This operator L_f naturally corresponds to two operators: $\sum_{k=0}^\infty c_k u_{\rho_k}^*$ in $\mathcal{A}$, and $\sum_{k=0}^\infty c_k U_{\rho_k}^*$ in $\mathcal{A}p$. Conversely, each operator in $\mathcal{A}$ or $\mathcal{A}p$ must have the form as above.

Proof of Lemma 6.7.1 Now assume that $\Phi : \Omega_0 \to \Omega$ is a holomorphic regular covering map and the conditions I–III hold. Given a unitary operator U in $\mathcal{V}^*(\Phi)$, write

$$Uh(z) = \sum_{k=0}^\infty c_k U_{\rho_k}^* h(z).$$

This operator gives rise to an $f \in l^2(G(\Phi))$:

$$f = \sum_{k=0}^\infty c_k \epsilon_{\rho_k}.$$

Since U is a unitary operator, $f^* * f = \epsilon_e$. Put

$$L_f = \sum_{k=0}^\infty c_k L_{\rho_k},$$

which holds on the linear span of all ϵ_{ρ_k}, a dense subspace of $l^2(G(\Phi))$. Then by Proposition 6.5.2, L_f is a unitary operator, and hence a bounded operator in $\mathcal{L}(G(\phi))$. Thus, the above reasoning shows that each unitary operator U in $\mathcal{V}^*(\Phi)$ induces a unitary operator L_f.

Condition I shows that for each $S \in \mathcal{V}^*(\phi)$, there exists a unique $f \in l^2(G(\phi))$ formally satisfying

$$S = \Theta(f) \triangleq \sum_{k=0}^\infty c_k U_{\rho_k}^*.$$

The above sum does not mean convergence in weak or strong operator topology. As done in Sect. 6.5 of this chapter, define a linear map Λ on $\mathcal{V}^*(\phi)$ as follows:

$$\Lambda : \Theta(f) \mapsto L_f;$$

that is,

$$\Lambda : \sum_k f(\rho_k) U_{\rho_k}^* \mapsto \sum_k f(\rho_k) L_{\rho_k}.$$

By condition I, Λ is well-defined and injective. In the above paragraph, it has been shown that Λ maps each unitary operator in $\mathcal{V}^*(\Phi)$ to another in $\mathcal{L}(G(\phi))$. Since a von Neumann algebra is the finite linear span of its unitary operators, Λ maps each operator in $\mathcal{V}^*(\Phi)$ into $\mathcal{L}(G(\phi))$.

By conditions II and III, one can verify that Λ is a *-homomorphism between the von Neumann algebras $\mathcal{V}^*(\Phi)$ and $\mathcal{L}(G(\phi))$. Since each U_{ρ_k} is in $\mathcal{V}^*(\Phi)$ and $\mathcal{V}^*(G(\Phi))$ is generated by

$$\{U_{\rho_k} : k \geq 0\},$$

$\mathcal{V}^*(G(\Phi))$ is a *-subalgebra of $\mathcal{V}^*(\Phi)$. Then by the discussions below Lemma 6.7.1, it is not difficult to show that $\Lambda\left(\mathcal{V}^*(G(\Phi))\right) = \mathcal{L}(G(\phi))$, and hence Λ is onto. Since Λ is also injective, Λ is an isomorphism from the von Neumann algebra $\mathcal{V}^*(\Phi)$ onto $\mathcal{L}(G(\phi))$. Note that $\Lambda^{-1}(\mathcal{L}(G(\phi))) = \mathcal{V}^*(G(\Phi))$, forcing $\mathcal{V}^*(\Phi) = \mathcal{V}^*(G(\phi))$. The proof of Lemma 6.7.1 is complete.   □

In Sect. 5.6 of this chapter, when studying thin Blaschke products B, we get the following property of B:

For each $z \in E$, there is an open neighborhood $U(U \subseteq E)$ of z that admits a complete local inverse, such that $\mathbb{C} - \overline{B^{-1} \circ B(U)}$ is connected.

Inspired by this property and the methods in Sect. 5.6 of this chapter, we give the following condition for the holomorphic regular (branched) covering map Φ:

There is an open neighborhood U such that $\mathbb{C} - \overline{\Phi^{-1} \circ \Phi(U)}$ is connected.

In this case, we say that Φ has *Runge's property*. Then as done in Sect. 5.6 of this chapter, by lifting those operators S in $\mathcal{V}^*(\Phi)$ to $\tilde{S}$ defined on $L^2(U)$(regarded as a subspace of $L^2(\Omega_0)$), one can verify I–III under the condition that Φ has the Runge's property in single-variable case.

Corollary 6.7.2 *Suppose $\Omega_0 \subseteq \mathbb{C}, \Omega \subseteq \mathbb{C}$, and $\Phi : \Omega_0 \to \Omega$ is a holomorphic regular covering map satisfying Runge's property. Then $\mathcal{V}^*(G(\Phi)) = \mathcal{V}^*(\Phi)$. That is, $\mathcal{V}^*(\Phi)$ is generated by U_ρ with $\rho \in G(\Phi)$.*

In the case of $\Omega_0 = \mathbb{D}$, we do not know whether all holomorphic covering maps $\phi : \mathbb{D} \to \Omega$ satisfy Runge's property. The following example shows that if $\Omega_0 \neq \mathbb{D}$, then there is an exception even if $d = 1$.

Example 6.7.3 Let Ω be the complex plane $\mathbb{C}$ minus discrete points. In particular, put $\Omega = \mathbb{C} - \{0, 1\}$, and let $\phi : \mathbb{D} \to \Omega$ be a holomorphic covering map, called the modular function. For the definition and more details of modular function, refer to [Ap, Chap. 2], [Og], also see [Mil, p. 48]. Let K be a small closed disk centered at a, and put $\Omega_0 = \mathbb{D} - \phi^{-1}(K)$. Then

$$\phi|_{\Omega_0} : \Omega_0 \to \Omega - K$$

is also a covering map with the same deck transformation group as ϕ. Write $\psi = (\phi - a)^{-1}|_{\Omega_0}$, a bounded covering map. By the geometric property of ϕ, for each point $w \in \mathbb{C}$, the closure of $\phi^{-1}(w)$ contains $\mathbb{T}$. This immediately implies that ψ does not satisfy Runge's property. For this map ψ, we do not know whether Condition I holds.

Stimulated by Example 6.7.3, we will go a bit beyond the theme of this section and study the restriction of a covering map.

For a bounded holomorphic covering map $\phi : \mathbb{D} \to \Omega$ and an open subset G of Ω, write $\phi^{-1}(G) = G_0$. Then $\phi|_{G_0} : G_0 \to G$ is also a regular covering map. Let $\mathcal{W}^*(\phi, G_0)$ be the von Neumann algebra generated by the multiplication operator M_ϕ on $L_a^2(G_0)$, and set $\mathcal{V}^*(\phi, G_0) \triangleq \mathcal{W}^*(\phi, G_0)'$. By the methods in this chapter, it is not difficult to prove that $\mathcal{V}^*(\phi, G_0)$ is unitarily isomorphic to $\mathcal{V}^*(\phi)$.

The following shows that in a different situation $\mathcal{V}^*(\phi, G_0)$ can be trivial.

Proposition 6.7.4 *Suppose $\phi : \mathbb{D} \to \Omega$ is a holomorphic covering map, and K is a compact subset of $\mathbb{D}$ whose interior is not empty. Then the von Neumann algebra $\mathcal{V}^*(\phi, \mathbb{D} - K)$ is trivial.*

Proof Suppose $\phi : \mathbb{D} \to \Omega$ is a holomorphic covering map, and K is a compact subset of $\mathbb{D}$ with an interior point a. Since K is compact, so is $\phi(K)$. Then there is a point $z_0 \in \mathbb{D}$ such that $w_0 = \phi(z_0) \notin \phi(K)$. Since ϕ is a covering map, there is an enough small disk V centered at w_0 such that $V \cap \phi(K) = \emptyset$ and

$$\phi^{-1}(V) = \bigsqcup_{i=0}^{\infty} V_i,$$

where $\phi|_{V_i} : V_i \to V$ are biholomorphic for all i. Then by the proof of Theorem 6.2.2, for each operator $S \in \mathcal{V}^*(\phi)$, there is a sequence $\{c_k\} \in l^2$ such that

$$Sh(z) = \sum_k c_k h \circ \rho_k(z) \rho_k'(z), h \in L_a^2(\mathbb{D} - K), z \in V_0, \tag{6.24}$$

where all ρ_k run over the deck transformation group $G(\phi)$. By Lemma 6.2.1, $\{\rho_j(a)\}$ is an interpolating sequence, forcing $\lim_{j \to \infty} |\rho_j(a)| = 1$. Also noting that there is an enough small disk Δ such that $a \in \Delta \subseteq K$, we conclude that $\{\rho_j(\Delta)\}$ tends to $\mathbb{T}$;

that is,

$$\lim_{j\to\infty} \min\{|\rho_j(z)| : z \in \overline{\Delta}\} = 1.$$

Since K is compact, only finitely many $\rho_j^{-1}(\Delta)$ has intersection with K, say

$$\rho_0^{-1}(\Delta), \rho_1^{-1}(\Delta), \cdots, \rho_n^{-1}(\Delta).$$

Suppose conversely there is an operator $S \in V^*(\phi)$ such that in (6.24) $c_{i_0} \neq 0$ for some $i_0 > n$. Set

$$f = \frac{1}{z-a} \in H^\infty(\mathbb{D} - K).$$

Then by (6.24),

$$Sf(z) = \sum_k c_k f \circ \rho_k(z) \rho_k'(z), z \in V_0. \tag{6.25}$$

One can show that the right-hand side of (6.25) is a holomorphic function in $\mathbb{D} - \phi^{-1}(\phi(a))$, and thus

$$Sf(z) = \sum_k c_k f \circ \rho_k(z) \rho_k'(z), \ z \in \rho_{i_0}^{-1}(\Delta) - \{\rho_{i_0}^{-1}(a)\}.$$

That is,

$$Sf - \sum_{k \neq i_0} c_k f \circ \rho_k(z)\rho_k'(z) = c_{i_0} f \circ \rho_{i_0} \rho_{i_0}'(z), \ z \in \rho_{i_0}^{-1}(\Delta) - \{\rho_{i_0}^{-1}(a)\}.$$

Observe that the left-hand side is holomorphic at $\rho_{i_0}^{-1}(a)$. However, since $f = \dfrac{1}{z-a}$, the right-hand side is unbounded in any neighborhood of $\rho_{i_0}^{-1}(a)$, which is a contradiction. Therefore,

$$c_j = 0, j \geq n.$$

Thus, it is shown that for any $S \in V^*(\phi)$, there is a finite sequence $\{c_j\}_{j=1}^n$ such that

$$Sh(z) = \sum_{k=1}^n c_k h \circ \rho_k(z)\rho_k'(z), z \in V_0.$$

Note that the coefficients are uniquely determined when h run over polynomials. Rewrite

$$Sp(z) = \sum_{k=1}^{n} c_k p \circ \rho_k(z) \rho_k'(z), \ p \in \mathbb{C}[z] \tag{6.26}$$

Therefore, $\dim V^*(\phi, \mathbb{D} - K) < \infty$.

Below we conclude that S equals a constant tuple of I. To see this, first consider the case that $G(\phi)$ is the infinite cyclic group generated by ρ. In the representation (6.26), put $\rho_k = \rho^j$, where j is largest possible one satisfying $c_k \neq 0$. Then by a simple calculation, the representation of S^2 must contain the term $U_{\rho^{2j}}$, and etc. Thus, $S, S^2, \cdots$, are linearly independent, which is a contradiction to the fact $\dim V^*(\phi, \mathbb{D} - K) < \infty$. Therefore, $V^*(\phi, \mathbb{D} - K)$ is trivial. In general, $G(\phi)$ is a free group, which can be dealt with in a similar way. The proof is complete. □

Remark 6.7.5 The condition on K can reduce to the following: $\mathbb{D} - K$ is biholomorphic to a multiply-connected domain whose "holes" contains an open disk. For example, if K equals the union of two segments in $\mathbb{D}$, then $\mathbb{D} - K$ is biholomorphic to $\mathbb{D}$ minus two disjoint smaller disks in $\mathbb{D}$, as is well known. In this case, $V^*(\phi, \mathbb{D} - K)$ is trivial.

It is also worthwhile to mention that if ϕ is a finite Blaschke product (a branched covering map), then by a careful choice of K, it can happen that $V^*(\phi, \mathbb{D} - K)$ is nontrivially abelian, and $\dim V^*(\phi, \mathbb{D} - K) = \text{order } \phi$. For details, see Remark 4.2.6.

A multi-variable version of Proposition 6.7.4 fails, see the following example.

Example 6.7.6 Set $\Phi = (p, q)$ with $p = z^4 w^4$ and $q = z^4 + w^4$. Write $\omega = \sqrt{-1}$, $\varepsilon = 10^{-2}$ and $z_0 = \frac{1}{3}$. Denote by E of all points $(z, w) \in \mathbb{C}^2$ such that $|z - \omega^j z_0| + |w - \omega^k z_0| \leq \varepsilon$ for some $(j, k) \in \mathbb{Z}^2$. Note that E is exactly the union of finitely many closed balls in $\mathbb{B}_2$. Put $\Omega_0 = \mathbb{B}_2 - E$, and consider the von Neumann algebra $V^*(\Phi)$ over $L_a^2(\Omega_0)$. Then one can show that $V^*(\Phi, \Omega_0)$ is a non-abelian finite dimensional von Neumann algebra generated by unitary operators U_ρ with the following forms:

$$U_\rho : f(z, w) \rightarrow f(\omega^j z, \omega^k w), f \in L_a^2(\Omega_0)$$

or

$$U_\rho : f(z, w) \rightarrow f(\omega^j w, \omega^k z), f \in L_a^2(\Omega_0)$$

where j and k are integers. Clearly, $V^*(\Phi, \Omega_0)$ is nontrivial.

To end this section, we provide the following conjecture.

Conjecture 6.7.7 *Given finite points $w_1, \cdots, w_k$ in $\mathbb{D}$, put*

$$\Omega = \mathbb{D} - \{w_1, \cdots, w_k\}.$$

Let $\phi : \mathbb{D} \to \Omega$ be a holomorphic covering map. If B is a finite Blaschke product with order $B \geq 2$, then $V^*(B \circ \phi)$ is a type II_1 factor. Furthermore, $V^*(B \circ \phi)$ is never $*$-isomorphic to any free group von Neumann factor $\mathcal{L}(F_n)(2 \leq n \leq \infty)$.

It may be of interest to consider the structures of $V^*(B \circ \phi)$ and $V^*(\phi \circ B)$.

Part II: In the Case of Weighted Bergman Spaces

In the following, new techniques will be introduced to generalize those results in Part I to the case of weighted Bergman spaces. In addition, a class of group-like von Neumann algebras are constructed, which are shown to be $*$-isomorphic to group von Neumann algebras.

6.8 Representation of Operators in $V_\alpha^*(\phi)$

This section first introduces some notations from [Huang2]. As done in Sect. 2.4 of Chap. 2, let $L_{a,\alpha}^2(\mathbb{D})$ denote the *weighted Bergman space*, which consists of all holomorphic functions f over $\mathbb{D}$ satisfying

$$\|f\|^2 \triangleq (\alpha + 1) \int_{\mathbb{D}} |f(z)|^2 (1 - |z|^2)^\alpha dA(z) < \infty.$$

When $\alpha = 0$, $L_{a,0}^2(\mathbb{D})$ is the usual Bergman space $L_a^2(\mathbb{D})$. In this section, the weighted Bergman space $L_{a,\alpha}^2(\mathbb{D})$ is always fixed; for any bounded holomorphic function ϕ over $\mathbb{D}$, M_ϕ denotes the multiplication operator on $L_{a,\alpha}^2(\mathbb{D})$. In what follows, let $\mathcal{W}_\alpha^*(\phi)$ be the von Neumann algebra generated by M_ϕ and write

$$V_\alpha^*(\phi) \triangleq \mathcal{W}_\alpha^*(\phi)',$$

the commutant algebra of $\mathcal{W}_\alpha^*(\phi)$. If $\alpha = 0$, we also write $\mathcal{W}^*(\phi)$ for $\mathcal{W}_0^*(\phi)$, and $V^*(\phi)$ for $V_0^*(\phi)$.

This section will present the weighted-Bergman-space versions of those results in Sect. 6.2 of this chapter, including the representation of operators in $V_\alpha^*(\phi)$. It is shown that $V_\alpha^*(\phi)$ is generated by some unitary operators U_ρ which arise from those members ρ in $G(\phi)$.

It is well-known that the reproducing kernel of $L_{a,\alpha}^2(\mathbb{D})$ at w is given by

$$K_w^\alpha(z) = \frac{1}{(1 - \overline{w}z)^{2+\alpha}},$$

and its normalized kernel is defined by

$$k_w^\alpha(z) \triangleq \frac{K_w^\alpha(z)}{\sqrt{K_w^\alpha(w)}} = \frac{(1 - |w|^2)^{\frac{2+\alpha}{2}}}{(1 - \overline{w}z)^{2+\alpha}},$$

where

$$(1 - \overline{w}z)^{2+\alpha} = \exp\big((2 + \alpha)\ln(1 - \overline{w}z)\big), \text{ with } \ln 1 = 0.$$

Note that for each $\rho \in \text{Aut}(\mathbb{D})$, the linear map

$$h \mapsto k_{\rho^{-1}(0)}^\alpha \, h \circ \rho$$

defines a unitary operator on $L_{a,\alpha}^2(\mathbb{D})$. In addition, from

$$\varphi_w(z) = \frac{w - z}{1 - \overline{w}z},$$

we get $k_w^\alpha = (-\varphi_w')^{\frac{2+\alpha}{2}}$. Now, *we can make the following convention: if $\alpha = 0$ mod 2, define*

$$U_\rho h = (\rho')^{\frac{2+\alpha}{2}} h \circ \rho, \; h \in L_{a,\alpha}^2(\mathbb{D}); \tag{6.27}$$

otherwise, put

$$U_\rho h = k_{\rho^{-1}(0)}^\alpha h \circ \rho, \; h \in L_{a,\alpha}^2(\mathbb{D}). \tag{6.28}$$

In both cases, U_ρ is a unitary operator and $U_{id} = I$. The difference between (6.27) and (6.28) is exactly a unimodular constant multiple.

As before, for a regular branched covering map $\phi : \mathbb{D} \to \Omega$, $G(\phi)$ denotes the deck transformation group of ϕ, and we write

$$G(\phi) = \{\rho_k : k = 0, 1, \cdots\}.$$

For each $w \in \mathbb{D}$, set

$$\pi_w : h \mapsto \{U_{\rho_k} h(w)\}, \; h \in L_{a,\alpha}^2(\mathbb{D}),$$

which will turn out to be a bounded operator from the weighted Bergman space to l^2 or $\mathbb{C}^n$. Our first aim is to establish a similar version of Theorem 6.2.2 as follows, due to Huang [Huang2].

Theorem 6.8.1 *Suppose $\phi : \mathbb{D} \to \Omega$ is a holomorphic regular branched covering map, and S is a unitary operator, which commutes with M_ϕ on the weighted*

Bergman space $L^2_{a,\alpha}(\mathbb{D})$. If $G(\phi)$ is infinite (or finite), then there is a unique operator $W : l^2 \to l^2$ (or $W : \mathbb{C}^n \to \mathbb{C}^n$) such that

$$W\pi_w(h) = \pi_w(Sh), \ h \in L^2_{a,\alpha}(\mathbb{D}), \ w \in \mathbb{D}.$$

This W is necessarily a unitary operator. Moreover, there is a unique square-summable sequence $\{c_k\}$ satisfying

$$Sh(z) = \sum_{k=0}^{\infty} c_k U_{\rho_k} h(z), \ h \in L^2_{a,\alpha}(\mathbb{D}), \ z \in \mathbb{D},$$

where ρ_k run over $G(\phi)$.

Recall that $\mathcal{E}_\phi$ is the critical value set of ϕ; that is,

$$\mathcal{E}_\phi = \{\phi(z) : \text{ there is a } z \in \mathbb{D} \text{ such that } \phi'(z) = 0\}.$$

In Sect. 6.2 of this chapter, it was mentioned that $\mathcal{E}_\phi$ is a discrete subset of $\mathbb{D}$. Before establishing Theorem 6.8.1, let us make an observation: *for each $\lambda \in \mathbb{D} - \phi^{-1}(\mathcal{E}_\phi)$,*

$$h \mapsto \{(1 - |\rho_j(\lambda)|^2)^{\frac{2+\alpha}{2}} h(\rho_j(\lambda))\}$$

defines a bounded invertible linear map from $\left[(\phi - \phi(\lambda))L^2_{a,\alpha}(\mathbb{D})\right]^{\perp}$ onto l^2. The reasoning is as follows. For a given $\lambda \in \mathbb{D} - \phi^{-1}(\mathcal{E}_\phi)$, Lemma 6.2.1 states that $\{\rho_k(\lambda)\}$ is an $H^\infty(\mathbb{D})$-interpolating sequence; equivalently, $\{\rho_k(\lambda)\}$ is a uniformly separated sequence. Let B denote the Blaschke product for $\{\rho_k(\lambda)\}$. By Proposition 2.1.17,

$$\left|\frac{\phi - \phi(\lambda)}{B}\right|$$

is bounded and bounded below, forcing

$$\left[(\phi - \phi(\lambda))L^2_{a,\alpha}(\mathbb{D})\right]^{\perp} = L^2_{a,\alpha}(\mathbb{D}) \ominus BL^2_{a,\alpha}(\mathbb{D}).$$

Then applying Proposition 2.4.5 gives our conclusion, as desired. Using this result and following the discussions in Sect. 6.2 of this chapter yield Theorem 6.8.1.

A similar version of Corollary 6.2.5 is stated as follows.

Proposition 6.8.2 *On the weighted Bergman space $L^2_{a,\alpha}(\mathbb{D})(\alpha > -1)$, the von Neumann algebra $\mathcal{V}^*_\alpha(\phi)$ equals the SOT-closure (or WOT-closure) of span $\{U_\rho : \rho \in G(\phi)\}$.*

We intend to generalize all results in Sects. 6.2–6.6 of this chapter to $L^2_{a,\alpha}(\mathbb{D})(\alpha > -1)$. However, there is some trouble here. In more detail, given

$G(\phi) = \{\rho_j\}$, even if we redefine all U_{ρ_j}, it seems difficult to tell whether

$$U_{\rho_j} U_{\rho_i} = U_{\rho_i \circ \rho_j}.$$

This shows that we can not directly apply the proofs in Sects. 6.2–6.6 of this chapter.

Next, we present a complete characterization for the commutativity of $\mathcal{V}_\alpha^*(\phi)$, which has nothing to do with the parameter α [Huang2].

Theorem 6.8.3 *Suppose $\phi : \mathbb{D} \to \Omega$ is a holomorphic regular branched covering map. Then $\mathcal{V}_\alpha^*(\phi)$ is abelian if and only if $G(\phi)$ is abelian. In particular, if ϕ is a holomorphic covering map, then the following are equivalent:*

(1) *the commutant algebra $\mathcal{V}_\alpha^*(\phi)$ is abelian;*
(2) *the fundamental group $\pi_1(\Omega)$ of Ω is abelian;*
(3) *the range Ω of ϕ is conformally isomorphic to the disk, annuli or the punctured disk.*

Before we give the proof of Theorem 6.8.3, some words are in order. For the above map ϕ, there is a clear description for when its deck transformation group $G(\phi)$ is abelian, in the language of holomorphic regular branched covering map as below. One can also refer to Remark 6.6.3.

Remark 6.8.4 The deck transformation group $G(\phi)$ is abelian if and only if one of the following holds:

(1) ϕ is a holomorphic covering map, and $\pi_1(\Omega)$ is abelian;
(2) there is a biholomorphic map $\psi : \mathbb{D} \to \Omega$ and a $\rho \in \mathrm{Aut}(\mathbb{D})$ such that $\phi(z) = \psi(\rho(z)^n)$ with $n \geq 2$.

Proof of Theorem 6.8.3 The proof is from [Huang2].

Assume that ϕ is a holomorphic regular branched covering map and rewrite $\Gamma = G(\phi)$, the deck transformation group of ϕ. By the proof of Theorem 6.6.2, Γ is abelian if and only if Γ is trivial or Γ is a cyclic group (finite or infinite). In this situation, write $\Gamma = \{\rho_j\}$ and denote its generator by τ. Then it is not difficult to show that each U_{ρ_j} equals a constant tuple of U_τ^n for some integer n. This, combined with Proposition 6.8.2, shows that $\mathcal{V}_\alpha^*(\phi)$ is abelian if $G(\phi)$ is abelian. The inverse direction is trivial.

The remaining part follows from Theorem 6.3.2. $\qquad\qquad\square$

It remains to deal with the case of orbifold Riemann domains. However, it is not easy to tell whether a similar version of Corollary 6.4.6 is true or not. Fortunately, we still have a generalized version of Theorem 6.6.2, stated as follows.

Theorem 6.8.5 *Suppose ϕ is a holomorphic universal covering of the orbifold Riemann surface (Ω, ν) with the cone set $\sum$. Then for each $\alpha > -1$, $\mathcal{V}_\alpha^*(\phi)$ is a type II_1 factor with the following exceptions:*

1. *Σ is empty and $\pi_1(\Omega)$ is abelian. In this case, Ω is conformally isomorphic to the disk, annuli or the punctured disk, and ϕ is a covering map.*

2. $\Sigma = \{w_0\}$ is a singlet and $\Omega \cong \mathbb{D}$. In this case, $\phi = \psi(\varphi_\alpha^n)(\alpha \in \mathbb{D})$, where ψ is a conformal map from $\mathbb{D}$ onto Ω, $\psi(0) = w_0$ and $n = \nu(w_0)$.

3. $\Sigma = \{w_0, w_1\}$ with $\nu(w_0) = \nu(w_1) = 2$ and $\Omega \cong \mathbb{D}$. In this case, $\mathcal{V}_\alpha^*(\phi)$ is a finite von Neumann algebra whose center is infinite dimensional.

Proof The proof comes from [Huang2].

In the proof of Theorem 6.6.2, it is shown that except for the three cases in Theorem 6.8.5, the deck transformation group $\pi_1(\Omega, \nu)$ is always an i.c.c. group. Then by the same discussions, one can show that $\mathcal{V}_\alpha^*(\phi)$ is a type II$_1$ factor. As for case (1) and (2), $\pi_1(\Omega, \nu)$ is a cyclic group, and then by Proposition 6.8.2 $\mathcal{V}_\alpha^*(\phi)$ is an abelian von Neumann algebra. Thus it remains to deal with case (3), where $\pi_1(\Omega, \nu)$ is isomorphic to $\mathbb{Z}_2 * \mathbb{Z}_2$.

In case (3), $\pi_1(\Omega, \nu)$ is generated by two members in Aut$(\mathbb{D})$: ρ and τ, with the relation $\rho^2 = e$ and $\tau^2 = e$, where $\rho^2 \triangleq \rho \circ \rho$, $\rho\tau \triangleq \rho \circ \tau$, and etc. Observe that for any positive integer n, $\{(\rho\tau)^n, (\tau\rho)^n\}$ is a finite conjugacy class in $\pi_1(\Omega, \nu)$. Soon we will see that *for each n, $\{(\rho\tau)^n, (\tau\rho)^n\}$ induces an operator $U_{(\rho\tau)^n} + cU_{(\tau\rho)^n}$ in the center of $\mathcal{V}_\alpha^*(\phi)$.*

For this, write $a = (\rho\tau)^n$ and consider the conjugacy class $\{a, a^{-1}\}$, where $a^{-1} = (\tau\rho)^n$. There must be a unique unimodular constant ξ_ρ satisfying

$$U_\rho U_a U_\rho^* = \xi_\rho U_{a^{-1}},$$

and thus

$$U_\rho U_{a^{-1}} U_\rho^* = \overline{\xi_\rho} U_a.$$

Then it follows that $U_a + \xi_\rho U_{a^{-1}}$ commutes with U_ρ. Since $U_a = U_{\rho\tau}^n$ and $U_a^* = U_{a^{-1}}$, we deduce that $U_a + \xi_\rho U_{a^{-1}}$ also commutes with $U_{\rho\tau}$. Note that for any $b \in \Gamma - \{e\}$, U_b is always a constant multiple of a finite product of U_ρ and $U_{\rho\tau}$, and hence $U_a + \xi_\rho U_{a^{-1}}$ commutes with U_b. By Proposition 6.8.2, the WOT-closure of all U_b equals $\mathcal{V}_\alpha^*(\phi)$, and hence $U_a + \xi_\rho U_{a^{-1}}$ lies in the center of $\mathcal{V}_\alpha^*(\phi)$. Thus in case (3) $\mathcal{V}_\alpha^*(\phi)$ has a nontrivial center of infinite dimension. The proof of Theorem 6.8.5 is complete.   □

6.9   The Structure of $\mathcal{V}_\alpha^*(\phi)$

In this section, all results on $L_a^2(\mathbb{D})$ in Sects. 6.2–6.6 of this chapter will be generalized to $L_{a,\alpha}^2(\mathbb{D})(\alpha > -1)$. In particular, we have the following theorem as the main result in this section [Huang2].

Theorem 6.9.1 *Assume that $\phi : \mathbb{D} \to \Omega$ is a bounded holomorphic regular branched covering map. Then the von Neumann algebras $\mathcal{V}_\alpha^*(\phi)$ is *-isomorphic to the group von Neumann algebra $\mathcal{L}(G(\phi))$.*

If $\phi : \mathbb{D} \to \Omega$ is a bounded holomorphic covering map, then $\mathcal{L}(G(\phi)) = \mathcal{L}(\pi_1(\Omega))$ since $G(\phi) \cong \pi_1(\Omega)$.

It is not difficult to see that once Theorem 6.9.1 is proved, all results in Sects. 6.2–6.6 of this chapter will be carried out without any difficulty. The main difficulty in proving Theorem 6.9.1 lies in the defect of the definition of U_ρ, see Sect. 6.2 of this chapter. Precisely, there is no evidence showing that the map $\rho \mapsto U_\rho$ is a unitary representation. We need the following claim from [Huang2].

Claim For each regular branched covering map $\phi : \mathbb{D} \to \Omega$, by assigning a suitable unimodular constant ξ_ρ to U_ρ, the map

$$\rho \mapsto \xi_\rho U_\rho, \rho \in G(\phi)$$

defines a unitary representation from $G(\phi)$ to unitary operators on $L^2_{a,\alpha}(\mathbb{D})$.

Before giving the proof of the claim, we will make an observation. For $\alpha = 0, 2, 4, \cdots$, by our convention (6.27)

$$\rho \mapsto U_\rho, \rho \in G(\phi)$$

is already a unitary representation. More generally, for $\alpha = 0, 1, 2, \cdots$, see [Lang1, pp. 185–187] for a unitary representation. Below, we will see that the claim holds in a general case.

Proof of the claim The proof is from [Huang2].

To prove the claim, it suffices to construct a representation π of $G(\phi)$, such that

$$\pi(\rho) = \xi_\rho U_\rho, \ \rho \in G(\phi),$$

where ξ_ρ are unimodular constants.

For this, note that the deck transformation $G(\phi)$ is always the free product of some cyclic groups, with each isomorphic to $\mathbb{Z}$ or $\mathbb{Z}_k$, see Remark 6.6.3. Therefore, there is a sequence of generators:

$$\rho_1, \cdots, \rho_n, \cdots,$$

where all $\rho_n \in G(\phi)$. We define each U_{ρ_n} as in (6.28). In general, for each member ρ in $G(\phi)$, let

$$\rho = \rho_{i_1}^{n_1} \cdots \rho_{i_k}^{n_k}, \quad n_i \in \mathbb{Z} - \{0\}, 1 \le i \le k \tag{6.29}$$

be its unique reduced form. Then set $\pi(\rho) \triangleq U_{\rho_{i_1}}^{n_1} \cdots U_{\rho_{i_k}}^{n_k}$. It is straightforward to check that π gives a unitary representation, and that $\pi(\rho)$ is a unimodular constant tuple of U_ρ. Therefore, the proof of the claim is complete.

In addition, we mention that such a representation π is not unique, since one may redefine each U_{ρ_n} by attaching any unimodular constant to it. $\qquad\qquad\square$

With the claim proved, by applying the discussion in Sects. 6.2–6.6 of this chapter one can give Theorem 6.9.1. In particular, if $\phi : \mathbb{D} \to \Omega$ is a holomorphic covering map, then $G(\phi) \cong \pi_1(\Omega)$. By Theorem 6.9.1, $\mathcal{V}^*(\Omega)$ is $*$-isomorphic to $\mathcal{L}(F_n)$, where n is the number of generators of $\pi_1(\Omega)$.

The following context comes from [Huang2]. Below, we will use the above claim to give a shorter proof of Theorem 6.9.1 in the case when ϕ is a holomorphic covering map and $\pi_1(\Omega)$ is not abelian. The approach presented below is different from the methods in Sects. 6.2–6.6 of this chapter.

Two lemmas will be established here. One is from [Jon, Exercise 3.4.6].

Lemma 6.9.2 *If $\mathcal{A}$ is a von Neumann algebra generated by the self-adjoint, multiplicative closed set E, and p is a (self-adjoint) projection in $\mathcal{A}$ or $\mathcal{A}'$, then pEp generates the von Neumann algebra $p\mathcal{A}p$.*

For the following lemma, see [Jon, Corollary 3.4.4].

Lemma 6.9.3 *Suppose $\mathcal{A}$ is a factor on the underlying Hilbert space $\mathcal{H}$, and $p \in \mathcal{A}$ is a self-adjoint projection. Then $p\mathcal{A}p$ is a factor on $p\mathcal{H}$, and so is $p\mathcal{A}'$. Moreover, the map $x \mapsto xp$ from $\mathcal{A}'$ to $\mathcal{A}'p$ is a $*$-isomorphism.*

We only discuss the case $\alpha = 0$, and the remaining is similar. As done below Lemma 6.7.1, let $L^2(\mathbb{D})$ be the Hilbert space consisting of all Lebesgue-measurable functions over $\mathbb{D}$, which are square integrable with respect to dA. Set

$$u_{\rho_k} f = f \circ \rho_k \, \rho_k', f \in L^2(\mathbb{D}),$$

and put $E = \{u_{\rho_k} : k = 0, 1 \cdots\}$. Clearly, E is a self-adjoint, multiplicative closed set, which generates a von Neumann algebra, say $\mathcal{A}$. Denote by p the orthogonal projection from $L^2(\mathbb{D})$ onto $L_a^2(\mathbb{D})$. Clearly, p is in the commutant of E, which implies that $p \in \mathcal{A}'$. By Lemma 6.9.2, the von Neumann algebra $\mathcal{A}p$ is generated by pEp; that is, $\mathcal{A}p$ is generated by $\{U_{\rho_k} : k = 0, 1, \cdots\}$, where

$$U_{\rho_k} = p \, u_{\rho_k}|_{pL^2(\mathbb{D})}, \quad k = 0, 1, \cdots.$$

Then by Proposition 6.8.2, $\mathcal{A}p = \mathcal{V}^*(\phi)$. Besides, by a similar discussion on [GHJ, pp. 143–145], one will get

$$\mathcal{A}p \overset{*}{\cong} \mathcal{A} \overset{*}{\cong} \mathcal{L}(G(\phi)).$$

For the sake of completeness, we provide the details for the discussion on [GHJ, pp. 143–145]. As pointed out in [GHJ, Chap. 3], there always exists a domain $\Delta \subseteq \mathbb{D}$ such that

(1) $\mathbb{D} - \cup_k \rho_k(\Delta)$ has null measure;
(2) $\rho_j(\Delta) \cap \rho_k(\Delta)$ has null measure for $j \neq k$.

Let dm be the restriction of the area measure dA on Δ. There is a unitary operator W from $L^2(\mathbb{D}, dA)$ onto $l^2(G(\phi)) \otimes L^2(\Delta, dm)$, which maps f to $\sum_k \epsilon_{\rho_k} \otimes f_{\rho_k}$, where ϵ_{ρ_k} is the characteristic function of $\{\rho_k\}$ and $f_{\rho_k}(z) = u_{\rho_k} f(z)$, $z \in \Delta$. By (6.23), we have

$$Wu_\rho^* = (L_\rho \otimes I)W. \tag{6.30}$$

Recall that for each $a \in G(\phi)$, $L_a : l^2(G) \to l^2(G)$ is defined by

$$L_a f(x) = f(a^{-1}x), x \in G(\phi), f \in l^2(G(\phi)).$$

Note that u_ρ is a natural extension of U_ρ, and both $\rho \mapsto u_{\rho*}$ and $\rho \mapsto L_\rho \otimes I$ are unitary representations (*this is exactly where the claim is applied*). Then (6.30) implies that the von Neumann algebra generated by $\{u_\rho : \rho \in G(\phi)\}$ is unitarily isomorphic to the von Neumann algebra generated by

$$\{L_\rho \otimes I : \rho \in G(\phi)\}.$$

That is, $\mathcal{A}$ is unitarily isomorphic to $\mathcal{L}(G(\phi)) \otimes I$. If $\pi_1(\Omega)$ is not abelian, then by Lemma 6.4.2 $\pi_1(\Omega)$ is a free group on two or more generators. In particular, $\pi_1(\Omega)$ is an i.c.c. group. In this case, since $G(\phi)$ is isomorphic to $\pi_1(\Omega)$, $\mathcal{L}(G(\phi))$ is a factor. Therefore, $\mathcal{A}$ is a factor. Then by Lemma 6.9.3 $\mathcal{A}p \overset{*}{\cong} \mathcal{A}$. Since $\mathcal{A} \overset{*}{\cong} \mathcal{L}(G(\phi))$, we have $\mathcal{V}^*(\phi) = \mathcal{A}p \overset{*}{\cong} \mathcal{L}(G(\phi))$.

Since $\mathcal{A}p = \mathcal{V}^*(\phi)$, $\mathcal{V}^*(\phi) \overset{*}{\cong} \mathcal{L}(G(\phi))$. This completes the proof of Theorem 6.9.1.   □

6.10 Group-Like von Neumann Algebras

This section defines a new class of von Neumann algebras, which turns out to be ∗-isomorphic to the group von Neumann algebra, see Corollary 6.10.6.

This section will adopt the convention (6.28); that is, for a holomorphic regular branched covering map ϕ and for each $\rho \in G(\phi)$,

$$U_\rho h = k_{\rho^{-1}(0)}^\alpha h \circ \rho, \; h \in L_{a,\alpha}^2(\mathbb{D}).$$

Below we will investigate the structure of $\mathcal{V}_\alpha^*(\phi)$ via a different approach. Note that Lemma 6.2.4(3) plays an important role in Sect. 6.5 of this chapter. Intuitively, it says that for any $S, T \in \mathcal{V}^*(\phi)$, ST exactly equals its "formal" product. Though in the case to be discussed as follows, this becomes difficult. Write

$$Sh(z) = \sum_{k=0}^{\infty} c_k U_{\rho_k} h(z), \; h \in L_a^2(\mathbb{D}), z \in \mathbb{D},$$

and

$$Th(z) = \sum_{k=0}^{\infty} c'_k U_{\rho_k} h(z), \ h \in L_a^2(\mathbb{D}), \ z \in \mathbb{D},$$

where $\{c_k\}$ and $\{c'_k\}$ are in l^2, and all ρ_k run over $G(\phi)$. Then ST still equals its formal product. Precisely, there is a sequence $\{d_k\} \in l^2$ satisfying

$$STh(z) = \sum_{k=1}^{\infty} d_k U_{\rho_k} h(z), \ h \in L_a^2(\mathbb{D}), \ z \in \mathbb{D},$$

The difference lies in the equations:

$$d_i = \sum_{\rho_j \circ \rho_k = \rho_i} \zeta(j,k) c_k c_j, \ i = 0, 1, \cdots$$

where $\zeta(j,k)$ are some unimodular constants.

A similar version for Lemma 6.2.4 is not obvious. We must develop new techniques to overcome this difficulty. The idea is to construct something like the group von Neumann algebra $\mathcal{L}(\Gamma)$ and its commutant $\mathcal{R}(\Gamma)$, where Γ is a countable discrete group. Below, Γ always denotes the deck transformation group $G(\phi)$ of ϕ, ϕ being a holomorphic regular branched map. Also, let e denote the identity in Γ.

Note that for each fixed pair (ρ, τ), there is a unimodular constant $\xi = \xi(\rho, \tau)$ satisfying

$$U_\rho U_\tau = \xi(\rho, \tau) U_{\tau \circ \rho}.$$

In particular, $U_\rho U_\tau 1 = \xi(\rho, \tau) U_{\tau \circ \rho} 1$. Then by convention (6.28), it is not difficult to check that

$$\xi(\rho, \tau) = \frac{|1 - \overline{\tau^{-1}(0)} \rho(0)|^{2+\alpha}}{(1 - \overline{\tau^{-1}(0)} \rho(0))^{2+\alpha}}.$$

Therefore $\xi(\rho, \rho^{-1}) = 1$, and then $U_\rho U_{\rho^{-1}} = I$. Also, $\xi(\rho, e) = \xi(e, \rho) = 1$.

Given $f, g \in l^2(\Gamma)$, define

$$f *_\alpha g(a) = \sum_{x \in \Gamma} \overline{\xi(x^{-1}a, x)} f(x) g(x^{-1}a).$$

Now for each $\rho \in \Gamma$, set

$$L_\rho^\alpha f = \epsilon_\rho *_\alpha f$$

and

$$R^\alpha_\rho f = f *_\alpha \epsilon_{\rho^{-1}}.$$

Let $\mathcal{L}_\alpha(\Gamma)$ be the WOT-closure of the span of all L^α_ρ, and $\mathcal{R}_\alpha(\Gamma)$ be the WOT-closure of that of all R^α_ρ. Both $\mathcal{L}_\alpha(\Gamma)$ and $\mathcal{R}_\alpha(\Gamma)$ are von Neumann algebras, and we call them *group-like von Neumann algebras*. In many cases, $\mathcal{L}_\alpha(\Gamma)$ is not equal to $\mathcal{L}(\Gamma)$, even if $\alpha = 0$.

Many results on $\mathcal{L}(\Gamma)$ and $\mathcal{R}(\Gamma)$ also hold for $\mathcal{L}_\alpha(\Gamma)$ and $\mathcal{R}_\alpha(\Gamma)$ [Con1, pp. 247–250]. For example, we have the following result [Huang2].

Proposition 6.10.1 *With $\mathcal{L}_\alpha(\Gamma)$ and $\mathcal{R}_\alpha(\Gamma)$ defined as above, then*

$$\mathcal{L}_\alpha(\Gamma)' = \mathcal{R}_\alpha(\Gamma) \quad \text{and} \quad \mathcal{R}_\alpha(\Gamma)' = \mathcal{L}_\alpha(\Gamma).$$

Proof The proof comes from [Huang2].

We have a trivial identity for ρ, σ and τ in Γ: $(U_\rho U_\sigma)U_\tau = U_\rho(U_\sigma U_\tau)$, which gives that

$$\xi(\rho, \sigma)\xi(\sigma \circ \rho, \tau) = \xi(\sigma, \tau)\xi(\rho, \tau \circ \sigma).$$

For simplicity, we abbreviate $*_\alpha$ by $*$ temporarily. By the above identity, it is easy to verify that

$$(\epsilon_\tau * \epsilon_\sigma) * \epsilon_\rho = \epsilon_\tau * (\epsilon_\sigma * \epsilon_\rho),$$

and hence

$$(\epsilon_\tau * p) * \epsilon_\rho = \epsilon_\tau * (p * \epsilon_\rho), p \in \mathcal{P},$$

where $\mathcal{P}$ denotes the linear span of all ϵ_ρ. By an argument on limit, the above also holds when p is replaced with any function f in $l^2(\Gamma)$. This immediately shows that L^α_τ commutes with R^α_ρ for all τ and ρ in Γ. Using this fact, by a similar argument in [Con1, p. 248, 249], one will get $\mathcal{L}_\alpha(\Gamma) = \mathcal{R}_\alpha(\Gamma)'$. Then by von Neumann bicommutant theorem, $\mathcal{R}_\alpha(\Gamma) = \mathcal{L}_\alpha(\Gamma)'$, as desired.

For completeness, we give the proof for $\mathcal{L}_\alpha(\Gamma) = \mathcal{R}_\alpha(\Gamma)'$. For this, we first make the following claim.

If $T \in \mathcal{R}_\alpha(\Gamma)'$, then there is an $f \in l^2(\Gamma)$ such that

$$Th = f * h, \ h \in l^2(\Gamma).$$

Similarly, if $S \in \mathcal{L}_\alpha(\Gamma)'$, then there is a $g \in l^2(\Gamma)$ such that

$$Sh = h * g, \ h \in l^2(\Gamma).$$

The proof of this claim comes from that of [Con1, Proposition 43.10]. In detail, assume that $T \in \mathcal{R}_\alpha(\Gamma)'$, and put $f = T\epsilon_e$. For any $\rho \in \Gamma$,

$$T\epsilon_\rho = TR^\alpha_{\rho^{-1}}\epsilon_e = R^\alpha_{\rho^{-1}}f = f * \epsilon_\rho,$$

and thus $Tp = f * p$, $p \in \mathcal{P}$, where $\mathcal{P} = span\{\epsilon_\rho : \rho \in \Gamma\}$. By taking a limit, for each $h \in l^2(G)$,

$$Th(\rho) = (f * h)(\rho), \quad \rho \in \Gamma,$$

and hence $Th = f * h$, $h \in l^2(G)$, as desired. The proof for the remaining part is similar. We emphasize that both f and g are unique in the claim.

Observe that

$$L^\alpha_\sigma R^\alpha_\rho = R^\alpha_\rho L^\alpha_\sigma, \quad \rho, \sigma \in \Gamma,$$

which gives

$$\mathcal{L}_\alpha(\Gamma) \subseteq \mathcal{R}_\alpha(\Gamma)'.$$

To show that $\mathcal{L}_\alpha(\Gamma) = \mathcal{R}_\alpha(\Gamma)'$, it remains to prove that

$$(\mathcal{L}_\alpha(\Gamma))'' \supseteq \mathcal{R}_\alpha(\Gamma)',$$

which reduces to show that if $T \in \mathcal{R}_\alpha(\Gamma)'$ and $S \in \mathcal{L}_\alpha(\Gamma)'$, then $TS = ST$. In fact, the above claim shows that there are two functions f and g in $l^2(\Gamma)$ such that $Th = f * h$ and $Sh = h * g$ for all $h \in l^2(\Gamma)$. Then by careful verification,

$$TSh = T(h * g) = f * (h * g) = (f * h) * g = STh$$

holds for $h = \epsilon_\rho$ with $\rho \in \Gamma$. Therefore, the above identities hold for any h in $\mathcal{P}$, which is dense in $l^2(\Gamma)$. Thus, $TS = ST$. The proof is complete. $\qquad\square$

Also, one can establish the finiteness of two von Neumann algebras: $\mathcal{L}_\alpha(\Gamma)$ and $\mathcal{R}_\alpha(\Gamma)$. Take $\mathcal{R}_\alpha(\Gamma)$ as an example. For each $h \in l^2(\Gamma)$, define R_h by setting

$$R_h g = g *_\alpha h,$$

where g satisfies $g *_\alpha h \in l^2(\Gamma)$. Clearly, each R_h is densely defined. Since $\mathcal{L}_\alpha(\Gamma)' = \mathcal{R}_\alpha(\Gamma)$, by the proof of Proposition 6.10.1 one can show that $\mathcal{R}_\alpha(\Gamma)$ consists of all R_h where h has the property that for any $g \in l^2(\Gamma)$, $g *_\alpha h \in l^2(\Gamma)$. As done in [Con1, Theorem 53.1], define a map $Tr : \mathcal{R}_\alpha(\Gamma) \to \mathbb{C}$ by

$$Tr : T \mapsto \langle T\epsilon_e, \epsilon_e \rangle.$$

It is easy to check that $Tr(1) = 1$ and Tr is a faithful map. Besides, for R_f and R_g in $\mathcal{R}_\alpha(\Gamma)$,

$$Tr(R_g R_f) = f *_\alpha g(e) = \sum_y \overline{\xi(y, y^{-1})} f(y) g(y^{-1}) = \sum_y f(y) g(y^{-1}).$$

From the above, it is easy to see that $Tr(R_g R_f) = Tr(R_f R_g)$, and then Tr is a faithful trace. By Theorem 2.5.3, $\mathcal{R}_\alpha(\Gamma)$ is finite.

Theorem 6.10.2 *Both $\mathcal{L}_\alpha(\Gamma)$ and $\mathcal{R}_\alpha(\Gamma)$ are finite von Neumann algebras for $\alpha > -1$.*

Furthermore, with a similar discussion as in [Con1, p. 301] one sees that *if the deck transformation group Γ is an i.c.c. group, then both $\mathcal{L}_\alpha(\Gamma)$ and $\mathcal{R}_\alpha(\Gamma)$ are type II_1 factors.*

Recall that in Sect. 6.2 of this chapter, the construction of the trace on $\mathcal{L}(\Gamma)$ is presented in order to give the trace on $V^*(\phi)$, thus establishing the finiteness of $V^*(\phi)$. This also works in the situation here: just let tr be the linear map which takes each operator $S \in V^*_\alpha(\phi)$ with the form $\sum_{k=0}^\infty c_k U_{\rho_k}$ to c_0. By using an elementary analysis in $\mathcal{L}_\alpha(G(\phi))$, one can check that tr defines a faithful trace over $V^*_\alpha(\phi)$. Therefore, $V^*_\alpha(\phi)$ is also a finite von Neumann algebra. Then by using the methods in Sects. 6.2–6.6 of this chapter, one can show a similar version of Theorems 6.3.2 and 6.4.1 as below [Huang2].

Theorem 6.10.3 *Suppose $\phi : \mathbb{D} \to \Omega$ is a holomorphic covering map. If Ω is not conformally isomorphic to the disk, annuli or the punctured disk, then $V^*_\alpha(\phi)$ is a type II_1 factor, and $W^*_\alpha(\phi)$ is a type II_∞ factor. Otherwise, $V^*_\alpha(\phi)$ is abelian, and hence is of type I.*

Note that Theorems 6.9.1 and 6.8.3 cover Theorem 6.10.3.

Furthermore, by applying Theorem 6.10.2, one can establish a similar version of Proposition 6.5.2.

Lemma 6.10.4 *Suppose f is in $l^2(\Gamma)$, then $f *_\alpha f^* = \epsilon_e$ if and only if $f^* *_\alpha f = \epsilon_e$.*

Here f^* is defined by $f^*(a) = \overline{f(a^{-1})}$, which is the same as that in [Con1, p. 250, Exercise 10].

Following the discussions in Sect. 6.5 of this chapter gives the following result, which is from [Huang2].

Theorem 6.10.5 *For each $\alpha > -1$, there is a natural $*$-isomorphism from $V^*_\alpha(\phi)$ onto $\mathcal{L}_\alpha(G(\phi))$.*

In Sect. 6.9 of this chapter, it is shown that the von Neumann algebra $V^*_\alpha(\phi)$ is $*$-isomorphic to the original group von Neumann algebra $\mathcal{L}(G(\phi))$. Therefore, we have the following consequence, due to Huang [Huang2].

Corollary 6.10.6 *For each $\alpha > -1$, $\mathcal{L}_\alpha(G(\phi))$ is $*$-isomorphic to $\mathcal{L}(G(\phi))$.*

Whenever $G(\phi)$ is an i.c.c. group, $\mathcal{L}_\alpha(G(\phi))$ and $\mathcal{L}(G(\phi))$ are type II factors in standard forms. Therefore, by Lemma 6.5.8 and Corollary 6.10.6, except for very restricted cases $\mathcal{L}_\alpha(G(\phi))$ are unitarily isomorphic to $\mathcal{L}(G(\phi))$. However, it seems not easy to give an intuitive and simple unitary isomorphism between them.

6.11   Weighted Bergman Spaces over the Upper Half Plane

This section shows that all results in Sects. 6.2–6.10 (except for Sect. 6.7) of this chapter, discussed on the unit disk $\mathbb{D}$, can be translated to the upper half plane Π.

In what follows, we shall adopt the notations in [Lang1], with a bit modification; and in this section, $i = \sqrt{-1}$, the imaginary unit. For each $\alpha > -1$, put

$$d\mu_\alpha(z) = (\alpha + 1)(2y)^{\alpha+2}\frac{dxdy}{\pi}, z = x + yi \in \Pi.$$

Let $L^2_{a,\alpha}(\Pi)$ denote the Bergman space over the upper plane Π, consisting of those holomorphic functions which are square-summable with respect to $d\mu_\alpha(z)$. Set

$$w = \frac{z - i}{z + i}, z \in \Pi,$$

which is a biholomorphic map from Π onto $\mathbb{D}$. Its inverse map is defined by

$$z = -i\frac{w + 1}{w - 1}, w \in \mathbb{D}.$$

Then put

$$T_\alpha f(w) = f\left(-i\frac{w + 1}{w - 1}\right)\left(\frac{-\sqrt{2}i}{w - 1}\right)^\alpha, f \in L^2_{a,\alpha}(\Pi)\, w \in \mathbb{D},$$

and it is not difficult to check that $T_\alpha : L^2_{a,\alpha}(\Pi) \to L^2_{a,\alpha}(\mathbb{D})$ is a unitary operator [Lang1, Chap. IX], whose inverse T^*_α is defined by

$$T^*_\alpha h(z) = h\left(\frac{z - i}{z + i}\right)\left(\frac{\sqrt{2}}{z + i}\right)^\alpha, h \in L^2_{a,\alpha}(\mathbb{D}), z \in \Pi.$$

Moreover, for each bounded holomorphic function ϕ over Π, there is a unimodular constant ξ such that

$$T^*_\alpha M_\phi T_\alpha = \xi M_{\phi\left(\frac{z-i}{z+i}\right)}.$$

This implies that studying $\mathcal{V}_\alpha^*(\phi)$ on $L_{a,\alpha}^2(\Pi)$ is equivalent to studying $\mathcal{V}_\alpha^*(\phi(\frac{z-i}{z+i}))$ on $L_{a,\alpha}^2(\mathbb{D})$. Observe that whenever ϕ is a holomorphic covering map, then $\phi(\frac{z-i}{z+i})$ is also a holomorphic covering map, and the case of holomorphic regular branched covering maps is similar. By a careful verification, all conclusions on the unit disk can be transferred to the upper plane.

6.12   Remarks on Chap. 6

This chapter is mainly based on Guo and Huang's papers [GH2] and [Huang2].

It has been a long decade since the beginning of studying reducing subspaces and commutants of a multiplication operator. However, not much is known about the structure of the related von Neumann algebras $\mathcal{V}^*(\phi)$ and $\mathcal{W}^*(\phi)$. One may refer to [AD] and [GH2] for "concrete" type II factors stemming from function spaces. There is an extensive literature on this line, see [AB, DW, Nor, Cow1, Cow2, Cow3, SWa, T1, T2, T3, T4] and [AC, ACR, AD, Cl, Cow1, Cu, JL, Ro, SZ1, SZ2, Zhu1, Zhu2].

Theorem 6.2.2, Lemma 6.2.4, Proposition 6.2.6, Corollaries 6.2.3 and 6.2.5 were first obtained in [GH2]. Theorems 6.3.2, 6.3.3 and Propositions 6.3.10, 6.3.11, 6.3.12 were shown by Guo and Huang [GH2]. Examples 6.3.5, 6.3.6 and 6.3.9 came also from [GH2].

Theorem 6.4.1, Proposition 6.4.9, Lemma 6.4.5, Corollary 6.4.6 and Example 6.4.3 are all from [GH2].

Theorem 6.5.4, Propositions 6.5.2, 6.5.3, Lemmas 6.5.1, 6.5.8 and Example 6.5.5 first appeared in [GH2]. Theorems 6.5.6, 6.6.2, Lemma 6.6.1 and Corollary 6.6.5 are proved by Guo and Huang [GH2].

Lemma 6.7.1, Proposition 6.7.4, Corollary 6.7.2, Examples 6.7.3 and 6.7.6 are new and due to Guo and Huang. All results from Sects. 6.8 to 6.11 of this chapter, concerning all theorems, propositions, corollaries and examples, come from [Huang2].

Chapter 7
Similarity and Unitary Equivalence

In this chapter, we will apply those methods developed in Chaps. 3–6 to study similarity and unitary equivalence of multiplication operators, defined on both the Hardy space and the Bergman space.

This chapter is divided into two parts: the first part (Sect. 7.1) focuses on similarity and unitary equivalence of multiplication operators defined on the Hardy space, and the counterparts on the Bergman space will be taken up in the latter part (Sects. 7.2 and 7.3). As will be seen, the treatises are quite different.

Throughout this chapter, for each bounded holomorphic function f, we rewrite the analytic Toeplitz operator T_f for M_f, multiplication operator with symbol f.

7.1 The Case of the Hardy Space

This section will deal with the unitary equivalence of two analytic Toeplitz operators on the Hardy space $H^2(\mathbb{D})$.

Recall that given a function f in $H^\infty(\mathbb{D})$, if for some λ in $\mathbb{D}$ the inner part of $f - f(\lambda)$ is a finite Blaschke product, then f is said to satisfy *Cowen's condition*. We restate Cowen-Thomson's theorem as follows (see Chap. 3).

Theorem 7.1.1 (Cowen-Thomson) *Suppose f is in $H^\infty(\mathbb{D})$ and satisfy Cowen's condition. Then there is a function $\tilde{f}$ in $H^\infty(\mathbb{D})$ and a finite Blaschke product B such that $f = \tilde{f} \circ B$ and $\{T_f\}' = \{T_B\}'$ holds on $H^2(\mathbb{D})$.*

Though Theorem 7.1.1 also holds on the Bergman space, see Theorem 3.1.1, our interest is focused on the Hardy space in this section. As pointed out in Chap. 3, this finite Blaschke product B in Theorem 7.1.1 is unique in the following sense: if there is a function h in $H^\infty(\mathbb{D})$ and a finite Blaschke product B_1 such that $f = h \circ B_1$ and $\{T_f\}' = \{T_{B_1}\}'$, then there is a Mobius map φ such that $B_1 = \varphi \circ B$.

© Springer-Verlag Berlin Heidelberg 2015
K. Guo, H. Huang, *Multiplication Operators on the Bergman Space*,
Lecture Notes in Mathematics 2145, DOI 10.1007/978-3-662-46845-6_7

An observation is in order. For a finite Blaschke product B, we have

$$\text{order } B = \dim \left(H^2(\mathbb{D}) \ominus BH^2(\mathbb{D}) \right). \tag{7.1}$$

Inspired by (7.1), for each inner function η, define *the order of η* as follows:

$$\text{order } \eta \triangleq \dim \left(H^2(\mathbb{D}) \ominus \eta H^2(\mathbb{D}) \right).$$

(As in [Pel, p. 722], this is called the degree of η.) Then for each inner function η, η is a finite Blaschke product if and only if order $\eta < \infty$. Note that T_η is an isometric operator on $H^2(\mathbb{D})$ for each inner function η, and it follows that for two inner functions η_1 and η_2, T_{η_1} is unitarily equivalent to T_{η_2} if and only if order $\eta_1 = $ order η_2.

In general, one has the following theorem, due to Cowen [Cow4].

Theorem 7.1.2 *Suppose that h is in $L^\infty(\mathbb{T})$ and η is an inner function of order $n(n \le \infty)$, then*

$$T_{h \circ \eta} \overset{unitary}{\cong} \bigoplus_n T_h,$$

where $\oplus_n T_h$ denotes the direct sum of n copies of T_h.

In the case of $h(z) = z$, Theorem 7.1.2 says that for each inner function η, on $H^2(\mathbb{D})$ T_η is unitarily equivalent to the direct sum of n copies of the Hardy space shift T_z, with $n = $ order η.

Proof The proof is due to Cowen. Since

$$n = \text{order } \eta = \dim(H^2(\mathbb{D}) \ominus \eta H^2(\mathbb{D})),$$

there is an orthonormal basis $\{w_k\}_{k=1}^n$ for $H^2(\mathbb{D}) \ominus \eta H^2(\mathbb{D})$. Define

$$U : \bigoplus_n H^2(\mathbb{D}) \longrightarrow H^2(\mathbb{D})$$

$$\bigoplus_{k=1}^n \left(\sum_{m=0}^\infty b_{k,m} z^m \right) \mapsto \sum_{k=1}^n \sum_{m=0}^\infty b_{k,m} \eta^m w_k.$$

where $\sum_{k,m=0}^\infty |b_{k,m}|^2 < \infty$. Then U is a unitary operator.

First consider a special case where h is a trigonometric polynomial. In this case, one can verify that

$$T_{h \circ \eta} U = U \left(\bigoplus_n T_h \right).$$

In general, let p_N be the N-th Cesaro mean of h. Then $\{p_N\}$ converges to h almost everywhere with respect to the arc-length measure and

$$\|p_N\|_\infty \leq \|h\|_\infty.$$

Thus, $\{p_N \circ \eta\}$ converges to $h \circ \eta$ in the weak*-topology in $L^\infty(\mathbb{T})$. Therefore, $\{T_{p_N \circ \eta}\}$ converges to $T_{h \circ \eta}$ in weak operator topology. Since

$$T_{p_N \circ \eta} U = U \bigoplus_n T_{p_N},$$

then letting $N \to \infty$ gives

$$T_{h \circ \eta} U = U \bigoplus_n T_h,$$

completing the proof. $\qquad\qquad\qquad\qquad\qquad\qquad\qquad\qquad\qquad\qquad\qquad$ □

Remark 7.1.3 Later, one will see that on the Bergman space T_B is similar to $\bigoplus_n T_z$, with $n = \text{order } B$, see [JL]. However, in most cases M_B is not unitarily equivalent to $\bigoplus_n T_z$.

The following is a simple application of Theorem 7.1.2.

Example 7.1.4 Given two inner functions η_1 and η_2 with the same order, we have

$$T_{\eta_1} \overset{\text{unitary equivalent}}{\cong} T_{\eta_2}.$$

Furthermore, applying Theorem 7.1.2 shows that for each $h \in L^\infty(\mathbb{T})$,

$$T_{h \circ \eta_1} \overset{\text{unitarily equivalent}}{\cong} T_{h \circ \eta_2}.$$

The following result is given by Cowen [Cow4, Theorem 2], with a bit modification.

Theorem 7.1.5 *Suppose both f and g are in $H^\infty(\mathbb{D})$ and satisfy Cowen's condition. Then the following are equivalent:*

(i) *T_f is unitarily equivalent to T_g;*
(ii) *T_f is similar to T_g;*
(iii) *There is a function $h \in H^\infty(\mathbb{D})$ and two finite Blaschke products B_1 and B_2 with the same order such that*

$$f = h \circ B_1 \quad \text{and} \quad g = h \circ B_2.$$

Proof (i) $\Rightarrow$ (ii) is straightforward. (iii) $\Rightarrow$ (i) follows directly from Example 7.1.4. It remains to show (ii) $\Rightarrow$ (iii).

By Theorem 7.1.1 there exist two functions $\tilde{f}$ and $\tilde{g}$ in $H^\infty(\mathbb{D})$ and two finite Blaschke products ϕ_1 and ϕ_2 satisfying

(1) $f = \tilde{f} \circ \phi_1$ and $g = \tilde{g} \circ \phi_2$.
(2) $\{T_f\}' = \{T_{\phi_1}\}'$ and $\{T_g\}' = \{T_{\phi_2}\}'$.

Since T_f is similar to T_g, there is a bounded invertible operator S such that $T_g = ST_f S^{-1}$, which induces a natural bijection

$$\{T_f\}'' \longrightarrow \{T_g\}''$$
$$V \mapsto SVS^{-1}.$$

Since $T_{\phi_1} \in \{T_{\phi_1}\}'' = \{T_f\}''$, and

$$\{T_g\}'' \subseteq \{T_\psi : \psi \in H^\infty(\mathbb{D})\}'' = \{T_\psi : \psi \in H^\infty(\mathbb{D})\},$$

there is a function $u \in H^\infty(\mathbb{D})$ satisfying

$$ST_{\phi_1}S^{-1} = T_u. \tag{7.2}$$

We claim that u is a finite Blaschke product. In fact, by a standard treatise of spectrum analysis and (7.2)

$$\sigma_e(T_u) = \sigma_e(T_{\phi_1}) = \mathbb{T}. \quad (\sigma_e \text{ denotes the essential spectrum})$$

Since

$$\sigma_e(T_u) = \bigcap_{0<r<1} \overline{\{u(z) : r < |z| < 1\}} = \mathbb{T},$$

it follows that each radial limit of u on $\mathbb{T}$(if exists) lies in $\{z : |z| = 1\}$, forcing u to be an inner function. By (7.2), T_u is Fredholm, and *Index* $T_u = $ *Index* T_{φ_1}. Therefore, u is a finite Blaschke product with order $u=$ order ϕ_1.

By a simple computation,

$$T_g = ST_f S^{-1}$$
$$= ST_{\tilde{f} \circ \phi_1} S^{-1}$$
$$= \tilde{f}(ST_{\phi_1} S^{-1})$$
$$= \tilde{f}(T_u)$$
$$= T_{\tilde{f} \circ u}.$$

The last and third identities hold firstly for analytic polynomials. Then by taking a limit in the weak operator topology, they also holds for an arbitrary function $\tilde{f}$ in

$H^\infty(\mathbb{D})$. Since $T_g = T_{\tilde{f} \circ u}$, $g = \tilde{f} \circ u$. Write $B_1 = \phi_1$ and $B_2 = u$, and thus

$$g = \tilde{f} \circ B_2 \quad \text{and} \quad f = \tilde{f} \circ B_1.$$

The proof is complete. $\qquad\qquad\qquad\qquad\qquad\qquad\qquad\qquad\qquad$ □

Remark 7.1.6 The statement of Theorem 7.1.5 rests on the structure of the Hardy space. However, on the Bergman space $L_a^2(\mathbb{D})$ one can show that under the same assumption of Theorem 7.1.5, the following are equivalent:

(*i*) T_f is unitarily equivalent to T_g;
(*ii*) There is a function $h \in H^\infty(\mathbb{D})$ and two finite Blaschke products B_1 and B_2 such that

$$f = h \circ B_1 \quad \text{and} \quad g = h \circ B_2,$$

and T_{B_1} is unitarily equivalent to T_{B_2}. The proof is just the same. Therefore, Theorem 7.2.1 (see next section) reduces to a very special case where both f and g are finite Blaschke products.

Now let us turns back to Theorem 7.1.5. In particular, if f and g are in $H^\infty(\overline{\mathbb{D}})$, then both f and g satisfy Cowen's condition. More specially, if both f and g are entire functions, then T_f is similar to T_g if and only if there is a unimodular constant c such that $g(z) = f(cz)$. The reasoning is as follows. Applying Theorem 7.1.5 and Corollary 3.1.2 shows that there is an integer n such that $f = \varphi(z^n)$ and $g = \varphi(c^n z^n)$, where φ is an entire function. It is not difficult to determine this integer n. In fact, if we write

$$f(z) = \sum_{k=0}^\infty c_k z^k, \quad z \in \mathbb{C},$$

then

$$n = \gcd\{k : k > 0, c_k \neq 0\}.$$

Example 7.1.7 Let p and q be polynomials, and then T_p is similar to T_q if and only if $q(z) = p(cz)$ for some unimodular constant c. Therefore, put $p(z) = 1 + z + z^2$ and $q(z) = 1 + 2z + z^2$. Then T_p is never similar to T_q.

For the problem of similarity, remarkable work is also due to Clark [Cla1, Cla2, Cla3, Cla4], Clark and Morrel [ClMo], who dealt with Toeplitz operators with rational symbol.

In more detail, in [ClMo] Clark and Morrel gave the following theorem on similarity.

Theorem 7.1.8 *Suppose $F(z)$ is a rational function which maps the unit circle onto a simple closed curve and the map $t \to F(e^{it})$ has winding number n*

$(n > 0)$ *for each interior point of $F(\mathbb{T})$. Suppose F is n-to-one in some annulus $\{z \in \mathbb{C} : s \leq |z| \leq 1\}$. Then T_F is similar to an analytic Toeplitz operator $T_{\tau(z^n)}$, where τ is the conformal map from the unit disk onto the interior of the curve $F(\mathbb{T})$.*

A special case of Theorem 7.1.8 was treated by Duren [Dur].

Now let F be a rational function without poles on the unit circle $\mathbb{T}$. Suppose σ_F is a compact set whose boundary $\partial\sigma_F$ contains $F(\mathbb{T})$, and n denotes winding number of the map $t \to F(e^{it})$ for each interior point λ of $F(\mathbb{T})$. Let τ denote the Riemann mapping function from the unit disk onto the interior of σ_F. As done in [Cla2], F is called to back up at $e^{i\theta}$ if the argument of $\tau^{-1}F(e^{it})$ is decreasing at $t = \theta$, and Σ denotes the set of all points where F backs up. Then we are ready to state Clark's Similarity Theorem.

Theorem 7.1.9 (Similarity Theorem) *If $\partial\sigma_F$ is an analytic curve in a neighborhood of $F(\mathbb{T})$, $n \geq 0$ and $F(\Sigma) \neq \partial\sigma_F$, then T_F is similar to $M_\tau \oplus V$, where M_τ is the multiplication operator by τ on $H^2(\mathbb{D}) \otimes \mathbb{C}^n$, and V is a normal operator whose spectrum is $F(\Sigma)$. Furthermore, V is compact and the spectral multiplicity of V of a spectral point λ equals the number of those points $e^{i\theta}$ where F backs up and $F(e^{i\theta}) = \lambda$.*

Clark also gave some delicate geometric characterization for similarity of Toeplitz operators, for the details see [Cla1, Cla3].

7.2 Unitary Equivalence on Analytic Multiplication Operators

This section elaborates on the unitary equivalence problem of analytic multiplication operators on the Bergman space.

As shown in Sect. 7.1, for two finite Blaschke products B_1 and B_2, T_{B_1} is unitarily equivalent to T_{B_2} on the Hardy space if and only if order $B_1 =$ order B_2. However, the case is different on the Bergman space. One can construct two finite Blaschke products B_1 and B_2 of the same order, but T_{B_1} is not unitarily equivalent to T_{B_2} on the Bergman space. For example, set $B_1(z) = z^3$ and $B_2(z) = z^2\varphi_a(z)$ with $a = \frac{1}{2}$. By Theorem 5.8.7, T_{B_1} has exactly three minimal reducing subspaces and T_{B_2} has exactly two. This implies that T_{B_1} is not unitarily equivalent to T_{B_2}.

The following problem received wide attention and interest:

When are two analytic multiplication operator defined on $L_a^2(\mathbb{D})$ unitarily equivalent?

Sun [Sun1] found that under a mild condition, the above problem reduces to a special case where the symbols of two analytic multiplication operator are both finite Blaschke products, which are necessarily of the same order. By analyzing function-theoretic properties of finite Blaschke products, he presented the following deep result. However, his proof in [Sun1] is complicated and unclear.

Theorem 7.2.1 (Sun) *Suppose both f and g are in $H^\infty(\mathbb{D})$ and satisfy Cowen's condition. The following are equivalent:*

(i) T_f is unitarily equivalent to T_g on $L_a^2(\mathbb{D})$;
(ii) There is a Mobius map $\psi : \mathbb{D} \to \mathbb{D}$ such that $g = f \circ \psi$.

A natural problem arises: when is an analytic multiplication operator, say T_h, unitarily equivalent to some unilateral weighted shift with finite multiplicity on the Bergman space. The following result provides a solution. Under the condition that h is a finite Blaschke product of order n, it was done in [SY]. In general, it was accomplished by Sun et al. [SZZ2]; a simpler version is presented in [GZ].

Theorem 7.2.2 (Sun-Zheng-Zhong) *For a function h in $H^\infty(\mathbb{D})$, T_h is unitarily equivalent to a nonzero unilateral weighted shift W of multiplicity n if and only if $h = c\varphi_\alpha^n$ for some $c \neq 0$ and $\alpha \in \mathbb{D}$, where $\varphi_\alpha(z) = \frac{\alpha - z}{1 - \bar{\alpha}z}$.*

Proof Assume that $h = c\varphi_\alpha^n$ for some $c \neq 0$ and $\alpha \in \mathbb{D}$. Define

$$U_\alpha : L_a^2(\mathbb{D}) \to L_a^2(\mathbb{D})$$

$$f \mapsto f \circ \varphi_\alpha \, \varphi_\alpha',$$

which proves to be a unitary operator. Then one can verify that

$$U_\alpha T_h = T_{h \circ \varphi_\alpha} U_\alpha.$$

Thus, $U_\alpha T_h = T_{cz^n} U_\alpha$, as desired.

On the inverse direction, assume that T_h is unitarily equivalent to a nonzero unilateral weighted shift W of multiplicity n. Without loss of generality, assume that $\|h\|_\infty = 1$. To finish the proof, we need to show that $h = c\varphi_\alpha^n$ for some unimodular constant c, with $\alpha \in \mathbb{D}$.

First, we show that h is a finite Blaschke product. The following reasoning is from [GZ]. Since T_h is subnormal, W is subnormal, and hence by a theorem in [Shi], the essential spectrum $\sigma_e(W)$ of W consists of a union of n circles centered at 0. Let $\mathcal{M}_\infty$ denote the maximal ideal space of $H^\infty(\mathbb{D})$. The Corona Theorem [Ga] shows that

$$\sigma_e(T_h) = h(\mathcal{M}_\infty - \mathbb{D}). \tag{7.3}$$

Note that by the connectedness of $\mathcal{M}_\infty - \mathbb{D}$ [Ga, Hof1], $\sigma_e(T_h)$ is connected, which gives that

$$\sigma_e(T_h) = \sigma_e(W) = \mathbb{T}. \tag{7.4}$$

Combing (7.3) with (7.4) gives that

$$\sigma_e(T_h) = \bigcap_{0 < r < 1} \overline{\{h(z); r < |z| < 1\}} = \mathbb{T}.$$

Thus, for each $\eta \in \mathbb{T}$, $\lim_{r \to 1^-} |h(r\eta)| = 1$ whenever $\lim_{r \to 1^-} |h(r\eta)|$ exists, forcing h to be an inner function. By Fredholmness of T_h, h is a finite Blaschke product. Below, rewrite $h = B$.

We can assume that W acts on the Bergman space, and note that

$$n = \dim(L_a^2(\mathbb{D}) \ominus WL_a^2(\mathbb{D})) = \dim(L_a^2(\mathbb{D}) \ominus BL_a^2(\mathbb{D})) = \operatorname{order} B.$$

Since W has n nonzero minimal orthogonal reducing subspaces, so does T_B. Then by the proof of Theorem 4.6.1, each local inverse of B extends to a member in $\operatorname{Aut}(\mathbb{D})$. By Theorem 6.6.2 $B = \psi(\rho^n)$, where ψ and ρ are in $\operatorname{Aut}(\mathbb{D})$. This reduces to the following case: $B = \psi(z^n)$. Thus it remains to show $\psi(z) = cz$ for some unimodular constant $c \in \mathbb{T}$.

Now assume conversely that there is a constant $c' \in \mathbb{T}$ and a nonzero number $a \in \mathbb{D}$ such that

$$B = c' \frac{a^n - z^n}{1 - \overline{a^n} z^n}.$$

Since T_B is unitarily equivalent to the unilateral weighted shift W of multiplicity n, then T_B^2 is unitarily equivalent to W^2, a unilateral weighted shift of multiplicity $2n$. Therefore, by similar discussions as above there exists a unimodular constant c'' and $b \in \mathbb{D}$ such that

$$B^2 = c'' \frac{b^{2n} - \tau(z)^{2n}}{1 - \overline{b^{2n}} \tau(z)^{2n}},$$

where $\tau \in \operatorname{Aut}(\mathbb{D})$. Write $\omega = \exp(\frac{\pi i}{n})$, the $2n$-th root of 1. We first show that $b = 0$. If $b \neq 0$, then all zeros of B in $\mathbb{D}$,

$$\tau^{-1}(b), \tau^{-1}(b\omega), \cdots, \tau^{-1}(b\omega^{2n-1}),$$

were simple zeros. However, a was a zero of B^2 with multiplicity 2, which is a contradiction. Thus, $b = 0$. In this case, $\tau^{-1}(0)$ is the only zero of B^2 with multiplicity $2n$. Since $n \geq 2$, this is a contradiction to the fact that a is a zero of B^2 with multiplicity 2. Therefore, $B(z) = cz^n$ for some unimodular constant c. The proof is complete. $\square$

7.3   Similarity of Analytic Toeplitz Operators

This section will deal with the similarity of analytic Toeplitz operators on the Bergman space.

As pointed out by Cowen and Douglas [Dou], a related natural question is: Is T_B on $L_a^2(\mathbb{D})$ similar to $T_z \otimes I_n$ on $L_a^2(\mathbb{D}) \otimes \mathbb{C}^n$, where $n = \operatorname{order} B$? This question was positively answered in [JL]. Below, by using Rudin's method [Ru1, Chap. 7], this section will establish a representation theorem of $L_a^2(\mathbb{D})$-functions, which is of independent interest, and from which one would obtain an immediate answer to the similarity problem. We state it as follows [GH1].

Theorem 7.3.1 *Suppose B is a finite Blaschke product of order n. Then there are bounded linear operators τ_i from $L_a^2(\mathbb{D})$ to $L_a^2(\mathbb{D})$ such that*

$$f(z) = \sum_{i=1}^{n} (\tau_i f)(B(z)) z^{i-1}, \ f \in L_a^2(\mathbb{D}).$$

The representation is unique; that is, if there are $f_1, \cdots, f_n$ in $L_a^2(\mathbb{D})$ satisfying

$$f(z) = \sum_{i=1}^{n} f_i(B(z)) z^{i-1}, \ z \in \mathbb{D}, \tag{7.5}$$

then $f_i = \tau_i f$ for $i = 1, \cdots, n$.

Proof The proof comes from [GH1].

Before presenting the proof, let us make an observation.

By proposition 4.2.5, the derivative of a finite Blaschke product never vanishes on $\mathbb{T}$. Thus, there exists a number $r \in (0, 1)$ such that B' never vanishes on $\overline{A_r}$, where

$$A_r = \{z \in \mathbb{C} : r < |z| < 1\}.$$

It is easy to check that there is a number $r'(r < r' < 1)$ satisfying

$$B^{-1}(\overline{A_{r'}}) \subseteq \overline{A_r}. \tag{7.6}$$

We first deal with uniqueness. It suffices to show that if there are $f_1, \cdots, f_n$ in $L_a^2(\mathbb{D})$ satisfying

$$\sum_{i=1}^{n} f_i(B(z)) z^{i-1} = 0, \ z \in \mathbb{D}, \tag{7.7}$$

then $f_i \equiv 0$ for all i.

In fact, from the above observation we can pick a number r_0 with $r' < r_0 < 1$. By (7.6), for each $w \in r_0 \mathbb{T}$,

$$B^{-1}(w) \subseteq B^{-1}(\overline{A_{r'}}) \subseteq \overline{A_r}.$$

Since B' never vanishes on $\overline{A_r}$, for each $w \in r_0\mathbb{T}$, $B^{-1}(w)$ consists of exactly n different points, denoted by $\beta_1(w), \cdots, \beta_n(w)$. Thus by (7.7),

$$\sum_{i=1}^{n} f_i(w)\beta_j(w)^{i-1} = 0, \ w \in r_0\mathbb{T} \text{ and } 1 \le j \le n.$$

Note that for each $w \in r_0\mathbb{T}$, $\left(\beta_j(w)^{i-1}\right)$ is just the Vandermonde matrix, and then the above equations force

$$f_1(w) = \cdots = f_n(w) = 0, \ w \in r_0\mathbb{T}.$$

Thus $f_i \equiv 0$ for $1 \le i \le n$, as desired.

Next, the proof for the existence is taken up as follows. By a similar argument as above, for any $w \in \overline{A_{r'}}$, there exist n distinct points $\beta_j(w) \in \overline{A_r}$ satisfying

$$B(\beta_j(w)) = w, j = 1, \cdots, n. \tag{7.8}$$

Moreover, since $B'(\beta_j(w)) \ne 0$, there is a positive number δ_w such that B maps each disk $O(\beta_j(w), \delta_w)$ biholomorphically onto some neighborhood of w. This implies the following.

Fact For any $w \in \overline{A_{r'}}$, there exists an $\varepsilon_w > 0$, and n different holomorphic functions $\beta_1, \cdots, \beta_n$ which are univalent on $O(w, \varepsilon_w)$ such that each of them satisfies (7.8), and $\beta_i(z) \ne \beta_j(z)$ for $z \in O(w, \varepsilon_w)$, $i \ne j$. In addition, the derivative of each β_j^{-1} are bounded on $O(w, \varepsilon_w)$.

As done in [Ru1, Chap. 7, pp. 169–172], for each polynomial P there are rational maps R_i in $A(\mathbb{D})$ such that

$$P(z) = \sum_{i=1}^{n} R_i(B(z))z^{i-1}. \tag{7.9}$$

To see (7.9), recall that B is a finite Blaschke product of order n, and write

$$B(z) = c \prod_{i=1}^{n} \frac{z - a_i}{1 - \overline{a}_i z}.$$

Multiplying the equation $w = B(z)$ by the dominator $\prod_{i=1}^{n}(1 - \overline{a}_i z)$ of B, we get

$$(c - aw)z^n = p_0(w) + p_1(w)z + \cdots + p_{n-1}(w)z^{n-1},$$

where $|a| < 1$ and $|c| = 1$, and each $p_i (0 \le i \le n - 1)$ is a polynomial of degree at most 1. Therefore, there are rational functions $\hat{R}_1, \cdots, \hat{R}_n$, with no poles on $\overline{\mathbb{D}}$ such that

$$z^n = \sum_{i=1}^{n} \hat{R}_i(B(z))z^{i-1}. \tag{7.10}$$

Multiplying (7.10) by z yields that

$$z^{n+1} = \hat{R}_n(B(z))z^n + \sum_{i=1}^{n-1} \hat{R}_i(B(z))z^i. \tag{7.11}$$

Then combining (7.10) with (7.11) shows that there are rational functions $\tilde{R}_1, \cdots, \tilde{R}_n$ in $A(\mathbb{D})$ such that

$$z^{n+1} = \sum_{i=1}^{n} \tilde{R}_i(B(z))z^{i-1}.$$

By induction, there are rational functions $\tilde{R}_{1,p}, \cdots, \tilde{R}_{n,p}$ in $A(\mathbb{D})$ such that

$$z^p = \sum_{i=1}^{n} \tilde{R}_{i,p}(B(z))z^{i-1}.$$

Therefore, (7.9) holds; that is, for each polynomial P, there are rational maps R_i (depending on P) in $A(\mathbb{D})$ such that

$$P(z) = \sum_{i=1}^{n} R_i(B(z))z^{i-1}. \tag{7.12}$$

As done in the proof of Theorem 7.4.1 in [Ru1], we shall first define τ_i on polynomials and τ_i are well-defined and linear. Precisely, in view of (7.12), put $\tau_i P = R_i$.

In what follows, we will use the fact to prove that τ_i are bounded with respect to the Bergman norm. Now rewrite (7.12) by

$$P(\beta_j(w)) = \sum_{i=1}^{n} R_i(w)\beta_j(w)^{i-1}, \ 1 \le j \le n \text{ and } w \in \overline{A_{r'}}.$$

By Cramer's rule, for each $k(1 \le k \le n)$

$$R_k(w) = \frac{\det V_k(P)(w)}{\det \left(\beta_j(w)^{i-1} \right)}, \ w \in \overline{A_{r'}}, \tag{7.13}$$

where $V_k(P)(w)$ denotes the matrix $\left(\beta_j(w)^{i-1}\right)$ whose k-th column is replaced with $(P(\beta_1(w)), \cdots, P(\beta_n(w)))^T$. The above discussion shows that the Vandermonde determinant $\det\left(\beta_j(w)^{i-1}\right)$ is continuous on $\overline{A_{r'}}$ and has no zero point. Then by (7.13), there is a constant $C > 0$ such that

$$\int_{A_{r'}} |R_k(w)|^2 dA(w) \leq C \int_{A_{r'}} \sum_{j=1}^n |P(\beta_j(w))|^2 dA(w). \tag{7.14}$$

Since all neighborhoods $O(w, \varepsilon_w)$ of $w(w \in \overline{A_{r'}})$ consist of an open cover of the compact set $\overline{A_{r'}}$, by Henie-Borel's theorem one can pick finitely many of them:

$$O(w_1, \varepsilon_1), \cdots, O(w_N, \varepsilon_N),$$

whose union contains $\overline{A_{r'}}$. Then by (7.14),

$$\int_{A_{r'}} |R_k(w)|^2 dA(w) \leq C \int_{\bigcup_{l=1}^N O(w_l,\varepsilon_l) \cap \overline{\mathbb{D}}} \sum_{j=1}^n |P(\beta_j(w))|^2 dA(w)$$

$$\leq C \sum_{l=1}^N \sum_{j=1}^n \int_{\beta_j(O(w_l,\varepsilon_l) \cap \overline{\mathbb{D}})} |P(z)|^2 |(\beta_j^{-1})'(z)|^2 dA(z)$$

$$\leq CN \int_{\mathbb{D}} nM|P(z)|^2 dA(z)$$

$$= CnNM \int_{\mathbb{D}} |P(z)|^2 dA(z),$$

where

$$M = \sup\{|(\beta_j^{-1})'(z)|^2 : z \in O(w_l, \varepsilon_l), 1 \leq j \leq n, 1 \leq l \leq N\} < \infty.$$

Besides, there is a numerical constant C' satisfying

$$\int_{\mathbb{D}} |f(w)|^2 dA(w) \leq C' \int_{A_{r'}} |f(w)|^2 dA(w), \ f \in L_a^2(\mathbb{D}),$$

and hence $\int_{\mathbb{D}} |R_k(w)|^2 dA(w) \leq C' \int_{A_{r'}} |R_k(w)|^2 dA(w)$. Combining this inequality with the above arguments shows that

$$\|\tau_k P\|^2 = \|R_k\|^2 \leq K\|P\|^2, \tag{7.15}$$

where $K = nMNCC' < \infty$ depends only on B.

Recall that every function f in $L_a^2(\mathbb{D})$ can be expressed as a limit of polynomials $\{p_m\}$ in the Bergman-norm. By boundedness of $\tau_k(1 \leq k \leq n)$ (see (7.15)), $\{\tau_k p_m\}_m$ are uniformly bounded in norm, and hence $\{\tau_k p_m\}_m$ is a normal family. Then it follows that there is a subsequence $\{p_{m_l}\}$ such that $\{\tau_k p_{m_l}\}_l$ converges to some holomorphic function f_k (in the Bergman space) uniformly on compact subsets of $\mathbb{D}$. Clearly, these functions f_k satisfy

$$f(z) = \sum_{k=1}^{n} f_k(B(z))z^{k-1}, \ z \in \mathbb{D}.$$

Then by uniqueness of the representation, f_k are independent of the choices of $\{p_m\}$ and its subsequence. Put $\tau_k f = f_k$ and the proof is complete. $\qquad \square$

Remark 7.3.2 The above theorem remains true if the Bergman space is replaced with the weighted Bergman spaces, Dirichlet space and Hardy spaces $H^p(\mathbb{D})(p \geq 1)$, and the proof is similar.

We have the following corollary.

Corollary 7.3.3 *Let B be a finite Blaschke product of order n. Then there is a bounded invertible operator S from $L_a^2(\mathbb{D})$ to $L_a^2(\mathbb{D}) \otimes \mathbb{C}^n$ satisfying*

$$ST_B = (T_z \otimes I_n)S.$$

Consequently, for two finite Blaschke products B_1 and B_2, T_{B_1} is similar to T_{B_2} on the Bergman space if and only if order $B_1 = $ order B_2.

Proof Assume the operators $\tau_1, \cdots, \tau_n$ are defined as in Theorem 7.3.1 and set $Sf = (\tau_1 f, \cdots, \tau_n f)$ for each $f \in L_a^2(\mathbb{D})$. By Theorem 7.3.1, it is easy to see that S has a bounded inverse.

Moreover, we have $ST_B = (T_z \otimes I_n)S$. In fact, for each $f \in L_a^2(\mathbb{D})$,

$$T_B f = B \sum_{i=1}^{n} \tau_i f(B)z^{i-1} = \sum_{i=1}^{n}(z\tau_i f)(B)z^{i-1},$$

and hence by uniqueness,

$$ST_B f = (z\tau_1 f, \cdots, z\tau_n f) = (T_z \otimes I_n)Sf.$$

That is, $ST_B = (T_z \otimes I_n)S$.

Given two finite Blaschke products B_1 and B_2, if T_{B_1} is similar to T_{B_2} on the Bergman space, then $T_{B_1}^*$ is similar to $T_{B_2}^*$. Therefore,

$$\dim \ker T_{B_1}^* = \dim \ker T_{B_2}^*,$$

forcing order $B_1 = $ order B_2. The proof is complete. $\qquad \square$

An observation is in order. Given two finite Blaschke products B_1 and B_2 with the same order, by Corollary 7.3.3 there is a bounded invertible operator $S : L_a^2(\mathbb{D}) \to L_a^2(\mathbb{D})$ such that $ST_{B_1} = T_{B_2}S$, and thus for each polynomial p,

$$ST_{p(B_1)} = T_{p(B_2)}S. \tag{7.16}$$

For a function $f \in H^\infty(\mathbb{D})$, let p_k be the k-th Cesaro mean of f. Note that $\{p_k\}$ are polynomials satisfying $\|p_k\|_\infty \le \|f\|_\infty$, and that $\{p_k\}$ converges to f uniformly on compact subsets of $\mathbb{D}$. Then $\{T_{p_k(B_i)}\}$ converges to $T_{f(B_i)}$ in strong operator topology for $i = 1, 2$. Replacing p with p_k in (7.16) and then taking limits, we get

$$ST_{f(B_1)} = T_{f(B_2)}S.$$

That is, $T_{f(B_1)}$ is similar to $T_{f(B_2)}$.

The above observation inspires the following theorem, whose proof is similar to that of (ii)$\Rightarrow$(iii) in Theorem 7.1.5.

Theorem 7.3.4 *Suppose both f and g are in $H^\infty(\mathbb{D})$ and satisfy Cowen's condition. Then T_f is similar to T_g on the Bergman space $L_a^2(\mathbb{D})$ if and only if there are two finite Blaschke products B_1 and B_2 with the same order and a function $h \in H^\infty(\overline{\mathbb{D}})$ such that*

$$f = h \circ B_1 \quad \text{and} \quad g = h \circ B_2.$$

In a special case when both f and g are in $H^\infty(\overline{\mathbb{D}})$, Jiang and Zheng [JZ] got the following by applying the technique of K-theory, and the approach diverges much from the one presented here.

Corollary 7.3.5 (Jiang-Zheng) *Suppose f and g are in $Hol(\overline{\mathbb{D}})$. Then T_f is similar to to T_g on the Bergman space $L_a^2(\mathbb{D})$ if and only if there are two finite Blaschke products B_1 and B_2 with the same order and a function $h \in Hol(\overline{\mathbb{D}})$ such that*

$$f = h \circ B_1 \quad \text{and} \quad g = h \circ B_2.$$

Remark 7.3.6 Jiang and Zheng also proved Corollary 7.3.5 on the weighted Bergman spaces $L_{a,\alpha}^2(\mathbb{D})$ for $\alpha > -1$. By using the proof of (ii)$\Rightarrow$(iii) in Theorem 7.1.5, Corollary 7.3.5 also applies to the weighted Bergman spaces, the Dirichlet space and some other reproducing kernel Hilbert spaces because a weaker version of Theorem 7.1.1 also holds on the Dirichlet space.

A natural question arises from the Bergman space: for two infinite Blaschke products, are the corresponding Toeplitz operators always similar? On the Hardy space, the answer is affirmative. However, on the Bergman space $L_a^2(\mathbb{D})$, one can find two infinite Blaschke products ψ_1 and ψ_2 such that T_{ψ_1} is not similar to T_{ψ_2}. In fact, for a Blaschke product ψ, $\psi L_a^2(\mathbb{D})$ is closed if and only if ψ is a product of finitely many interpolating Blaschke products, see [Ho2, DS]. Then it is not difficult to construct two infinite Blaschke products ψ_1 and ψ_2 such that $\psi_1 L_a^2(\mathbb{D})$ is closed, while $\psi_2 L_a^2(\mathbb{D})$ is not closed; in this case, T_{ψ_1} is not similar to T_{ψ_2}, as desired.

At the end of Sect. 7.1 of this chapter, we see that if both f and g are entire functions, then T_f is similar to T_g on $H^2(\mathbb{D})$ if and only if there is a unimodular constant c such that $g(z) = f(cz)$. This result also holds on weighted Bergman spaces and the Dirichlet space, with the same discussion.

Example 7.3.7 If both f and g are polynomials, then T_f is similar to T_g on the Bergman space if and only if there is a unimodular constant c such that $g(z) = f(cz)$. For example, write $p_1(z) = 1 + 2z^3 + z^5$, $p_2(z) = -1 - 2z^3 - z^5$, Then T_{p_1} is never similar to T_{p_2}.

Also, there is some remaining questions: which kind of operator may lies in $\{T_B\}'$ where B is a given finite Blaschke product. To be precise, the first question asks whether there is a nonzero compact operator in $\{T_B\}'$? By Corollary 3.1.7, the zero operator is the only compact operator commuting with T_B. This statement holds not only on the Bergman space, but also on the Hardy space.

With regard to another question, compare the commutant algebra $\{T_B\}'$ defined on the Hardy space with $\{T_B\}'$ defined on the Bergman space, where B is a finite Blaschke product. As mentioned in [CW], those operators lying in these two different algebras have very close forms, and these two algebras share some similar properties and close connections, also see [Cow1, Cow2, Cow3, Cow4]. The following theorem shows that there is a great difference between these two commutant algebras. Precisely, for a finite Blaschke product B, T_B is itself a pure isometric operator on the Hardy space, and $T_B \in \{T_B\}'$. However, the case is different on the Bergman space, as illustrated by the following result due to Guo.

Theorem 7.3.8 *If B is a finite Blaschke product and T_B is defined on $L_a^2(\mathbb{D})$, then there is no pure isometry in $\{T_B\}'$.*

Proof Let B be a finite Blaschke product. Assume conversely that S is a pure isometry in $\{T_B\}'$. Note that $I = S^{*n}S^n$ for each positive integer n,

$$T_B^* T_B = T_B^* S^{*n} S^n T_B = S^{*n} T_B^* T_B S^n,$$

and hence

$$I - T_B^* T_B = S^{*n}(I - T_B^* T_B)S^n. \tag{7.17}$$

Note that $I - T_B^* T_B$ is compact, and that $\{S^n\}$ converges to 0 in the weak operator topology. Then for each $h \in L_a^2(\mathbb{D})$,

$$\limsup_{n\to\infty} \|S^{*n}(I - T_B^* T_B)S^n h\| \leq \limsup_{n\to\infty} \|(I - T_B^* T_B)S^n h\| = 0,$$

which, combined with (7.17), gives

$$I - T_B^* T_B = 0.$$

In particular, $\langle (I - T_B^* T_B)1, 1 \rangle = 0$. That is,

$$\int_{\mathbb{D}} (1 - |B(z)|^2) dA(z) = 0.$$

Since $1 - |B(z)|^2$ is a continuous, nonnegative function, it follows that $|B| \equiv 1$, forcing B to be a constant. This is a contradiction. Thus $\{T_B\}'$ contains no pure isometry.   □

Applying Cowen-Thomson's theorem(=Theorem 3.1.1) immediately gives the following corollary.

Corollary 7.3.9 *If φ is a function in $H^\infty(\mathbb{D})$ which satisfy Cowen's condition, then there is no pure isometry in $\{T_\varphi\}'$ defined on $L_a^2(\mathbb{D})$.*

7.4 Remarks on Chap. 7

As mentioned in the beginning, this chapter elaborates on similarity and unitary equivalence of multiplication operators, defined on the Hardy space and the (weighted) Bergman spaces. Some ideas developed in this book are intended to give a new approach for the results listed in this chapter.

Theorem 7.3.1 comes from [GH1]. Theorem 7.3.8 and Corollary 7.3.9 are new and due to Guo. This chapter mainly comes from [Cow4, GH1, JZ, Sun1, SZZ1]. For more materials on this line, the reader can refer to the literatures [Cow1, Cow3, Cow4, CW, GZ].

In the framework of Hilbert module for the multi-variable operator theory [DPa], similarity and unitary equivalence of operator tuples can be studied by module approach [CG, Guo1, Guo2, Guo3, Guo4, Guo5, Guo6], etc. As for the study of commutants of operator tuples, refer to [GW1, GW2, GW3, GW4]. A geometric investigation in classification of operators goes back to [CoD, Dou].

Chapter 8
Algebraic Structure and Reducing Subspaces

In preceding chapters, we investigated reducing subspaces of analytic multiplication operators and the related von Neumann algebras generated by these multiplication operators whose symbols range over finite Blaschke products, thin Blaschke products and covering Blaschke products. In most interesting situations, multiplication operators on function spaces are essentially normal. This chapter is firstly devoted to discussion of algebraic structure of general essentially normal operators. Then we apply these results to the study of algebraic structure and reducing subspaces of multiplication operators, and the related von Neumann algebras generated by these operators in the cases of both single variable and multi-variable.

8.1 Algebraic Structure of Essentially Normal Operators

The material of this section mainly comes from [Con2] and [Gil2]. In this section, we center our attention on Behncke's result concerning with the algebraic structure of essential normal operators, and then present some problems and conjectures which show a link between function theory and operator theory.

Before going on, some notations are in order. Throughout this chapter, $\mathcal{H}$ always denotes a Hilbert space, and $\mathcal{K}(\mathcal{H})$ denotes the algebra of all compact operators on $\mathcal{H}$. For any two operators A and B in $B(\mathcal{H})$, define

$$[A, B] \triangleq AB - BA.$$

An operator T is called *essential normal* if $[T^*, T] = T^*T - TT^* \in \mathcal{K}(\mathcal{H})$. As done in Sect. 2.5 in Chap. 2, for a collection $\mathcal{E}$ of projections, let $\bigvee_{P \in \mathcal{E}} P$ denote the orthogonal projection onto the closed space spanned by the ranges of P, where P run over $\mathcal{E}$. Similarly, for a collection of subspaces $\{\mathcal{H}_\alpha : \alpha \in \Lambda\}$, $\bigvee_{\alpha \in \Lambda} \mathcal{H}_\alpha$ denotes the closed subspace spanned by $\{\mathcal{H}_\alpha : \alpha \in \Lambda\}$.

© Springer-Verlag Berlin Heidelberg 2015
K. Guo, H. Huang, *Multiplication Operators on the Bergman Space*,
Lecture Notes in Mathematics 2145, DOI 10.1007/978-3-662-46845-6_8

Behncke [Be1] obtained the following result, which goes back to a paper of Suzuki [Su], where the case of $S - S^*$ being compact is discussed. We also call the reader's attention to [Con2, p. 159, Theorem 5.4].

Theorem 8.1.1 (Behncke) *If $S \in B(\mathcal{H})$ and S is essential normal, then we have the decomposition*

$$\mathcal{H} = \mathcal{H}_0 \oplus \mathcal{H}_1 \oplus \cdots,$$

where

(1) each $\mathcal{H}_n$ reduces S;
(2) $S_0 \triangleq S|_{\mathcal{H}_0}$ is a maximal normal operator;
(3) for any $n \geq 1$, $S_n \triangleq S|_{\mathcal{H}_n}$ is irreducible and essentially normal.

The decomposition is unique in the sense that if S_i and $\mathcal{H}_i (i \geq 0)$ are replaced with S_i' and $\mathcal{H}_i'$, which satisfy (1)–(3), and both S_0 and S_0' are maximal, then after reordering $\mathcal{H}_i' (i \geq 1)$ there is a unitary operator U commuting with S such that

$$U^* P_{\mathcal{H}_i} U = P_{\mathcal{H}_i'} \quad \text{and} \quad U^* S_i U|_{\mathcal{H}_i'} = S_i', \ i \geq 0.$$

Here, by saying S_0 is maximal, we mean that there is no subspace $\mathcal{K}_0 \supsetneq \mathcal{H}_0$ such that $\mathcal{K}_0$ reduces S and $S|_{\mathcal{K}_0}$ is normal.

Proof Because of its length, the proof is divided into two parts.

Existence This part comes from [Con2, p. 159].

Let $\mathcal{A}$ denote the von Neumann algebra generated by S and consider the ideal $\mathcal{F} \triangleq \mathcal{A} \cap \mathcal{K}(\mathcal{H})$ of $\mathcal{A}$. Write

$$[\mathcal{F}\mathcal{H}] \triangleq \overline{\text{span}\{Ah : A \in \mathcal{F}, \ h \in \mathcal{H}\}},$$

and clearly [$\mathcal{F}$ is invariant for $\mathcal{A}$. Since $\mathcal{A}$ is self-adjoint, both $[\mathcal{F}\mathcal{H}]$ and $[\mathcal{F}\mathcal{H}]^\perp$ are reducing for $\mathcal{A}$. Write $\mathcal{H}_0 = [\mathcal{F}\mathcal{H}]^\perp$, and denote $S|_{\mathcal{H}_0}$ by S_0.

By the essential normality of S, $[S^*, S] \in \mathcal{A} \cap \mathcal{K}(\mathcal{H})$. Then for any $h \in \mathcal{H}_0$, $[S^*, S]h \in \mathcal{F}\mathcal{H}_0$. On the other hand, $\mathcal{H}_0$ reduces $\mathcal{A}$ as well as S, and thus

$$[S^*, S]h \in \mathcal{H}_0 = [\mathcal{F}\mathcal{H}]^\perp, \ h \in \mathcal{H}_0,$$

forcing $[S^*, S]h = 0$. That is, $[S_0^*, S_0] = 0$, i.e. S_0 is normal.

By definition, $\mathcal{F}|_{\mathcal{H}_0^\perp}$ is a C^*-algebra of compact operators. By [Ar2, Theorem 1.4.5] there is a sequence $\{\mathcal{H}_n\}_{n \geq 1}$ of reducing subspaces for $\mathcal{F}$ such that

$$\mathcal{H}_0^\perp = \mathcal{H}_1 \oplus \mathcal{H}_2 \oplus \cdots,$$

and $\mathcal{F}|_{\mathcal{H}_n} = \mathcal{K}(\mathcal{H}_n)$ for $n \geq 1$. Since $\mathcal{H}_n = [\mathcal{F}\mathcal{H}_n]$ and $\mathcal{F}$ is an ideal of $\mathcal{A}$, $\mathcal{H}_n$ reduces $\mathcal{A}$, and hence $\mathcal{H}_n$ reduces S. Write $S_n = S|_{\mathcal{H}_n}$, and it is easy to see that

$$\mathcal{A}|_{\mathcal{H}_n} = \mathcal{W}^*(S_n),$$

where $\mathcal{W}^*(S_n)$ denotes the von Neumann algebra generated by S_n. The essential normality of $S_n(n \geq 1)$ follows directly from that of S. In addition, the irreducibility of $S_n(n \geq 1)$ follows from that of $\mathcal{K}(\mathcal{H}_n)$ and the fact

$$\mathcal{W}^*(S_n) = \mathcal{A}|_{\mathcal{H}_n} \supseteq \mathcal{F}|_{\mathcal{H}_n} = \mathcal{K}(\mathcal{H}_n). \tag{8.1}$$

It can be required that all $\mathcal{H}_n(n \geq 1)$ satisfy $\dim \mathcal{H}_n \geq 2$. Otherwise, replace $\mathcal{H}_0$ with

$$\bigoplus_{k;\, \dim \mathcal{H}_k=1} \mathcal{H}_k \oplus \mathcal{H}_0,$$

and then in the new decomposition

$$\mathcal{H} = \mathcal{H}_0 \oplus \mathcal{H}_1 \oplus \cdots,$$

we have $\dim \mathcal{H}_n \geq 2$ for all $n \geq 1$. In this case, put $S_0 \triangleq S|_{\mathcal{H}_0}$, which is a normal operator.

It remains to show that S_0 is maximal. For this, assume conversely that there is a subspace $\mathcal{K}_0 \supsetneq \mathcal{H}_0$ such that $\mathcal{K}_0$ reduces S and $S|_{\mathcal{K}_0}$ is normal. Then $S|_{\mathcal{K}_0 \ominus \mathcal{H}_0}$ is normal. Since

$$P_{\mathcal{K}_0 \ominus \mathcal{H}_0} \sum_{i \geq 1} P_{\mathcal{H}_i} = P_{\mathcal{K}_0 \ominus \mathcal{H}_0}(I - P_{\mathcal{H}_0}) \neq 0,$$

there is a positive integer j such that $P_{\mathcal{K}_0 \ominus \mathcal{H}_0} P_{\mathcal{H}_j} \neq 0$. By the minimality of $P_{\mathcal{H}_j}$ and Lemma 2.5.5, there is a projection $P' \in \mathcal{A}'$ satisfying

$$P_{\mathcal{H}_j} \sim P' \leq P_{\mathcal{K}_0 \ominus \mathcal{H}_0}.$$

Thus, there is an operator $V \in \mathcal{A}'$ satisfying

$$V^*V = P_{\mathcal{H}_j} \quad \text{and} \quad VV^* = P'. \tag{8.2}$$

Note that the linear map

$$\mathcal{W}^*(S_j) \longrightarrow \mathcal{W}^*(S|_{P'\mathcal{H}})$$

$$T \longmapsto VTV^*$$

defines a $*$-isomorphism. Since $S|_{\mathcal{K}_0 \ominus \mathcal{H}_0}$ is normal and $P'\mathcal{H}$ is a subspace of $\mathcal{K}_0 \ominus \mathcal{H}_0$ that reduces S, then $S|_{P'\mathcal{H}}$ is normal, and thus both $\mathcal{W}^*(S|_{P'\mathcal{H}})$ and $\mathcal{W}^*(S_j)$ are abelian. Observe that by (8.1) $\mathcal{W}^*(S_j) \supseteq \mathcal{K}(\mathcal{H}_j)$, and then we get $\dim \mathcal{H}_j = 1$, which is a contradiction. Therefore, S_0 is maximal.

The proof for the existence of the desired decomposition is complete.

Uniqueness   The following proof is due to Guo and Huang.

Assume that there is a family of reducing subspaces $\mathcal{H}'_i$ satisfying such conditions as in Theorem 8.1.1. Let $P_{\mathcal{H}_n}$ and $P_{\mathcal{H}'_n}$ denote the orthogonal projections onto $\mathcal{H}_n$ and $\mathcal{H}'_n$, respectively, and rewrite

$$P_n = P_{\mathcal{H}_n} \quad \text{and} \quad Q_n = P_{\mathcal{H}'_n}.$$

Note that for $n \geq 1$, P_n and Q_n are minimal projections in $\mathcal{A}'$.

First, assume that $\mathcal{A}'$ is a finite von Neumann algebra. In this case, we provide the proof of the uniqueness as follows. First we recall a fact, see [Con1, Proposition 48.5b].

Fact   For a finite von Neumann algebra $\mathcal{B}$, two projections E and F are equivalent if and only if there is a unitary operator U in $\mathcal{B}$ satisfying $U^*EU = F$. In this case, $E^\perp \sim F^\perp$.

Since $\sum Q_n = I$, there is some integer $i_1 (i_1 \geq 0)$ such that $P_1 Q_{i_1} \neq 0$. By the proof of Corollary 2.5.6, Q_{i_1} is equivalent to the projection P_1 in $\mathcal{A}'$. Then there is a unitary operator $W \in \mathcal{A}'$ satisfying $W^* P_1 W = Q_{i_1}$. Soon we will see that there is some integer $i_2(i_2 \neq i_1)$ such that Q_{i_2} is equivalent to P_2 in $\mathcal{A}'$. For this, note that $WQ_{i_1}W^* = P_1$, which gives that

$$P_2 \sum_{n \neq i_1} WQ_n W^* = P_2(I - P_1) \neq 0.$$

By similar reasoning as above, there is an integer i_2 such that $WQ_{i_2}W^* \sim P_2$ in $\mathcal{A}'$. Since $Q_{i_2} \sim WQ_{i_2}W^*$, then $Q_{i_2} \sim P_2$, as desired. This procedure can be repeated. Eventually one gets a sequence $\{i_n\}$ of $\mathbb{Z}_+$ such that each $P_n \sim Q_{i_n}$ for $n \geq 1$. This implies

$$\sum_{n \geq 1} P_n \sim \sum_{n \geq 1} Q_{i_n}. \tag{8.3}$$

Furthermore, we claim that for each $n \geq 1$, $i_n \neq 0$. Otherwise, assume that there is an integer $k(k \geq 1)$ such that $P_k \sim Q_0$. Then by a similar discussion as below (8.2), there is a $*$-isomorphism between $\mathcal{W}^*(S_k)$ and $\mathcal{W}^*(S|_{Q_0})$. However, $\mathcal{W}^*(S_k)$ is not abelian, but $\mathcal{W}^*(S|_{Q_0})$ is abelian. This is a contradiction. Thus, $i_n \neq 0$ for each $n \geq 1$.

Below, it will be shown that $\{i_n : n \geq 1\} = \mathbb{Z}_+ - \{0\}$. For this, write

$$\Lambda \triangleq \mathbb{Z}_+ - \{0\} - \{i_n : n \geq 1\}$$

and assume conversely that Λ is not empty. Then by (8.3) and the above fact,

$$P_{\mathcal{H}_0} \sim \sum_{n \in \Lambda} Q_{i_n},$$

which implies that $S|_{\mathcal{H}_0}$ is unitarily equivalent to $S|_{\bigvee_{n \in \Lambda} \mathcal{H}'_n}$, and hence both $S|_{\mathcal{H}_0}$ and $S|_{\bigvee_{n \in \Lambda} \mathcal{H}'_n}$ are normal operators. Pick a nonzero integer $k \in \Lambda$, $S|_{\mathcal{H}'_k}$ is normal because $\mathcal{H}'_k$ reduces S. Therefore $S|_{H'_0 \oplus \mathcal{H}'_k}$ is also normal, which gives a contradiction to the maximality of S'_0.

Since $\{i_n : n \geq 1\} = \mathbb{Z}_+ - \{0\}$, by the afore-mentioned fact we get $P_0 \sim Q_0$. Now write $i_0 = 0$, and $P_n \sim Q_{i_n}$ holds for all $n \geq 0$. That is, for each $n \geq 0$ there exists an operator $V_n \in \mathcal{A}'$ satisfying

$$V_n^* V_n = P_n \text{ and } V_n V_n^* = Q_{i_n}.$$

Define $U = \sum_{n \geq 0} V_n^*$, which proves to be a well-defined unitary operator in $\mathcal{A}'$, satisfying

$$U^* P_n U = Q_{i_n}, \; n \geq 0.$$

Then it is straightforward to verify that $U^* S_n U|_{\mathcal{H}'_{i_n}} = S|_{\mathcal{H}'_{i_n}}$ for $n \geq 0$. We are done in the case of $\mathcal{A}'$ being finite.

In general, the proof of uniqueness can be handled as below. For this, we will discuss the structure of $\mathcal{A}'$. By Corollary 2.5.6, for any two minimal projections P_i and $P_j(i, j \geq 1)$, either $P_i \perp P_j$ or $P_i \sim P_j$. Therefore, there is a partition $\bigwedge$ consisting of pairwise disjoint subsets $E_k(k \geq 1)$ of $\{1, 2, \cdots\}$ such that

(1) $\bigsqcup_{k \geq 1} E_k = \{1, 2, \cdots\}$;
(2) if $j, l \in E_k$, then $P_j \sim P_l$.
(3) for any two integers j and l belonging to different E_k, P_j is never equivalent to P_l.

Clearly, $\mathcal{A}'|_{\bigvee_{i \in E_k} P_i \mathcal{H}}$ is a homogenous von Neumann algebra. Recall that a von Neumann algebra is *homogeneous* if there is a family of orthogonal abelian projections that are mutually equivalent and whose sum is the identity. Since for each $j \geq 1$, P_j is a minimal projection in $\mathcal{A}'$, by

$$P_j \mathcal{A}' P_j = \mathbb{C} P_j, \; j \geq 1,$$

which, combined with Theorem 2.5.2 yields that

$$\mathcal{A}'|_{\bigvee_{i\in E_k} P_i \mathcal{H}} \overset{unitarily\ isomorphic}{\cong} M_{n_k}(\mathbb{C}) \otimes I_{\mathcal{H}_{i_k}}, \tag{8.4}$$

where $n_k = \sharp E_k$ and $i_k \in E_k$. For $n_k = \infty$, we regard $M_{n_k}(\mathbb{C})$ as $B(l^2)$, the algebra of all linear bounded operators on l^2.

By Corollary 2.5.7, $\bigvee_{i\in E_k} P_i$ lies in the center $Z(\mathcal{A})$ of $\mathcal{A}$ for each k, and then $P_0 \in Z(\mathcal{A})$ because

$$P_0 = I - \sum_{k\geq 1} \bigvee_{i\in E_k} P_i.$$

Therefore, by (8.4)

$$\mathcal{A}' \overset{unitarily\ isomorphic}{\cong} \mathcal{A}'|_{\mathcal{H}_0} \oplus \bigoplus_k M_{n_k}(\mathbb{C}) \otimes I_{\mathcal{H}_{i_k}}.$$

That is,

$$\mathcal{A}' \overset{unitarily\ isomorphic}{\cong} \mathcal{W}^*(S_0)' \oplus \bigoplus_k M_{n_k}(\mathbb{C}) \otimes I_{\mathcal{H}_{i_k}}, \tag{8.5}$$

and thus

$$\mathcal{A} \overset{unitarily\ isomorphic}{\cong} \mathcal{W}^*(S_0) \oplus \bigoplus_k I_{n_k}(\mathbb{C}) \otimes B(\mathcal{H}_{i_k}). \tag{8.6}$$

Here, it is required that all $\dim \mathcal{H}_{i_k} \geq 2$; that is, $\dim \mathcal{H}_i \geq 2$ for $i \geq 1$. In fact, if there were some i_k such that $\dim \mathcal{H}_{i_k} = 1$, then clearly $\mathcal{H}_0$ could be enlarged to be some $\mathcal{K}_0$ with $S|_{\mathcal{K}_0}$ being normal. This would be a contradiction to the maximality of S_0. By (8.5), the remaining is an easy exercise.   □

The proof of uniqueness in Theorem 8.1.1 immediately derives a consequence.

Corollary 8.1.2 *Suppose S is essential normal, and denote by $\mathcal{A}$ the von Neumann algebra generated by S. Then both $\mathcal{A}$ and $\mathcal{A}'$ are of type* I.

Proof Suppose S is essential normal, and let $\mathcal{A}$ denote the von Neumann algebra generated by S. By (8.6), $\mathcal{A}$ is the direct sum of $\mathcal{W}^*(S_0)$ and some type I von Neumann algebras. Since S_0 is normal, the von Neumann algebra $\mathcal{W}^*(S_0)$ generated by S_0 is abelian, and hence $\mathcal{W}^*(S_0)$ is of type I. It is known that the direct sum of type I von Neumann algebras is of type I [Con1], and thus $\mathcal{A}$ is of type I. Then by Proposition 2.5.4, $\mathcal{A}'$ is of type I. The proof is complete.   □

It is of interest to consider Theorem 8.1.1 and its corollary in the setting of concrete operators in reproducing kernel Hilbert spaces, specifically, multiplication operators on classical function spaces such as the Hardy space, the Bergman space and etc.

Example 8.1.3 Let $\mathcal{H}$ denote the Bergman space $L_a^2(\mathbb{D})$ or the Hardy space $H^2(\mathbb{D})$, and define $S = M_{z^n}$ for some positive integer n. Denote by $\mathcal{A}$ the von Neumann algebra generated by S, and clearly $\mathcal{A}' = \{S, S^*\}'$.

In the case of $\mathcal{H}$ being the Bergman space over $\mathbb{D}$, it is not difficult to verify that S has exactly n minimal reducing spaces: $M_0, M_1, \cdots, M_{n-1}$, where

$$M_i = \overline{span}\{z^{nk-1-i} : k = 1, 2, \cdots\}, \ 0 \le i \le n - 1.$$

Writing $\mathcal{H}_j = M_{j-1}$, $1 \le j \le n$, we have the decomposition of $\mathcal{H}$ as in Theorem 8.1.1. By Proposition 2.5.1(e), each von Neumann algebra equals its norm closure of the linear span of its projections. Then it follows that $\mathcal{A}'$ is the linear span of $\{P_{\mathcal{H}_j} : 1 \le j \le n\}$, and hence $\mathcal{A}'$ is abelian.

If $\mathcal{H} = H^2(\mathbb{D})$, the Hardy space, then $\mathcal{A}'$ is not abelian whenever $n \ge 2$. This is true in a general setting where $S = M_B$ with B a Blaschke product satisfying order $B \ge 2$. We call the reader's attention to Example 2.6.6. Precisely, let S be a pure isometry defined on a Hilbert space $\mathcal{N}$, i.e. $\bigcap_n S^n \mathcal{N} = 0$. Set $\mathcal{N}_0 = \mathcal{N} \ominus S\mathcal{N}$, and then $\mathcal{A}'$ is $*$-isomorphic to $B(\mathcal{N}_0)$.

Assume that $\mathcal{H} = L_a^2(\mathbb{D})$ or $H^2(\mathbb{D})$. For a function f in the disk algebra $A(\mathbb{D})$, put $S = M_f$. If f is a polynomial, it is easy to see that $[S, S^*]$ is a compact operator. Since each function in $A(\mathbb{D})$ can be uniformly approximated by polynomials on $\overline{\mathbb{D}}$, it follows that $[S, S^*]$ is compact for each $f \in A(\mathbb{D})$. In this case, Corollary 8.1.2 shows that both $\mathcal{A}$ and $\mathcal{A}'$ are type I von Neumann algebras. Similar results also holds on the weighted Bergman spaces.

In some situation, the space $\mathcal{H}_0$ in Theorem 8.1.1 does appear. For this, just consider the case of Example 8.1.3 where $\mathcal{H} = H^2(\mathbb{D})$ and $S = M_B$, B being a finite Blaschke product. Nevertheless, sometimes $\mathcal{H}_0$ degenerates. For example, given a nonconstant function f in $A(\mathbb{D})$ consider the multiplication operator M_f acting on a weighted Bergman space $L_{a,\alpha}^2(\mathbb{D})$ with $\alpha > -1$. In this case, M_f is essential normal; that is, M_f satisfies the assumption of Theorem 8.1.1. Besides, one can show that $M_f^* M_f h - M_f M_f^* h = 0$ if and only if $h = 0$, which implies that the space $\mathcal{H}_0$ vanishes. In this situation, let $\mathcal{A}$ denote the von Neumann algebra generated by M_f, and by (8.5)

$$\mathcal{A}' \overset{\text{unitarily isomorphic}}{\cong} \bigoplus_k M_{n_k}(\mathbb{C}) \otimes I_{\mathcal{H}_{i_k}}. \tag{8.7}$$

where $1 \le n_k \le \infty$ and $\mathcal{H}_{i_k}$ are some Hilbert spaces. This provides a rough structure of $\mathcal{A}'$ that is closely related to the structure of all reducing subspaces for M_f, as illustrated in Sect. 2.6 of Chap. 2.

It is meaningful to pose a condition for all n_k in (8.7) to be 1 on a fixed reproducing Hilbert space, and this is equivalent to the abelian property of $\mathcal{A}'$. Inspired by this, we raise the following problem on the Bergman space.

Suppose $\phi \in H^\infty(\mathbb{D})$ and M_ϕ denotes the multiplication operator defined on the Bergman space $L_a^2(\mathbb{D})$. Put $\mathcal{V}^*(\phi) = \{M_\phi, M_\phi^*\}'$. If $\mathcal{V}^*(\phi)$ is a type I von Neumann algebra, then is $\mathcal{V}^*(\phi)$ abelian?

Note that Douglas-Putinar-Wang's theorem (Theorem 4.2.1) follows directly from an affirmative answer to the above problem even in the case of ϕ being a member of $A(\mathbb{D})$. Such an affirmative answer would also imply the following.

Conjecture 8.1.4 If f is in the disk algebra $A(\mathbb{D})$, then the von Neumann algebra $\mathcal{V}^*(f)$ defined on $L_a^2(\mathbb{D})$ is abelian.

Some words are in order. Cowen-Thomson's theorem (Theorem 3.1.1) states that if ϕ lies in Cowen's class, then there exists a finite Blaschke product B and an H^∞-function ψ such that $\phi = \psi(B)$ and $\{M_\phi\}' = \{M_B\}'$ holds on the Bergman space $L_a^2(\mathbb{D})$. In particular, $\mathcal{V}^*(\phi) = \mathcal{V}^*(B)$. Then by Douglas-Putinar-Wang's theorem, for each finite Blaschke product B the von Neumann algebra $\mathcal{V}^*(B)$ is abelian, and so is $\mathcal{V}^*(\phi)$. Therefore, if ϕ belongs to Cowen's class, then $\mathcal{V}^*(\phi)$ is abelian.

On the other hand, to the best of our knowledge it is not known whether each function in $A(\mathbb{D})$ belongs to Cowen's class, for details see Sect. 3.1 in Chap. 3. Thus Conjecture 8.1.4 still remains open, and an operator-theoretic approach toward it will be inspiring and interesting.

In [Gil2], an operator A is called *primary* if $\mathcal{W}^*(A)$ is a factor. Gilfeather presented a generalization of Theorem 8.1.1 as follows [Gil2].

Theorem 8.1.5 (Gilfeather) Let A be an operator on a Hilbert space $\mathcal{H}$ and $p(z, \bar{z})$ be a non-commutative complex polynomial such that $p(A, A^*)$ is a compact operator. Then there exists a unique sequence of central projections $\{P_i\}_{i=0}^n (n \leq \infty)$ in $\mathcal{W}^*(A)$ so that

$$A = A_0 \oplus \left(\bigoplus_{i=1}^n A_i \right),$$

where $A_0 \triangleq AP_0\mathcal{H}$ satisfies $p(A_0, A_0^*) = 0$, and $\mathcal{H} \ominus P_0\mathcal{H}$ is separable, $A_i \triangleq AP_i\mathcal{H}(i \geq 1)$ are primary operators with $p(A_i, A_i^*)$ compact and nonzero.

Before the proof of Theorem 8.1.5 is presented, some words are in order. In Gilfeather's paper [Gil2] main interest is focused on special cases of Theorem 8.1.5 where $p(z, \bar{z})$ is one of the following: (1) $p(z, \bar{z}) = z - \bar{z}$; (2) $p(z, \bar{z}) = z\bar{z} - \bar{z}z$; (3) $p(z, \bar{z}) = z\bar{z}z - \bar{z}z^2$; (4) $p(z, \bar{z}) = 1 - \bar{z}z$; (5) $p(z, \bar{z}) = z - z\bar{z}z$. As mentioned in [Gil2], case (1) was studied by Brodskii and Livsic [BL], also concerned in Suzuki's original work. Cases (2) and (3) were studied by Behncke [Be1, Be2]; and case (3) by Brown. Case (4) was studied by Gilfeather in his thesis [Gil1], also by Sz. Nagy and Foias in the case of A being a contraction.

Also, Gilfeather gave more detailed descriptions for the algebraic structure of the von Neumann algebra $\mathcal{W}^*(A)$ and its commutant $\mathcal{V}^*(A)$. One of them is the following result from [Gil2, Proposition 3]. Its proof is similar to that of [Su, Proposition 2], as mentioned in [Gil2].

Proposition 8.1.6 *Let A be a primary operator and $p(z, \bar{z})$ is a non-commutative polynomial such that $p(A, A^*)$ is compact and nonzero. Then $\mathcal{V}^*(A)$ is a type I von Neumann algebra.*

Gilfeather also get some consequences [Gil2].

Corollary 8.1.7 *Let A be an operator and $p(z, \bar{z})$ is a non-commutative polynomial such that $p(A, A^*)$ is compact. Then $\mathcal{W}^*(A)$ is type I if and only if the operator A_0 in Theorem 8.1.5 generates a type I von Neumann algebra.*

The following presents special cases of independent interest, see [Gil2].

Corollary 8.1.8 *Let A be an operator such that $p(A, A^*)$ is compact. Then $\mathcal{W}^*(A)$ is a type I von Neumann algebra if p has one of the following forms:*

(1) $p(z, \bar{z}) = z - \bar{z}$;
(2) $p(z, \bar{z}) = \bar{z}z - z\bar{z}$;
(3) $p(z, \bar{z}) = 1 - \bar{z}z$.

Note that case (2) of Corollary 8.1.8 covers Corollary 8.1.2.

Some comments on Theorem 8.1.5 and Corollary 8.1.8 is in order. Consider the cases of (2) and (3) in Corollary 8.1.8. Now B denotes a finite Blaschke product and M_B is the corresponding multiplication operator on the Bergman space. Both $M_B^* M_B - M_B M_B^*$ and $I - M_B^* M_B$ are non-negative compact operators; furthermore, it is not difficult to verify the trivialness of their kernels. By the spectral theorem, there is a sequence $\{\lambda_n\}$ of distinct positive numbers satisfying

$$I - M_B^* M_B = \sum_n \lambda_n E_n,$$

where E_n denotes the eigenvector space $\{h \in L_a^2(\mathbb{D}) : (I - M_B^* M_B)h = \lambda_n h\}$ for each n. Then a question is naturally raised:

$$\text{is } E_n \text{ of rank one for each } n?$$

If the answer is affirmative, it will lead to Douglas-Putinar-Wang's theorem that was mentioned below Conjecture 8.1.4. The reasoning is as follows. To prove the von Neumann algebra $\mathcal{V}^*(B)$ is abelian, it suffices to show any two operators T_1 and T_2 in $\mathcal{V}^*(B)$ commutes. Since T_1 commutes with $I - M_B^* M_B$, T_1 commutes with E_n for each n and so does T_2. By the assumption each E_n is of rank one, and write

$$E_n = e_n \otimes e_n, \ n \geq 1.$$

Note that $\{e_k : k \geq 1\}$ is an orthonormal basis and T_1 commutes with each E_n. Therefore T_1 is diagonal with respect to $\{e_k : k \geq 1\}$ and so does T_2. Then it is clear that T_1 commutes with T_2 and $V^*(B)$ is abelian. That is, we would obtain Douglas-Putinar-Wang's theorem alternatively if the answer for the above problem were affirmative. The above discussion applies to the case $B = z^N$ for a positive integer N. Since

$$I - M_{z^N}^* M_{z^N} = \sum_n \frac{N}{N + n + 1} e_n \otimes e_n,$$

where $e_n = \sqrt{n+1}\, z^n$, $n = 0, 1, \cdots$, we see that the von Neumann algebra $V^*(z^N)$ is abelian. Besides, it is of interest to find more applications of Theorem 8.1.5 in function theory.

Now turn back to the theory of von Neumann algebras. Recall that given a family $\mathcal{F}$ of operators in a von Neumann algebra $\mathcal{A}$, the *central support* of $\mathcal{F}$ is defined to be

$$\inf\{ C : C \text{ is a projection in } Z(\mathcal{A}) \text{ satisfying } AC = A, A \in \mathcal{F}\}.$$

In particular, the central support of a single operator A in $\mathcal{A}$ is defined to be

$$\inf\{ C : C \text{ is a projection in } Z(\mathcal{A}) \text{ satisfying } AC = A\}.$$

The following result is from [Gil2].

Proposition 8.1.9 *Let $\mathcal{A}$ be a von Neumann algebra such that $\mathcal{K} \cap \mathcal{A}$ has the central support I. Then the lattice of projections in the center $Z(\mathcal{A})$ of $\mathcal{A}$ is atomic; that is, each nonzero projection P in $Z(\mathcal{A})$ majorizes a nonzero minimal projection in $Z(\mathcal{A})$.*

Proof The proof is from [Gil2].

Suppose P is a nonzero projection in $Z(\mathcal{A})$. Since $\mathcal{K} \cap \mathcal{A}$ has the central support I, it is not difficult to see that there is an operator T in $\mathcal{K} \cap \mathcal{A}$ such that $PT \neq 0$. Write $T_0 = PTT^*P$, and then

$$PT_0 = T_0 \text{ and } T_0^* = T_0.$$

By the spectral decomposition, we deduce that $E = PE$ for each nonzero spectral projection E of T_0. By Proposition 2.5.1 $E \in W^*(T_0) \subseteq \mathcal{A}$. Since T_0 is compact, each spectral projection E of T_0 is finite dimensional and then there exists a nonzero minimal projection E_1 in $\mathcal{A}$ such that $E_1 \leq E$.

Now let Q be the central support of E_1. To finish the proof, it suffices to show that Q is a minimal projection in $Z(\mathcal{A})$ majorized by P. By [Con1, Proposition 43.7], Q is the orthogonal projection onto $[Ah : h \in E_1, A \in \mathcal{A}]$. Since $E_1 \leq P$ and $P \in Z(\mathcal{A})$, we have $Q \leq P$.

It remains to show that Q is minimal. Now consider any nonzero projection $R \in Z(\mathcal{A})$ with $R \leq Q$. First observe that $RE_1 \neq 0$. In fact, if $RE_1 = 0$, then $(Q - R)E_1 = E_1$, then by the definition of central support, $Q \leq Q - R$, forcing $R = 0$, which is a contradiction. Thus, we get $RE_1 \neq 0$ as desired. Since $RE_1 = RE_1^2 = E_1RE_1$ and E_1 is minimal, we have $RE_1 = E_1$. Since Q is the central support of E_1, $Q \leq R$. Also, we have $R \leq Q$, forcing $R = Q$, as desired. $\square$

To end this section, we give the proof of Theorem 8.1.5, which comes from [Gil2].

Proof of Theorem 8.1.5 First we describe the subspace $\mathcal{H} \ominus P_0\mathcal{H}$. Write

$$w(A, A^*) = \prod_{1 \leq i \leq n} A^{k_i} A^{*m_i},$$

where k_i and m_i are non-negative integers and $n \in \mathbb{Z}_+$. In this case, we say $w(A, A^*)$ is a *word* in A and A^*. Denote by M the subspace of $\mathcal{H}$ generated by

$$\{w(A, A^*)x : w(A, A^*) \text{ is a word in } A \text{ and } A^* \text{ and } x \in p(A, A^*)\mathcal{H}\}.$$

Since the range of a compact operator is separable and $p(A, A^*)$ is compact, $p(A, A^*)\mathcal{H}$ is separable, and so is M. By definition, M is invariant under both A and A^*. That is, M is reducing for A. Set $Q \triangleq P_M$, the orthogonal projection onto M. Then we have $Q \in W^*(A)' \equiv V^*(A)$. Besides, for any T in $V^*(A)$, and $y = w(A, A^*)p(A, A^*)z$ with $z \in \mathcal{H}$,

$$Ty = Tw(A, A^*)p(A, A^*)z = w(A, A^*)p(A, A^*)Tz \in M.$$

Therefore, M is invariant under $V^*(A)$, and thus by von Neumann bi-commutant theorem

$$Q = P_M \in V^*(A)' = W^*(A).$$

Therefore, $Q \in Z(\mathcal{A})$.

Now put $P_0 = I - Q$, $A_0 = A|_{P_0\mathcal{H}}$. Soon one will see that $p(A_0, A_0^*) = 0$. For this, note that for any $x \in P_0\mathcal{H}$, $x = P_0x = (I - Q)x$, and thus

$$p(A_0, A_0^*)x = Qp(A_0, A_0^*)x = Qp(A_0, A_0^*)(I - Q)x = Q(I - Q)p(A_0, A_0^*)x = 0.$$

The third identity follows from that $Q \in Z(\mathcal{A})$. Therefore, $p(A_0, A_0^*) = 0$. In particular, M equals the subspace of $\mathcal{H}$ generated by

$$\{w(A_Q, A_Q^*)x : w(A, A^*) \text{ is a word in } A_Q \text{ and } A_Q^* \text{ and } x \in p(A_Q, A_Q^*)Q\mathcal{H}\},$$

where A_Q stands for $A|_{Q\mathcal{H}}$.

By [Con1, Proposition 43.8],

$$W^*(A)|_{QH} = W^*(A|_{QH}) \text{ and } Z(A)|_{QH} = Z(A|_{QH}).$$

As discussed in the above paragraph, the identity operator I_Q on QH is the central projection of the family of all operators with the following form:

$$w(A_Q, A_Q^*)p(A_Q, A_Q^*),$$

where $w(A_Q, A_Q^*)$ is a word in A_Q and A_Q^*. Since each $w(A_Q, A_Q^*)p(A_Q, A_Q^*)$ is compact, I_Q equals the central support of $\mathcal{K} \cap W^*(A_Q)$. By Proposition 8.1.9, the lattice of projections in $Z(A_Q)$ is atomic. By Zorn's lemma, there is a maximal family $\{\widetilde{P_i}\}$ of mutually orthogonal minimal projections in $Z(A_Q)$. By maximality, we have

$$(\text{SOT}) \sum \widetilde{P_i} = I_Q.$$

Also note that QH is separable, and thus $\{\widetilde{P_i}\}$ is a countable family, denoted by $\{\widetilde{P_i}\}_{i=1}^n (n \leq \infty)$. Since $Z(A_Q) = Z(A)|_{QH}$, there are projections $\{Q_i\}_{i=1}^n$ in $Z(A)$ such that $Q_i|_{QH} = \widetilde{P_i}$. Put

$$P_i = Q_i Q \equiv Q Q_i Q,$$

and then $P_i|_{QH} = \widetilde{P_i}$. Also, it is easy to see that $\{P_i\}_{i=1}^n$ is a family of mutually orthogonal minimal projections in $Z(A)$ satisfying

$$\sum_i P_i = Q.$$

Since P_i is minimal, $A|_{P_i H} \equiv A_i$ is primary. The compactness of $p(A_i, A_i^*)$ follows from the identity $p(A_i, A_i^*) = p(A, A^*)|_{P_i H}$. It remains to show that for each $i \geq 1$, $p(A_i, A_i^*) \neq 0$. For this, assume conversely that there is some $j \geq 1$ such that $p(A_j, A_j^*) = 0$. Then for any word w, $w(A_i, A_i^*)p(A_j, A_j^*) = 0$, from which it follows that

$$P_j x = 0, \quad x \in w(A, A^*)p(A, A^*)\mathcal{H}.$$

Thus $P_j|_M = 0$, and hence $P_j Q = 0$. Since $P_j \leq Q, P_j = 0$, which is a contradiction. This finishes the proof of Theorem 8.1.5. $\square$

8.2 Algebraic Structure and Reducing Subspaces

This section addresses some operator-theoretic considerations for reducing subspaces of multiplication operators defined on spaces of holomorphic functions.

The following theorem is quite useful, see [Ar2, Lemma 1.4.1].

Theorem 8.2.1 *Suppose that $\mathcal{A}$ is a C^*-algebra of compact operators and P is a projection in $\mathcal{A}$. Then P is minimal if and only if for each $A \in \mathcal{A}$, $PAP = \mathbb{C}P$.*

Before going on, it is worthwhile to mention that the results in this section apply to both single variable case and multi-variable case, as we will see. But we state them in single-variable version for convenience. In this section, $\mathcal{H}$ will denote the underlying Hilbert space consisting of some holomorphic functions. Below, for simplification, let ϕ temporarily be a bounded holomorphic function on $\mathbb{D}$ and $\mathcal{W}^*(\phi)$ denotes the von Neumann algebra generated by M_ϕ which acts on $\mathcal{H}$. Also, we make this assumption:

$$\mathcal{W}^*(\phi) \cap \mathcal{K}(\mathcal{H}) \neq \{0\}.$$

The identity $\mathcal{W}^*(\phi) \cap \mathcal{K}(\mathcal{H}) \neq \{0\}$ can happen. For example, if ϕ is a function in the disk algebra $A(\mathbb{D})$, then either on $H^2(\mathbb{D})$ or on $L_a^2(\mathbb{D})$, $\mathcal{W}^*(\phi)$ contains a nonzero compact operator $M_\phi^* M_\phi - M_\phi M_\phi^*$, see Example 8.1.3. But it is not always easy to judge whether $\mathcal{W}^*(\phi) \cap \mathcal{K}(\mathcal{H}) \neq \{0\}$ holds. For instance, if ϕ satisfies the assumption in Theorem 3.1.1, then from the geometric property of ϕ it is difficult to judge whether $\mathcal{W}^*(\phi) \cap \mathcal{K}(\mathcal{H}) \neq \{0\}$ holds. Actually, we still have $\mathcal{W}^*(\phi) \cap \mathcal{K}(\mathcal{H}) \neq \{0\}$ in this case since by Theorem 3.1.1 and von Neumann bicommutant theorem, there is a finite Blaschke product B such that $\mathcal{W}^*(\phi) = \mathcal{W}^*(B)$ and

$$M_B^* M_B - M_B M_B^* \in \mathcal{W}^*(\phi) \cap \mathcal{K}(\mathcal{H}).$$

The following result is due to Guo. Let $[\mathcal{W}^*(\phi)\xi]$ be the closure of $\mathcal{W}^*(\phi)\xi$.

Theorem 8.2.2 *Suppose E is a minimal projection in $\mathcal{W}^*(\phi)$ and ξ is a unit vector in $E\mathcal{H}$. Then the closed subspace $[\mathcal{W}^*(\phi)\xi]$ is a minimal reducing subspace for M_ϕ.*

Proof Write $K = [\mathcal{W}^*(\phi)\xi]$ and let P_K denote the orthogonal projection onto K. Note that K is a reducing subspace for M_ϕ. To prove that K is minimal, it suffices to show that for any nonzero projection $P \in \mathcal{W}^*(\phi)'$, if $P \leq P_K$, then $P = P_K$.

To see this, write $Q = P - \langle P\xi, \xi\rangle P_K$. Clearly, $Q \in \mathcal{W}^*(\phi)'$. Then for any $S, T \in \mathcal{W}^*(\phi)$, we have

$$\langle QS\xi, T\xi\rangle = \langle QSE\xi, TE\xi\rangle$$

$$= \langle T^*QSE\xi, E\xi\rangle$$

$$= \langle EQT^*SE\xi, \xi\rangle$$

$$= \langle QET^*SE\xi, \xi\rangle.$$

Since E is a minimal projection in $\mathcal{W}^*(\phi)$, by Theorem 8.2.1 there is some constant λ such that $ET^*SE = \lambda E$. Therefore,

$$\langle QS\xi, T\xi \rangle = \langle Q\lambda E\xi, \xi \rangle = \lambda \langle Q\xi, \xi \rangle = 0.$$

That is, for any $S, T \in \mathcal{W}^*(\phi)$, $\langle QS\xi, T\xi \rangle = 0$, forcing $QS\xi = 0$, and hence $Q|_K = 0$. Then $Q = QP_K = 0$, which immediately gives

$$P = \langle P\xi, \xi \rangle P_K.$$

Thus, $P = P_K$, completing the proof. $\qquad\qquad\qquad\qquad\qquad\qquad\qquad\qquad\qquad\qquad\Box$

Proposition 8.2.3 *Suppose E a minimal projection in $\mathcal{W}^*(\phi)$ and $\dim E\mathcal{H} \geq 2$. For two orthogonal unit vectors ξ_1 and ξ_2 in E, write*

$$H_i = [\mathcal{W}^*(\phi)\xi_i], \quad i = 1, 2.$$

Then $H_1 \perp H_2$.

Proof This follows directly from that

$$\langle S\xi_1, T\xi_2 \rangle = \langle SE\xi_1, TE\xi_2 \rangle = \langle ET^*SE\xi_1, \xi_2 \rangle = \lambda \langle \xi_1, \xi_2 \rangle = 0, \quad S, T \in \mathcal{W}^*(\phi),$$

where λ is some constant. $\qquad\qquad\qquad\qquad\qquad\qquad\qquad\qquad\qquad\qquad\qquad\Box$

Combining Theorem 8.2.2 with Proposition 8.2.3 gives the following result, due to Guo.

Corollary 8.2.4 *Suppose E a minimal projection in $\mathcal{W}^*(\phi)$ and $\xi \in E\mathcal{H}$. Then $[\mathcal{W}^*(\phi)\xi] \cap E\mathcal{H} = \mathbb{C}\xi$.*

Let $\mathcal{E}$ be a family of projections. Recall that $\bigvee_{E \in \mathcal{E}} E$ denote the orthogonal projection onto the closed space spanned by the ranges of E, where E run over $\mathcal{E}$. Combing Corollary 8.2.4 with Proposition 2.6.4, one gets the following.

Corollary 8.2.5 *Let $\mathcal{E}$ denotes the set of all minimal projections in $\mathcal{W}^*(\phi)$. Suppose $\bigvee_{E \in \mathcal{E}} E = I$. Then each reducing subspace for M_ϕ is the direct sum of some minimal reducing subspaces, which must have the form $[\mathcal{W}^*(\phi)\xi]$ for some $\xi \in E\mathcal{H}$, where $E \in \mathcal{E}$. In this case, both $\mathcal{W}^*(\phi)$ and its commutant $\mathcal{V}^*(\phi)$ are type I von Neumann algebras.*

Theorem 8.2.2, Proposition 8.2.3, Corollaries 8.2.4 and 8.2.5 win in general. In more detail, one can replace M_ϕ with a tuple $\mathbf{T}$ of bounded linear operators, $\mathcal{W}^*(\phi)$ with $\mathcal{W}^*(\mathbf{T})$, the von Neumann algebra generated by $\mathbf{T}$. Then the above mentioned results still hold. The proofs are just the same.

Combing Corollaries 2.5.6, 2.5.8 with Corollary 2.5.2 yields the following.

Corollary 8.2.6 *Let $\mathcal{E}$ denotes the set of all minimal projections in a von Neumann algebra $\mathcal{A}$ and suppose*

$$\bigvee_{E \in \mathcal{E}} E = I.$$

Then there is a family $\{\Lambda_i\}$ of subsets of $\mathcal{E}$ such that

$$\sum_i \sum_{E \in \Lambda_i} E = I.$$

1. each Λ_i consists of pairwisely orthogonal, mutually equivalent projections in $\mathcal{A}$;
2. if E' and E'' lie in different Λ_i, then E' is not equivalent to E'';
3. $\sum_i \sum_{E \in \Lambda_i} E = I$.

Consequently, the von Neumann algebra $\mathcal{A}$ is $*$-isomorphic to

$$\bigoplus_i M_{n_i}(\mathbb{C}),$$

where n_i denotes the cardinality of $\{\Lambda_i\}$, allowed to be infinity.

By a simple application of Corollary 8.2.6, one can give the structure of $\mathcal{W}^*(\phi)$ in Corollary 8.2.5. However, in practice we usually put $\mathcal{A} = \mathcal{V}^*(\phi)$, and then Corollary 8.2.6 proves useful in characterizing the structure of the von Neumann algebra $\mathcal{V}^*(\phi)$, where ϕ is a bounded holomorphic function either in one variable or in multi-variables, as will be illustrated in the next section.

8.3 Monomial Case

In the following two sections, we confine our attentions to concrete examples of the von Neumann algebra $\mathcal{V}^*(\Phi)$, where Φ denotes a tuple of bounded holomorphic functions defined on a domain in $\mathbb{C}^d$.

The main focus of this section is on the study of $\mathcal{V}^*(\Phi)$ where Φ is a monomial. Throughout this section, z and w are adopted to denote a single complex variable, and $\mathbf{z}$ denotes variables in $\mathbb{C}^d$.

Let us begin with a relatively easy example.

Example 8.3.1 Put $p(z, w) = zw$. As follows, we will consider the von Neumann algebra $\mathcal{V}^*(p)$ defined on the Bergman space $L_a^2(\mathbb{D}^2)$, and the reducing subspaces of M_p. Since the study of reducing subspaces of M_p is in some sense equivalent to that of the lattice of projections in $\mathcal{V}^*(p)$, it suffices to centre our attention on the investigation of reducing subspaces of M_p.

Before we present the description for reducing subspaces of M_p, some notations are in order. For each $\alpha \in \mathbb{Z}_+^2$ with $\alpha = (m, n)$, set $\overline{\alpha} = (n, m)$. Write

$$[\alpha] = \{\beta : \beta \in \mathbb{Z}_+^2, \beta - \alpha = k(1, 1) \ or \ \beta - \overline{\alpha} = k(1, 1) \ for \ some \ k \in \mathbb{Z}\},$$

and put $\mathbf{z}^\alpha = z^m w^n$. Write $E_\alpha = span\{\mathbf{z}^\alpha, \mathbf{z}^{\overline{\alpha}}\}$, and put

$$P_\alpha = P_{E_\alpha}.$$

Concerning $\mathbb{Z}_+^2$, define the partial order $\leq$ by setting

$$\alpha \leq \beta \ if \ \alpha_1 \leq \beta_1 \ and \ \alpha_2 \leq \beta_2,$$

where $\alpha = (\alpha_1, \alpha_2)$ and $\beta = (\beta_1, \beta_2)$. For each α, if $(1, 1) \leq \alpha$ fails, then either $\alpha_1 = 0$ or $\alpha_2 = 0$. In this case, it is easy to see that for each nonzero subspace M_α of E_α, the reducing subspace $[M_\alpha]_p$ generated by M_α is the direct sum of $p^k M_\alpha (k \geq 0)$. That is, $[M_\alpha]_p$ is the closure of the linear span of

$$\{\mathbf{z}^\beta : \beta \in [\alpha]\}.$$

In general, one can show that each nonzero closed subspace M is reducing for M_p if and only if M has the form:

$$M = \oplus_\alpha [M_\alpha]_p,$$

where either $\alpha_1 = 0$ or $\alpha_2 = 0$ for any α, and M_α is a subspace of E_α. To see this, the "if" part is now clear. The "only if" part follows from a fact which will be demonstrated in the next paragraph: for any $f \in M$, the reducing subspace $[f]$ generated by f contains $P_\alpha f$, because $P_\alpha \in \mathcal{W}^*(p)$.

The remaining part will prove that $P_\alpha \in \mathcal{W}^*(p)$ for all $\alpha \in \mathbb{Z}_+^2$. For this, recall that

$$\{e_n = \sqrt{n + 1} z^n, n = 0, 1, \cdots\}$$

is an orthonormal basis of $L_a^2(\mathbb{D})$. By direct computations, we have

$$M_{z^k} e_n = \sqrt{\frac{n + 1}{n + k + 1}} e_{n+k}, n \geq 0,$$

and

$$M_{z^k}^* e_n = \sqrt{\frac{n - k + 1}{n + 1}} e_{n-k}, n \geq k.$$

Thus,

$$M_{z^k}^* M_{z^k} e_n = \frac{n+1}{n+k+1} e_n, n \geq k. \tag{8.8}$$

Since $\{e_n(z) e_m(w)\}_{n,m=0}^{\infty}$ is an orthonormal basis of $L_a^2(\mathbb{D}^2)$, then by (8.8)

$$M_p^{*k} M_p^k e_n(z) e_m(w) = \frac{n+1}{n+k+1} \frac{m+1}{m+k+1} e_n(z) e_m(w).$$

Observe that $(n', m') \in \{(n, m), (m, n)\}$ if and only if

$$\frac{n+1}{n+k+1} \frac{m+1}{m+k+1} = \frac{n'+1}{n'+k+1} \frac{m'+1}{m'+k+1}, \ k = 1, 2, \cdots \tag{8.9}$$

To see (8.9), let us consider the bounded holomorphic function

$$h(z) = \frac{n+1}{n+1+z} \frac{m+1}{m+1+z} - \frac{n'+1}{n'+1+z} \frac{m'+1}{m'+1+z}$$

defined on the right half plane $\{z \in \mathbb{C} : Re\, z > 0\}$. Since $h(k) = 0$ for $k = 1, 2, \cdots$, then by Lemma 2.1.6 h is identically zero. That is,

$$\frac{n+1}{n+1+z} \frac{m+1}{m+1+z} = \frac{n'+1}{n'+1+z} \frac{m'+1}{m'+1+z}.$$

Therefore, either $(n', m') = (n, m)$ or $(n', m') = (m, n)$. By (8.9) and spectrum decomposition, it follows that $\mathcal{W}^*(p)$ contains P_α for each $\alpha \in \mathbb{Z}_+^2$, as desired.

By Corollary 8.2.6, one can show that $\mathcal{V}^*(p)$ is $*$-isomorphic to the direct sum of countably many $M_2(\mathbb{C}) \bigoplus \mathbb{C}$.

The following example may be a bit more complicated. However, the idea is similar as in Example 8.3.1.

Example 8.3.2 Put $q = z^2 w$. Below, we will study the reducing subspaces of M_q which is defined on the Bergman space $L_a^2(\mathbb{D}^2)$. As done in the above example, set $e_n = \sqrt{n+1} z^n, n = 0, 1, \cdots$. By direct computations, we have

$$M_q^{*k} M_q^k e_n(z) e_m(w) = \frac{n+1}{n+2k+1} \frac{m+1}{m+k+1} e_n(z) e_m(w).$$

Observe that

$$\frac{n+1}{2} = m'+1 \ \text{and} \ \frac{n'+1}{2} = m+1$$

or $(n', m') = (n, m)$ if and only if

$$\frac{n+1}{n+2k+1}\frac{m+1}{m+k+1} = \frac{n'+1}{n'+2k+1}\frac{m'+1}{m'+k+1}, \quad k = 1, 2, \cdots. \qquad (8.10)$$

This relation naturally gives an equivalence $\sim$ on $\mathbb{Z}_+^2$: $(n', m') \sim (n, m)$ if and only if (8.10) holds.

Put

$$[\alpha] = \{\beta \in \mathbb{Z}_+^2 : \beta - \alpha' = k(2, 1) \ for \ some \ \alpha' \sim \alpha \ and \ k \in \mathbb{Z}\}.$$

Again, set $E_\alpha = span\{\mathbf{z}^\alpha, \mathbf{z}^{\overline{\alpha}}\}$, and put

$$P_\alpha = P_{E_\alpha}.$$

Then all P_α belong to $\mathcal{W}^*(q)$.

For each α, if $(2, 1) \leq \alpha$ does not hold, then either $\alpha_1 = 0, 1$ or $\alpha_2 = 0$. For these α, if M_α is a nonzero subspace of E_α, then the reducing subspace $[M_\alpha]_p$ generated by M_α is the direct sum of $p^k M_\alpha (k \geq 0)$; that is, $[M_\alpha]_p$ is spanned by

$$\{\mathbf{z}^\beta : \beta \in [\alpha]\}.$$

Furthermore, each nonzero subspace M is reducing for M_p if and only if M has the following form

$$M = \bigoplus_\alpha [M_\alpha]_p,$$

where each α dissatisfies $(2, 1) \leq \alpha$ and M_α is a subspace of E_α. Also by applying Corollary 8.2.6, one can show that $V^*(q)$ is *-isomorphic to the direct sum of countably many $M_2(\mathbb{C}) \oplus \mathbb{C}$.

In general, a similar approach leads to the conclusion that if q is a monomial, then $V^*(q)$ defined on $L_a^2(\mathbb{D}^2)$ is a non-abelian type I von Neumann algebra. In particular, if $q = z^k w^l$ with $k, l \geq 1$, then $V^*(q)$ is *-isomorphic to the direct sum of countably many $M_2(\mathbb{C}) \oplus \mathbb{C}$. The remaining case is almost trivial. Observe that in all cases, the center of $V^*(q)$ is nontrivial.

It is worthy to point out that the above approach can be applied to deal with $V^*(q)$ defined on $L_a^2(\mathbb{D}^d)$ with $(d \geq 1)$ where q is a monomial.

As following, the ideas in Examples 8.3.1 and 8.3.2 enable us to go a bit further.

Example 8.3.3 Let $\mathcal{H}$ denote a Hilbert space with the orthogonal basis $\{\mathbf{z}_\alpha : \alpha \in \mathbb{Z}_+^d\}$. Put

$$e_\alpha \triangleq \frac{\mathbf{z}^\alpha}{\|\mathbf{z}^\alpha\|}, \quad \alpha \in \mathbb{Z}_+^d,$$

and then $\{e_\alpha : \alpha \in \mathbb{Z}_+^d\}$ is an orthonormal basis of $\mathcal{H}$. Naturally, it is required that each polynomial q defines a bounded multiplication operator M_q on $\mathcal{H}$. Fix $\alpha_0 \in \mathbb{Z}_+^d$ and put

$$p(\mathbf{z}) = \mathbf{z}^{\alpha_0}.$$

The main focus is still on the reducing subspaces of M_p and the von Neumann algebra $\mathcal{V}^*(p)$ on $\mathcal{H}$.

Let $\lambda(k, \alpha)$ denote the coefficients satisfying

$$M_p^{*k} M_p^k e_\alpha = \lambda(k, \alpha) e_\alpha, k \geq 0, \alpha \in \mathbb{Z}_+^d,$$

and put

$$\lambda(k, \alpha) = 0, \ k \geq 0, \ \alpha \in \mathbb{Z}^d - \mathbb{Z}_+^d.$$

We make the following assumption: for any α and β in $\mathbb{Z}_+^d$, $\lambda(k, \alpha) = \lambda(k, \beta)$ holds for all $k \in \mathbb{Z}_+$ if and only if $\lambda(k, \alpha - \alpha_0) = \lambda(k, \beta - \alpha_0)$ holds for all $k \in \mathbb{Z}_+$, provided that either $\alpha - \alpha_0$ or $\beta - \alpha_0$ belongs to $\mathbb{Z}_+^d$. Though this assumption does not hold in general, it is valid in some cases such as in the setting of the Bergman space over $\mathbb{D}^d$.

This assumption naturally gives a classification of $\mathbb{Z}_+^d$. Precisely, let $\tilde{\alpha}$ denote the set of all members β in $\mathbb{Z}_+^d$ such that $\lambda(k, \alpha) = \lambda(k, \beta)$ for all $k \in \mathbb{Z}_+$. Two observations are in order. Given α and α' in $\mathbb{Z}_+^d$, we have

$$M_p^k e_\alpha = \sqrt{\lambda(k, \alpha)} e_{\alpha + k\alpha_0}$$

and

$$M_p^k e_{\alpha'} = \sqrt{\lambda(k, \alpha')} e_{\alpha' + k\alpha_0}$$

Thus if $\alpha' \in \tilde{\alpha}$, then $\sqrt{\lambda(k, \alpha)} = \sqrt{\lambda(k, \alpha')}$ for all $k \in \mathbb{Z}_+$. Similarly,

$$M_p^{*k} e_\alpha = \sqrt{\lambda(k, \alpha - k\alpha_0)} e_{\alpha - k\alpha_0}$$

and

$$M_p^{*k} e_{\alpha'} = \sqrt{\lambda(k, \alpha' - k\alpha_0)} e_{\alpha' - k\alpha_0}$$

Then $\sqrt{\lambda(k, \alpha - k\alpha_0)} = \sqrt{\lambda(k, \alpha' - k\alpha_0)}$ if $\alpha' \in \tilde{\alpha}$.

With these observations, we can determine the reducing subspaces as follows. As done in Example 8.3.1, there is a partial order $\leq$ on $\mathbb{Z}_+^d$. Let $\mathcal{E}$ denote the set of all members α in $\mathbb{Z}_+^d$ which dissatisfy $\alpha_0 \leq \alpha$. Write

$$E_\alpha = \overline{span\{\mathbf{z}^\beta : \beta \in \tilde{\alpha}\}},$$

and rewrite

$$P_\alpha = P_{E_\alpha}.$$

Given a subspace M_α of E_α with $\alpha \in \mathcal{E}$, let $[M_\alpha]_p$ denote the closure of the linear span of $p^k M_\alpha (k \geq 0)$. By observations in last paragraph, each $[M_\alpha]_p$ is a reducing subspaces for M_p. Besides, by spectrum decomposition $P_\alpha \in \mathcal{W}^*(p)$ for each $\alpha \in \mathbb{Z}_+^d$. Then one can show that each nonzero subspace M is reducing for M_p if and only if M has the form:

$$M = \bigoplus_\alpha [M_\alpha]_p,$$

where M_α is a subspace of E_α for each $\alpha \in \mathcal{E}$.

In Example 8.3.3, we make the following assumption: for any α and β in $\mathbb{Z}_+^d$, $\lambda(k, \alpha) = \lambda(k, \beta)$ holds for all $k \in \mathbb{Z}_+$ if and only if $\lambda(k, \alpha - \alpha_0) = \lambda(k, \beta - \alpha_0)$ holds for all $k \in \mathbb{Z}_+$, provided that either $\alpha - \alpha_0$ or $\beta - \alpha_0$ belongs to $\mathbb{Z}_+^d$. However, even the Hardy space $H^2(\mathbb{D}^d)$ dissatisfies this assumption.

In fact, it can be replaced with a weaker one: for any α and β in $\mathbb{Z}_+^d$, $\lambda(k, \alpha) = \lambda(k, \beta)$ holds for all $k \in \mathbb{Z}_+$ if and only if $\lambda(k, \alpha + \alpha_0) = \lambda(k, \beta + \alpha_0)$ for all $k \in \mathbb{Z}_+$. Then we still have $P_\alpha \in \mathcal{W}^*(p)$ for each $\alpha \in \mathbb{Z}_+^d$. But there is some difference here. In more detail, there is probably some $\beta \in \tilde{\alpha}$ satisfying $\alpha_0 \leq \beta$, but $\alpha_0 \leq \alpha$ does not hold. In this case, note that $M_p^k M_p^{*k} P_\alpha \in \mathcal{W}^*(p)$ for all positive integers. There are possibly two or more orthogonal projections, say P_α' and P_α'' satisfying

$$P_\alpha = P_\alpha' + P_\alpha'',$$

and $P_\alpha', P_\alpha'' \in \mathcal{W}^*(p)$. By some subtle modification of the set $\mathcal{E}$ defined in Example 8.3.3, one can apply those ideas to characterizing the reducing subspaces of M_p.

Some comments on Examples 8.3.1 and 8.3.2 are in order. The von Neumann algebra $\mathcal{V}^*(z^k w^l)(k, l \geq 1)$ is closely related to those reducing subspaces for $M_{z^k w^l}$, which is firstly considered in [LZ], and completely characterized in [SL], both on the unweighted and weighted Bergman spaces over $\mathbb{D}^2$. However, the approach is quite different from those displayed in Examples 8.3.1 and 8.3.2.

Precisely, write $p = z^k w^l$ where $k, l \in \mathbb{Z}_+$. Lu and Zhou [LZ], Shi and Lu [SL], showed that on the weighted Bergman spaces $L_{a,\alpha}^2(\mathbb{D}) \otimes L_{a,\alpha}^2(\mathbb{D})(\alpha > -1)$, each nonzero reducing subspace for M_p always contains a minimal reducing subspace, whose form was explicitly characterized in [LZ] and [SL], but it will be omitted here for its complicity. Their results indeed imply that $V^*(p)$ is of type I. To see this, recall that a von Neumann algebra $\mathcal{A}$ is of type I if it is discrete; that is, for each nonzero central projection Z in $\mathcal{A}$, there is a nonzero abelian projection P in $\mathcal{A}$ satisfying $P \le Z$. Since a minimal projection is always abelian, it immediately follows that the von Neumann algebra $V^*(p)$ is of type I on $L^2_{a,\alpha}(\mathbb{D}) \otimes L^2_{a,\alpha}(\mathbb{D})$ $(\alpha > -1)$, where $p = z^k w^l$, $k, l \in \mathbb{Z}_+$. If either $0 \le a \le k - 1$ or $0 \le b \le l - 1$ holds, write

$$L_{a,b} = \overline{span\,\{z^a w^b p^n : n \in \mathbb{Z}_+\}}.$$

It is clear that $L_{a,b}$ is a reducing subspace for M_p. The following result is interesting [SL].

Theorem 8.3.4 *Let $\alpha > -1$ and $\alpha \ne 0$. Assume that $p = z^k w^l$, $k, l \in \mathbb{Z}_+$ and $k \ne l$. Then all $L_{a,b}$ defined as above are the only minimal reducing subspaces for M_p on $L^2_{a,\alpha}(\mathbb{D}) \otimes L^2_{a,\alpha}(\mathbb{D})$.*

While on the unweighted Bergman space $L^2_a(\mathbb{D}^2)$, it is not the case; it usually has more minimal reducing subspaces than the weighted case, see [SL, LZ] for the details, and also refer to Example 8.3.1. It appears that the study of reducing subspaces of a multiplication operator rests heavily on the structure of the underlying function space.

Recall that on the Bergman space over $L^2_a(\mathbb{D}^2)$, if q is a monomial, then $V^*(q)$ is never abelian, see Example 8.3.2. The next example is of interest because it provides on $L^2_a(\mathbb{B}_2)$ and $H^2(\mathbb{B}_2)$ some abelian von Neumann algebras which are induced by a multiplication operator defined by a special monomial.

Example 8.3.5 Let $\| \cdot \|_{\partial \mathbb{B}_d}$ and $\| \cdot \|_{\mathbb{B}_d}$ denote the norm on the Hardy space $H^2(\mathbb{B}_d)$ and the Bergman space $L^2_a(\mathbb{B}_d)$, respectively. By [Ru2, Proposition 1.4.9],

$$\|\mathbf{z}^\alpha\|^2_{\partial \mathbb{B}_d} = \frac{(d-1)!\alpha!}{(d-1+|\alpha|)!},$$

and

$$\|\mathbf{z}^\alpha\|^2_{\mathbb{B}_d} = \frac{d!\alpha!}{(d+|\alpha|)!}.$$

Now put $p = z^t (t \ge 1)$. Then by direct computations on $L^2_a(\mathbb{B}_d)$ we have

$$M_p^{*k} M_p^k \mathbf{z}^\alpha = \frac{(d + |\alpha|)!(\alpha_1 + kt)!}{(d + |\alpha| + kt)!\alpha_1!} \mathbf{z}^\alpha, \quad k = 1, 2 \cdots.$$

Now consider the special case of $d = 2$. By using Stirling's formula

$$\Gamma(n+1) = n! \sim \sqrt{2\pi n}(\frac{n}{e})^n \quad (n \to \infty),$$

one can prove that

$$\frac{(d + |\alpha|)!(\alpha_1 + kt)!}{(d + |\alpha| + kt)!\alpha_1!} = \frac{(d + |\beta|)!(\beta_1 + kt)!}{(d + |\beta| + kt)!\beta_1!}, \quad k = 1, 2 \cdots$$

if and only if $\alpha = \beta$. Therefore, $\mathcal{W}^*(q)$ contains all orthogonal projections P_α onto $\mathbb{C}z^\alpha$. It is then easy to verify that each operator in $\mathcal{V}^*(p)$ is diagonal with respect to the orthogonal basis $\{z^\alpha : \alpha \in \mathbb{Z}_+^d\}$, and thus $\mathcal{V}^*(p)$ is abelian. Furthermore, for each $\alpha = (\alpha_1, \alpha_2)$ with $\alpha_1 < t$, let M_α denote the reducing subspace spanned by

$$\{p^n z^\alpha : n = 0, 1, \cdots\},$$

and $Q_\alpha \triangleq P_{M_\alpha}$. It is not difficult to verify that these mutually orthogonal projections Q_α span the von Neumann algebra $\mathcal{V}^*(p)$, where all α run over $\mathbb{Z}_+^2$. Therefore, $\dim \mathcal{V}^*(p) = \infty$.

Similarly, the von Neumann algebra $\mathcal{V}^*(p)$ defined on $H^2(\mathbb{B}_2)$ is abelian.

It is natural to take more considerations on multiplication operators induced by polynomials, but this is not easy in general. By developing complicated techniques, Dan and Huang showed the following result, see [DH, Theorem 1.1].

Theorem 8.3.6 *Put* $p(z, w) = z^k + w^l$ *where* $k, l \geq 1$. *Then the von Neumann algebra* $\mathcal{V}^*(p)$ *defined on* $L_a^2(\mathbb{D}^2)$ *is of type I. Furthermore,* $\mathcal{V}^*(p)$ *is* *-isomorphic to*

$$\bigoplus_{i=1}^{m} M_2(\mathbb{C}) \oplus \left(\bigoplus_{i=1}^{m'} \mathbb{C} \right),$$

where $m = \frac{\delta^2 - \delta}{2}$ *and* $m' = kl - \delta^2 + 2\delta$ *with* $\delta = GCD(k, l)$.

The following corollary [DH] is straightforward, which completely characterizes the commutativity of $\mathcal{V}^*(p)$ in an algebraic way.

Corollary 8.3.7 *If* $p(z, w) = z^k + w^l$ *with* $k, l \geq 1$, *then the center* $Z(p)$ *of* $\mathcal{V}^*(p)$ *is nontrivial; and* $\mathcal{V}^*(p)$ *is abelian if and only if* $GCD(k, l) = 1$. *In this case,* $\mathcal{V}^*(p)$ *equals its center* $Z(p)$.

Also, Dan and Huang completely characterized all minimal reducing subspaces for M_p where $p(z, w) = z^k + w^l$; for the details, see [DH]. Furthermore, Wang, Dan and Huang [WDH] consider reducing subspaces of multiplication operator M_p defined on $L_a^2(\mathbb{D}^2)$, where $M_p = \alpha z^k + \beta w^l$, $\alpha, \beta \in \mathbb{C}$. It essentially reduces to studying M_{p_α}, where $p_\alpha = z^k + \alpha w^l$, $\alpha \in (0, 1]$. Then it is shown that for $\alpha \in (0, 1)$, $\mathcal{V}^*(p_\alpha)$ is *-isomorphic to $\mathcal{V}^*(z^k) \otimes \mathcal{V}^*(w^l)$, acting on $L_a^2(\mathbb{D}) \otimes L_a^2(\mathbb{D})$, and hence $\mathcal{V}^*(p_\alpha)$ is abelian. It is worthwhile to mention an interesting fact that there is always a polynomial p such that $\mathcal{V}^*(p)$ is trivial on $L_a^2(\mathbb{D}^d)(d \geq 2)$, as shown in [WDH]. There is some chance that this fact wins in general; that is, there is probably a polynomial p such that $V^*(p)$ on $L_a^2(\Omega)$ is trivial, where Ω is a bounded domain in $\mathbb{C}^d$ with $d \geq 2$. In the case of $d = 1$, it is trivially true. Recently, Guo and Wang generalize the above results to a general situation. A unilateral weighted shift A is said to be simple if its weight sequence $\{\alpha_n\}$ satisfies $\nabla^3(\alpha_n^2) \neq 0$ for all $n \geq 2$, where ∇ is the backward difference operator defined by $\nabla[f](n) = f(n) - f(n-1)$. It is shown that if A and B are two simple unilateral weighted shifts, then $A \otimes I + I \otimes B$ is reducible if and only if A and B are unitarily equivalent. Also, a furthermore consideration is done for the reducing subspaces of $A^k \otimes I + I \otimes B^l$, and these results are applied to the study of reducing subspaces of multiplication operators $M_{z^k + \alpha w^l}$ on general function spaces, see [GuoW].

Concerning results in [SL, LZ, DH] and [WDH], some comments are in order, which will illustrate how Corollary 8.2.6 works. In fact, given a multiplication operator M_ϕ, the assumption in Corollary 8.2.6 says there is a family $\{M_i\}$ of minimal reducing subspaces of M_ϕ whose closed linear span equals the whole space. In this case, one can pick a subfamily of $\{M_i\}$ whose members are mutually orthogonal. In practice, such a subfamily is not necessarily unique and it can be obtained in this way. First, by observation one can get a family of mutually orthogonal reducing subspaces. This can be handled if ϕ is not complicated; for example, either ϕ is a monomial or $\phi = z + w$. It is possible that some of them are minimal while others are not. Then by further investigation one may get a more subtle decomposition of the whole space in terms of minimal reducing subspaces. This job, however, can be challenging as well as determining which of them are unitarily equivalent and which are not. After this, by applying Corollary 8.2.6 one can get clear structures of reducing subspaces for M_ϕ and of the von Neumann algebra $V^*(\phi)$. It is reasonable to assume that for a large class of polynomials p, $V^*(p)$ satisfies the assumption in Corollary 8.2.6, and thus the above approach is applicable though it can be never a piece of cake.

To end this section, it is worthwhile to mention Curto, Muhly and Yan's result, which, though, is not quite along this line. Precisely, let $\mathcal{J}$ be an ideal in $\mathbb{C}[z, w]$ and $[\mathcal{J}]$ be the closure of $\mathcal{J}$ in the Hardy space $H^2(\mathbb{D}^2)$. Let $C^*(\mathcal{J})$ be the unital C^*-algebra generated by $M_z|_{[\mathcal{J}]}$ and $M_w|_{[\mathcal{J}]}$. It is shown in [CMY] that if $\mathcal{J}$ is a homogenous algebra, then the commuting pair $(M_z|_{[\mathcal{J}]}, M_w|_{[\mathcal{J}]})$ is essentially doubly commuting and $C^*(\mathcal{J})$ is of type I. Recall that two commuting pair (T_1, T_2) defined on a Hilbert space $\mathcal{H}$ is called essentially doubly commuting if $[T_i^*, T_j] \in \mathcal{K}(\mathcal{H})$ for $1 \leq i < j \leq 2$.

8.4 More Examples in Multi-variable Case

In studying the commutant algebra $V^*(\Phi)$, it is natural to raise the following.

Question 8.4.1 Suppose $\Phi = (\phi_1, \cdots, \phi_n)$ and M_Φ acts on the Bergman space $L_a^2(\Omega)$, where Ω is a Reinhardt domain in $\mathbb{C}^d$. If $d > n$, is $\dim V^*(\Phi) = \infty$? If $d < n$, under what conditions do we have $V^*(\Phi) = \mathbb{C}I$?

The case $d = n$ may be of most interest. However, little is known about the case of $d \geq 2$. In addition, it is worthwhile to mention that if $\Omega = \mathbb{D}^d$, then one can construct a single polynomial P such that $V^*(P) = \mathbb{C}I$, see [WDH].

In this section more examples will be provided, and all discussions below are based on Bergman spaces.

Example 8.4.2 Suppose that $\Phi(z, w) = (\phi_1(z), \phi_2(w))$, where ϕ_1 and ϕ_2 are bounded holomorphic functions over $\mathbb{D}$. Then on $L_a^2(\mathbb{D}^2)$ $V^*(\Phi)$ is $*$-isomorphic to $V^*(\phi_1) \otimes V^*(\phi_2)$, where $V^*(\phi_1)$ and $V^*(\phi_2)$ are defined over $L_a^2(\mathbb{D})$.

For example, write $\Phi = (\phi_1(z), \phi_2(w))$ where $\phi_j : \mathbb{D} \to \Omega_j (j = 1, 2)$ are two bounded holomorphic covering maps. Noting

$$G(\Phi) = \{(\rho, \sigma); \rho \in G(\phi_1), \sigma \in G(\phi_2)\},$$

one can show that both $V^*(\Phi)$ and $V^*(\phi_1) \otimes V^*(\phi_2)$ are $*$-isomorphic to the group von Neumann algebra $\mathcal{L}(G(\Phi))$. Before continuing, we need a notion. Given two groups G_1 and G_2, $G = G_1 \times G_2$ is called the external direct product of G_1 and G_2 if its multiplication is defined by

$$(a_1, a_2) \cdot (b_1, b_2) \triangleq (a_1 b_1, a_2 b_2), \ (a_1, a_2) \in G, \ (b_1, b_2) \in G.$$

Note that $G(\Phi)$ is exactly the external direct product of $G(\varphi_1)$ and $G(\varphi_2)$. If one of $G(\varphi_j)$ is non-abelian, then by Lemma 6.4.2 $G(\Phi)$ is an i.c.c. group, which differs from any group arising from orbifold domain, see Sect. 6.6 of Chap. 6. In particular, $G(\Phi)$ is distinct from all free groups.

Furthermore, each operator S in $V^*(\Phi)$ has the following form: there is a unique vector $\{c_k\}$ in l^2 such that

$$Sh(z, w) = \sum_{k=0}^{\infty} c_k h \circ \rho_k(z, w) \det \frac{\partial \rho_k(z, w)}{\partial(z, w)}, \ h \in L_a^2(\mathbb{D}^2), \ (z, w) \in \mathbb{D}^2,$$

where ρ_k runs over $G(\Phi)$, and $\det \frac{\partial \rho_k(z,w)}{\partial(z,w)}$ denotes the determinant of the Jacobian matrix $\frac{\partial \rho_k(z,w)}{\partial(z,w)}$. By applying the ideas of Sects. 6.5 and 6.6 in Chap. 6, one can obtain the uniqueness of the coefficients c_k.

To the best of our knowledge, it is not known whether $V^*(\Phi)$ over $L_a^2(\mathbb{D}^2)$ always has this structure $V^*(\phi_1) \otimes V^*(\phi_2)$ where $\phi_1, \phi_2 \in H^\infty(\mathbb{D})$ if Φ is a holomorphic covering map on $\mathbb{D}^2$.

The next example shows that on the Bergman space, even if $\dim V^*(\Phi) < \infty$, $V^*(\Phi)$ is not necessarily abelian. However, not a single example of $V^*(\Phi)$ is known to be non-abelian in single variable case if we require $\dim V^*(\Phi) < \infty$. We call the reader's attention to Conjectures 4.4.8 and 8.1.4.

Example 8.4.3 Write $p = z^2 + w^2$ and $q = z^2 w^2$, and put $\Phi = (p, q)$. Let Ω be the unit ball $\mathbb{B}_2$ or the bidisk $\mathbb{D}^2$. There are exactly 8 members in $G(\Phi)$: $\rho_0, \cdots, \rho_7$, which are defined by

$$\rho_0(z, w) = (z, w), \ \rho_1(z, w) = (-z, w), \ \rho_2(z, w) = (z, -w), \ \rho_3(z, w) = (-z, -w);$$

and

$$\rho_4(z, w) = (w, z), \ \rho_5(z, w) = (-w, z), \ \rho_6(z, w) = (w, -z), \ \rho_7(z, w) = (-w, -z).$$

Each ρ_j naturally defines a unitary operator U_j on $L_a^2(\Omega)$:

$$U_j f = f \circ \rho_j, f \in L_a^2(\Omega).$$

By the ideas in Chaps. 4 or 5, one can show that the von Neumann algebra $\mathcal{V}^*(\Phi)$ is generated by $\{U_j : 0 \le j \le 7\}$. Since $U_1 U_4 \ne U_4 U_1$, $\mathcal{V}^*(\Phi)$ is not abelian.

Furthermore, $\mathcal{V}^*(\Phi)$ is $*$-isomorphic to the group von Neumann algebra $\mathcal{L}(G(\Phi))$. By a careful look, one can verify that $G(\Phi)$ is isomorphic to the dihedral group D_4 defined by

$$D_4 = \{\rho^i \tau^j : i, j \in \mathbb{Z}; \ \rho^2 = \tau^4 = e, \rho\tau = \tau^{-1}\rho\},$$

where e denotes the identity of the group. Therefore, $\mathcal{V}^*(\Phi)$ is $*$-isomorphic to $\mathcal{L}(D_4)$.

In general, *the dihedral group D_n* is defined to be the group generated by ρ and τ satisfying

$$\rho^2 = \tau^2 = (\rho\tau)^n = e,$$

see [DF]. It is of interest to raise a natural question: for each integer $n \ge 4$, is there a family Φ of polynomials such that $\mathcal{V}^*(\Phi)$ is $*$-isomorphic to $\mathcal{V}^*(D_n)$ on some Bergman space?

Chapter 6 investigated the von Neumann algebra $\mathcal{V}^*(\phi)$ generated by a single multiplication operator M_ϕ where ϕ is a covering map. There it was shown that the structure of $\mathcal{V}^*(\phi)$ has close connection with the deck transformation group. Such a phenomena is not alone, see as follows.

Example 8.4.4 Let S_3 denote permutation group for $\{1, 2, 3\}$, and there are exactly 6 members in S_3, denoted by $\sigma_0, \cdots, \sigma_5$. Write

$$p_1 = z_1 + z_2 + z_3,$$

$$p_2 = z_1 z_2 + z_2 z_3 + z_3 z_1,$$

$$p_3 = z_1 z_2 z_3,$$

and put $\Phi = (p_1, p_2, p_3)$. Let Ω be the unit ball $\mathbb{B}_3$ or the polydisk $\mathbb{D}^3$. There are exactly 6 members in $G(\Phi)$: $\rho_0, \cdots, \rho_5$, which are defined by

$$\rho_j(z_1, z_2, z_3) = (z_{\sigma_j(1)}, z_{\sigma_j(2)}, z_{\sigma_j(3)}), j = 0, \cdots, 5.$$

Each ρ_j naturally defines a unitary operator U_j on $L_a^2(\Omega)$:

$$U_j f = f \circ \rho_j, f \in L_a^2(\Omega).$$

By applying the ideas in Chap. 5, one can prove that the von Neumann algebra $V^*(\Phi)$ is generated by $\{U_j : 0 \leq j \leq 5\}$, and thus $V^*(\Phi)$ is $*$-isomorphic to $\mathcal{L}(S_3)$. Note that $V^*(\Phi)$ is not abelian, and $\dim V^*(\Phi) = 6 < \infty$.

Of the non-abelian finite groups, the best known are perhaps the permutation groups $S_n (n \geq 3)$, and the only smallest non-abelian finite group is S_3. This indicates the following.

Conjecture 8.4.5 *Let* Φ *be a family of polynomials. If the von Neumann algebra* $V^*(\Phi)$, *defined on the Bergman space* $L_a^2(\Omega)$, *has finite dimension and* $\dim V^*(\Phi) < 6$, *then* $V^*(\Phi)$ *is abelian. In this case, it is generated by those* U_ρ *defined by*

$$U_\rho f = f \circ \rho \, J\rho, f \in L_a^2(\Omega),$$

where ρ *are members in* $\mathrm{Aut}(\Omega)$ *satisfying* $\phi \circ \rho = \phi$ *for each* ϕ *in* Φ, *and* $J\rho$ *denote the determinants of the Jocobian matrices of* ρ.

Inspired by Example 8.4.4, we consider the permutation groups $S_n (n \geq 3)$. Write

$$\phi_1 = \sum_{1 \leq j \leq n} z_j,$$

$$\phi_2 = \sum_{1 \leq i < j \leq n} z_i z_j,$$

$$\cdots$$

and $\phi_n = z_1 z_2 \cdots z_n$.

Proposition 8.4.6 *Set* $\Omega = \mathbb{B}_n$ *or* $\mathbb{D}^n$, *and write* $\Phi = (\phi_1, \cdots, \phi_n)$, *where* ϕ_i *are defined as above. Then the deck transformation group* $G(\Phi)$ *of* Φ *is isomorphic to* S_n, *and the von Neumann algebra* $V^*(\Phi)$ *on* $L_a^2(\Omega)$ *is* $*$-*isomorphic to* $\mathcal{L}(S_n)$.

Proof For each permutation σ of $\{1, 2, \cdots, n\}$, define a unitary operator U_σ by

$$U_\sigma f(z_1, \cdots, z_n) = f(z_{\sigma(1)}, \cdots, z_{\sigma(n)}), f \in L_a^2(\Omega).$$

Since for $1 \leq i \leq n$, U_σ commutes with M_{ϕ_i} and $M_{\phi_i}^*$, $U_\sigma \in \mathcal{V}^*(\Phi)$. Then it is easy to verify that

$$\sharp S_n \leq \dim \mathcal{V}^*(\Phi).$$

Write $s_n = \sharp S_n$, and as follows we will show that there is a neighborhood V_1 of some point $w \in \Omega$ such that

$$\Phi^{-1} \circ \Phi(V_1) = \bigsqcup_{i=1}^{s_n} V_i, \tag{8.11}$$

where all $\Phi|_{V_i} : V_i \to \Phi(V_1)$ are biholomorphic maps.

If (8.11) holds, then by using methods of the proof for Lemma 5.2.2, one have $\dim \mathcal{V}^*(\Phi) \leq s_n = \sharp S_n$, forcing $\dim \mathcal{V}^*(\Phi) = \sharp S_n = s_n$. Since each permutation $\sigma \in S_n$ naturally defines a member in $G(\Phi)$, then $\sharp G(\Phi) \geq s_n$. In order to show $s_n \geq \sharp G(\Phi)$, note that each member $\rho \in G(\Phi)$ naturally defines a unitary operator U_ρ, and all $U_\rho(\rho \in G(\Phi))$ generates a von Neumann algebra, denoted by $\mathcal{V}^*(G(\Phi))$. Clearly,

$$\mathcal{V}^*(\Phi) \supseteq \mathcal{V}^*(G(\Phi)),$$

forcing

$$s_n = \dim \mathcal{V}^*(\Phi) \geq \dim \mathcal{V}^*(G(\Phi)) = \sharp G(\Phi) \geq s_n.$$

Therefore, $\mathcal{V}^*(\Phi) = \mathcal{V}^*(G(\Phi))$ and $\sharp G(\Phi) = s_n$. Thus, $G(\Phi) \cong S_n$, which implies that $\mathcal{V}^*(\Phi)$ is $*$-isomorphic to $\mathcal{L}(S_n)$.

Below, to finish the proof we must show (8.11).

For $n = 1$ it is trivial. Now consider the case of $n = 2$. One can pick $(w_1, w_2) \in \Omega$ such that $w_1 \neq w_2$, and then

$$\det \frac{\partial(\phi_1, \phi_2)}{\partial(z_1, z_2)} \bigg|_{(w_1, w_2)} = \begin{vmatrix} 1 & 1 \\ w_2 & w_1 \end{vmatrix} \neq 0.$$

The function equations

$$\begin{cases} z_1 + z_2 = w_1 + w_2, \\ z_1 z_2 = w_1 w_2 \end{cases}$$

have exactly two solutions: $z = (w_1, w_2)$ or $z = (w_2, w_1)$. Thus, there is a neighborhood V of (w_1, w_2) such that

$$\Phi^{-1} \circ \Phi(V) = V' \bigsqcup V,$$

where both $\Phi|_V$ and $\Phi|_{V'}$ are biholomorphic maps onto $\phi(V)$. Therefore, in the case of $n = 2$ we obtain (8.11).

In general, this can be handled by induction. To make it clear, rewrite Ω_d for Ω, where d is the dimension of Ω. Also, rewrite

$$P_1^d = z_1 + \cdots + z_d,$$

$$P_2^d = \sum_{1 \le i < j \le d} z_i z_j, \cdots,$$

and

$$P_d^d = z_1 z_2 \cdots z_d.$$

By induction, assume that for $n = k(k > 1)$ there is a point $w' = (w_1, \cdots, w_k)$ in Ω_k such that $w_1 \cdots w_k \ne 0$ and for each $\lambda \in \Phi^{-1}(\Phi(w'))$,

$$\det \frac{\partial(P_1^k, \cdots, P_k^k)}{\partial(z_1, \cdots, z_k)}(\lambda) \ne 0; \tag{8.12}$$

also, $\sharp \Phi^{-1}(\Phi(w')) = \sharp S_k$.

Write $z = (z', z_{k+1})$ where $z' = (z_1, \cdots, z_k)$, and set $w = (w', 0)$ with $w \in \Omega_{k+1}$. Consider the following $k + 1$ equations

$$P_j^{k+1}(z) = P_j^{k+1}(w), \ j = 1, \cdots, k + 1. \tag{8.13}$$

In particular, the equation $P_{k+1}^{k+1}(z) = P_{k+1}^{k+1}(w)$ gives

$$z_1 \cdots z_{k+1} = 0.$$

By symmetry of the variables $z_1, \cdots, z_{k+1}$, we may assume that $z_{k+1} = 0$. In this case, (8.13) is reduced to k equations

$$P_j^k(z') = P_j^k(w'), \ j = 1, \cdots, k, \tag{8.14}$$

which have exactly $\sharp S_k$ solutions. Then the equations (8.13) have exactly $\sharp S_{k+1}$ solutions. After some computation, one get

$$\det \frac{\partial(P_1^{k+1}, \cdots, P_{k+1}^{k+1})}{\partial(z_1, \cdots, z_{k+1})}(w) = w_1 \cdots w_k \det \frac{\partial(P_1^k, \cdots, P_k^k)}{\partial(z_1, \cdots, z_k)}(w') \ne 0.$$

Similarly, rewriting $z = w^*$ in (8.13), i.e.

$$P_j^{k+1}(w^*) = P_j^{k+1}(w), \ j = 1, \cdots, k + 1,$$

we also have

$$\frac{\partial(P_1^{k+1},\cdots,P_{k+1}^{k+1})}{\partial(z_1,\cdots,z_{k+1})}(w^*) \neq 0.$$

This observation implies that there is a neighborhood W_1 of w such that

$$\Phi^{-1} \circ \Phi(W_1) = \bigsqcup_{i=1}^{s_{k+1}} W_i,$$

where all $\Phi|_{W_i} : W_i \to \Phi(W_1)$ are biholomorphic maps. Each W_i is an enough small neighborhood of a point in $\Phi^{-1} \circ \Phi(w)$. Consequently, there is a sub-domain V_1 of W_1 such that all points $(z_1,\cdots,z_{k+1})$ in V_1 satisfying $z_1 \cdots z_{k+1} \neq 0$. Pick an arbitrary point w'' of V_1, and for each $\lambda \in \Phi^{-1}(\Phi(w''))$,

$$\det \frac{\partial(P_1^{k+1},\cdots,P_{k+1}^{k+1})}{\partial(z_1,\cdots,z_k)}(\lambda) \neq 0.$$

The induction is complete. Then putting

$$V_i = (\Phi|_{W_i})^{-1}(V_1), \ i = 1,\cdots,k+1,$$

gives (8.11), as desired. The proof is complete.   □

Before continuing, let us introduce a notion, called the analytic Hilbert module. Now let Ω be a domain in $\mathbb{C}^d$, and denote by $A(\Omega)$ those holomorphic functions on Ω which can be continuously extended to $\overline{\Omega}$. Recall that a Banach space X contained in $Hol(\Omega)$ is called *an analytic Hilbert module* [CG, Guo1, Guo2, Guo3, Guo4, GW2, GW3, GW4] if the following hold:

(i) $1 \in X$ and for each $\lambda \in \Omega, f \mapsto f(\lambda)$ defines a bounded linear functional;
(ii) If $f \mapsto f(\lambda)$ defines a bounded linear functional for some $\lambda \in \mathbb{C}^d$, then $\lambda \in \Omega$;
(iii) the polynomial ring is dense in X and for each polynomial $p, pf \in X$ provided that $f \in X$.

By a submodule of X, we mean a closed subspace invariant under the coordinate operators $M_{z_j}, j = 1,\cdots,d$.

Write $\Phi \triangleq (\phi_1,\cdots,\phi_d)$, and let M_Φ denote the tuple $(M_{\phi_1},\cdots,M_{\phi_d})$. We have the following.

Proposition 8.4.7 *Suppose* $L_a^2(\Omega)$ *is an analytic Hilbert module over a domain* Ω *in* $\mathbb{C}^d$, $\phi_1,\cdots,\phi_d$ *are in* $A(\Omega)$, *and* Φ *has no zero on the boundary of* Ω *and each zero of* Φ *is a regular point. If* $1 \leq m = \dim L_a^2(\Omega) \ominus \Phi L_a^2(\Omega) < \infty$, *then we have*

(1) If M is a reducing subspace of M_Φ satisfying $\dim M \ominus \Phi M = 1$, then P_M belongs to $Z(\mathcal{V}^*(\Phi))$;

(2) $\dim \mathcal{V}^*(\Phi) \leq m$.

Proof The idea of the proof comes from [GH1]. The reader can alternatively refer to Chap. 4.

We claim that there are exactly m zeros of Φ. To see this, let N be the closure of $\Phi L_a^2(\Omega)$. Clearly, N is a submodule of finite codimension. Since $L_a^2(\Omega)$ is an analytic Hilbert module, by Guo [Guo1, Theorem 3.1] or Chen and Guo [CG, Corollary 2.2.6]

$$m = \operatorname{codim} N = \sum_{\lambda \in Z(N)} \dim N_\lambda,$$

where $N_\lambda = \{q \in \mathbb{C}[z_1, \cdots, z_d]; q(D)f|_\lambda = 0, \forall f \in N\}$. Here, D denotes the differential operator $(\frac{\partial}{\partial z_1}, \cdots, \frac{\partial}{\partial z_d})$, see [CG]. Now fix a point λ in the zero set $Z(N)$ defined by

$$Z(N) = \{z \in \Omega; h(z) = 0, h \in N\}.$$

Since $\Phi \in N$ and each zero point of Φ is a regular point,

$$\det\left(\frac{\partial \phi_i}{\partial z_j}\Big|_{z=\lambda}\right) \neq 0,$$

which guarantees that $z_i \notin N_\lambda (1 \leq i \leq d)$. Noting that N_λ is invariant under the partial operators $\frac{\partial}{\partial z_1}, \cdots, \frac{\partial}{\partial z_d}$, we get $N_\lambda = \mathbb{C}$ for all $\lambda \in Z(N)$, and hence $\dim N_\lambda = 1$. Thus there are exactly m different zeros of Φ in Ω.

Since these zeros are regular, there is a neighborhood Δ of 0 in $\mathbb{C}^d$ such that on Δ there exist m holomorphic branches of Φ^{-1}, say $\psi_1, \cdots, \psi_m$ satisfying

$$\Phi^{-1}(\Delta) = \bigsqcup_{k=1}^m \psi_k(\Delta).$$

For simplicity, let $\frac{\partial \psi_k}{\partial z}$ denote the Jacobian matrix of ψ_k. By applying the approaches in Sects. 4.3 and 4.4 of Chap. 4, we make the following assertion: If M_1 and M_2 are two reducing subspaces of M_Φ and $U : M_1 \to M_2$ is a unitary operator commuting with M_{ϕ_i} for each i, then there exists an $m \times m$ numerical unitary matrix W such that

$$W\begin{pmatrix} f(\psi_1(w)) \det \dfrac{\partial \psi_1}{\partial z}(w) \\ \vdots \\ f(\psi_m(w)) \det \dfrac{\partial \psi_m}{\partial z}(w) \end{pmatrix} = \begin{pmatrix} Uf(\psi_1(w)) \det \dfrac{\partial \psi_1}{\partial z}(w) \\ \vdots \\ Uf(\psi_m(w)) \det \dfrac{\partial \psi_m}{\partial z}(w) \end{pmatrix} \qquad (8.15)$$

holds for any f in M_1 and w in Δ. Now put

$$\mathcal{L}_{M_1,\Delta} = span\left\{ \begin{pmatrix} f(\psi_1(w)) \det \dfrac{\partial \psi_1}{\partial z}(w) \\ \vdots \\ f(\psi_m(w)) \det \dfrac{\partial \psi_m}{\partial z}(w) \end{pmatrix} : f \in M_1, w \in \Delta \right\} \subseteq \mathbb{C}^m.$$

Following similar arguments as in Sect. 4.4 of Chap. 4, one can show that for each reducing subspace M' of M_Φ,

$$\dim \mathcal{L}_{M' \ominus \Phi M',\Delta} = \dim \mathcal{L}_{M',\Delta} = \dim M' \ominus \Phi M'.$$

Then applying the proof of Theorem 4.4.1 leads to the conclusion (1) in Proposition 8.4.7.

Now set $M_1 = M_2 = L_a^2(\Omega)$. By a simple application of (8.15),

$$\dim \mathcal{V}^*(\Phi) \le m.$$

Therefore, we get (2) as desired. The proof is complete.   □

8.5 Remarks on Chap. 8

As mentioned in Sect. 8.1 of this chapter, the study of the algebraic structure of non-selfadjoint operators began with Brodskii and Livsic's work [BL], also concerned in Suzuki's original work [Su]. They considered bounded linear operators A whose imaginary parts are compact. In the case of A being essential normal, it was attacked by Behncke [Be1, Be2]. More general results were obtained by Gilfeather [Gil1, Gil2]. Some special cases were also considered by Brown, Sz. Nagy and Foias.

In this chapter, the materials of Sect. 8.1 mainly comes from [Con2] and [Gil2]; Sect. 8.2 is based on a manuscript of Guo; due to Guo and Huang, Sect. 8.3 concerns with reducing subspaces of multiplication operators in multi-variable cases. Examples 8.4.2, 8.4.3, 8.4.4, 8.3.1, 8.3.2, 8.3.3, 8.3.5 and Proposition 8.4.6 are given by Huang. Proposition 8.4.7 is obtained by Guo in [Guo5]. Theorem 8.3.4 is given by Shi and Lu in [SL]. Both Theorem 8.3.6 and Corollary 8.3.7 are established by Dan and Huang in [DH].

Appendix A
Berezin Transform

In this context, we mainly present a property of the Berezin transform. The following material comes mainly from [Str].

In what follows, let Ω be a domain in $\mathbb{C}^d$ and $\mathcal{H}$ be a reproducing kernel Hilbert space of holomorphic functions on Ω; that is, for every $\lambda \in \Omega$, the evaluation functional

$$E_\lambda : f \rightarrow f(\lambda), f \in \mathcal{H}$$

is bounded. Here, write $z = (z_1, \cdots, z_d)$ and $w = (w_1, \cdots, w_d)$, which denote the variables in Ω. For any multi-index $I = (i_1, \cdots, i_d) \in \mathbb{Z}_+^d$, z^I denotes

$$z_1^{i_1} \cdots z_d^{i_d}.$$

It is known that for each $w \in \Omega$, there exists a unique vector $K_w \in \mathcal{H}$ such that

$$f(w) = \langle f, K_w \rangle, \quad f \in \mathcal{H}.$$

This K_w is called the reproducing kernel of $\mathcal{H}$ at w. Define

$$k_w = \frac{K_w}{\|K_w\|},$$

called the normalized reproducing kernel at w. An important property of the reproducing kernel is that: if U is an open subset of Ω, then $span\{K_w; w \in U\}$ is dense in $\mathcal{H}$.

For each $S \in B(H)$, the *Berezin transform* $\tilde{S}$ of S is defined by

$$\tilde{S}(w) = \langle Sk_w, k_w \rangle, \quad w \in \Omega.$$

© Springer-Verlag Berlin Heidelberg 2015

K. Guo, H. Huang, *Multiplication Operators on the Bergman Space*,
Lecture Notes in Mathematics 2145, DOI 10.1007/978-3-662-46845-6

The following theorem gives a one-to-one correspondence between bounded linear operators on $\mathcal{H}$ and their Berezin transforms on Ω, see [Str, Theorem 2.2].

Theorem A.1 (Stroethoff) *Let $\mathcal{H}$ be a reproducing kernel Hilbert space of holomorphic functions on a domain Ω with $\Omega \subseteq \mathbb{C}^d$, and $S, T \in B(\mathcal{H})$. Then $S = T$ if and only if there is an open subset U of Ω such that*

$$\tilde{S}(w) = \tilde{T}(w), \ w \in U.$$

Following [Str], for a domain V in $\mathbb{C}^d$, write

$$V^* = \{(\overline{w_1}, \cdots, \overline{w_d}) : (w_1, \cdots, w_d) \in V\}.$$

Clearly, V^* is also a domain. The following comes from [Str, Lemma 2.3], which only concerns with single-variable case.

Lemma A.2 *Let U be an open subset of Ω. If h is a holomorphic function on $\Omega \times \Omega^*$ and $h(z, \bar{z}) = 0$, $z \in U$, then h is identically zero on $\Omega \times \Omega^*$.*

Proof The proof is from [Str].

Let U be an open subset of Ω. Assume that h is a holomorphic function on $\Omega \times \Omega^*$ and $h(z, \bar{z}) = 0$, $z \in U$. Without loss of generality, we may assume that $0 \in U$ and U is a polydisk centered at 0. Then $U \times U^*$ is also a polydisk, and we have the following expansion of h:

$$h(z, w) = \sum_{I, J \in \mathbb{Z}_+^d} c_{I,J} z^I w^J, \ (z, w) \in U \times U^*.$$

Note that $h(z, \bar{z}) = 0$, $z \in U$. That is,

$$h(z, \bar{z}) = \sum_{I, J \in \mathbb{Z}_+^d} c_{I,J} z^I \bar{z}^J = 0, \ z \in U.$$

Then for any $I, J \in \mathbb{Z}_+^d$,

$$\frac{\partial^{|I|+|J|}}{\partial z^I \partial \bar{z}^J} h \big|_{(0,0)} = 0.$$

and hence $c_{I,J} = 0$ for all I and J, forcing $h = 0$ on $U \times U^*$. Therefore by the uniqueness theorem, h is identically zero on $\Omega \times \Omega^*$.   □

In the proof of Lemma A.2, it is crucial to show $c_{I,J} = 0$. An alternative approach is as follows. Pick an enough small $\varepsilon > 0$, and integrate $\bar{z}^J z^I h(z, \bar{z})$ on $(\varepsilon \mathbb{T}) \times \cdots \times (\varepsilon \mathbb{T})$. Then one gets $c_{I,J} = 0$, as desired.

Now we are ready for the proof of Theorem A.1.

Proof of Theorem A.1 The proof is from [Str].

One direction is straightforward. To finish the proof, it is enough to prove that if there is an open subset U of Ω such that

$$\tilde{S}(w) = 0, \ w \in U.$$

then we must have $S^\bullet = 0$.

To see this, define

$$h(z, w) = \langle SK_{\overline{w}}, K_z \rangle, (z, w) \in \Omega \times \Omega^*.$$

Since for fixed variables $w = (w_1, \cdots, w_n)$, h is holomorphic in z and vice versa, then h is holomorphic in each variable (w_i or z_i) separately. The Hartogs theorem [Hor, Theorem 2.2.8] states that a function defined on a domain is holomorphic if it is holomorphic in each variable separately, and then h is a holomorphic function in (z, w). Since $\tilde{S}(w) = \langle Sk_w, k_w \rangle = 0$ for $w \in U$,

$$\langle SK_w, K_w \rangle = 0, \ w \in U,$$

and thus $h(w, \overline{w}) = 0$, $w \in U$. Then by Lemma A.2 h is identically zero on $\Omega \times \Omega^*$. In particular,

$$\langle SK_z, K_w \rangle = 0, z, w \in U.$$

Since $span\{K_w; w \in U\}$ is dense in $\mathcal{H}$, for each fixed $z \in U$, $SK_z = 0$. This shows that S is zero on the dense subspace $span\{K_z; z \in U\}$ of $\mathcal{H}$, forcing $S = 0$. The proof is complete. $\qquad\square$

Appendix B
Nordgren's Results on Reducing Subspaces

The following result is due to Nordgren [Nor], which gives a sufficient condition for an analytic Toeplitz operator to have no nontrivial reducing subspaces.

Theorem B.1 (Nordgren) *Let T_ϕ be an analytic Toeplitz operator on $H^2(\mathbb{D})$. If there is a Borel subset E of $\mathbb{T}$ such that*

(1) $m(E) > 0$;
(2) $\phi(E)$ *and* $\phi(\mathbb{T} - E)$ *are disjoint;*
(3) the restriction $\phi|_E$ *is one to one,*

then T_ϕ *has no nontrivial reducing subspace.*

If $\phi \in A(\mathbb{D})$, then there is no confusion on the conditions (1)–(3). Otherwise, ϕ restricted on $\mathbb{T}$ may be not well-defined on a set Z of Lebesgue measure zero. In this case, one can assign the value 0 to ϕ on Z, and thus the conditions (1)–(3) make sense.

Proof of Theorem B.1 The proof is from [Nor].

Note that the essential range Λ of ϕ on $\mathbb{T}$ is a closed set, which is invariant if the values of ϕ is changed on a set of Lebesgue measure zero. On the other hand, by Lusin's theorem [Hal1, p. 242], there is a compact subset E_0 of E such that $m(E_0) > 0$ and $\phi|_{E_0}$ is continuous. Since $\phi|_E$ is one to one, without loss of generality, we may assume that E is itself compact and $\phi(E)$ is compact. Therefore, $\phi|_E : E \to \phi(E)$ is a homeomorphism, and then the image of every Borel subset of E is a Borel subset of the complex plane. In this case, we will show that T_ϕ has no nontrivial reducing subspace.

First we will show that for any reducing subspace M of T_ϕ, M reduces T_{χ_F} for any Borel subset F of E. To see this, consider the measure $m\phi^{-1}$ defined on Borel subsets of Λ. Recall that for any measurable function g on Λ, we have

$$\int_\Lambda g(z)d(m\phi^{-1})(z) = \int_\mathbb{T} g \circ \phi(z)dm(z),$$

© Springer-Verlag Berlin Heidelberg 2015
K. Guo, H. Huang, *Multiplication Operators on the Bergman Space*,
Lecture Notes in Mathematics 2145, DOI 10.1007/978-3-662-46845-6

in the sense that if either integral exists, then so does the other, and these two integrals are equal, see [Hal1, Sect. 39, Theorem C]. Since $\phi|_E$ is a homeomorphism, for any Borel subset F of E the characteristic function $\chi_{\phi(F)}$ is a bounded Borel function. Then there exists a uniformly bounded sequence of polynomials $\{p_n(z, \bar{z})\}$, which converges to $\chi_{\phi(F)}$ almost everywhere on Λ with respect to the measure $m\phi^{-1}$. Therefore, $\{p_n(\phi, \bar{\phi})\}$ converges to $\chi_{\phi(F)}(\phi)$ almost everywhere on $\mathbb{T}$ with respect to m. Note that $\phi(E)$ and $\phi(\mathbb{T} - E)$ are disjoint and $F \subseteq E$, and then

$$\chi_{\phi(F)}(\phi) = \chi_F.$$

Therefore, $\{p_n(\phi, \bar{\phi})\}$ converges to χ_F almost everywhere with respect to m on $\mathbb{T}$. Now it is easy to see that $T_{p_n(\phi, \bar{\phi})}$ converges weakly to T_{χ_F}, which implies that each closed subspace that reduces all $T_{p_n(\phi, \bar{\phi})}$ also reduces T_{χ_F}. Thus, if M is a reducing subspace of T_ϕ, then M reduces T_{χ_F} for any Borel subset F of E.

Below, to finish the proof, it suffices to show that if M is a nonzero reducing subspace of T_ϕ, then $M = H^2(\mathbb{D})$. To see this, we assume that f is a nonzero function in M. Clearly, for any Borel subset F of E,

$$T_{\chi_F} f \in M.$$

For each bounded Borel function h supported on E, there is always a sequence of bounded functions $\{h_n\}$ converging uniformly to h, where each h_n is the linear span of $\chi_F(F \subseteq E)$. This immediately gives that $T_h f \in M$. For each $g \in M^\perp$, we have

$$\langle g, T_h f \rangle = 0.$$

That is, $\int_E h\bar{g}f dm = 0$ for any bounded Borel function h supported on E, forcing $gf = 0$ a.e. on E. By Riesz's theorem [Hof2, p. 51], any function in $H^2(\mathbb{D})$ can vanish only on a set of null measure, and so is f. Therefore, $g = 0$ a.e. on E. Since E has positive measure, by Riesz's theorem g is the zero function. By the arbitrariness of g, $M = H^2(\mathbb{D})$. The proof is complete.   □

In addition, Nordgren provided a sufficient condition for an analytic Toeplitz operator to have nontrivial reducing subspaces, on this line also refer to [Ba].

Theorem B.2 ([Nor]) *Let T_ϕ be an analytic Toeplitz operator on the Hardy space. If $\phi = \varphi \circ \eta$, where $\varphi \in H^\infty(\mathbb{D})$ and η is an inner function different from the Möbius map, then T_ϕ has nontrivial reducing subspaces.*

In the case of ϕ being an inner function, a description of the reducing subspaces of T_ϕ was well-known [Hal2].

Proof Assume that η is an inner function different from the Möbius map. We first show that on the Hardy space M_η has a nontrivial reducing subspace. Set $N = H^2(\mathbb{D}) \ominus \eta H^2(\mathbb{D})$, and then $\dim N \geq 2$. Pick a nonzero proper closed subspace M of N, and denote by $\tilde{M}$ the closed span of $\eta^k N(k \geq 0)$, which are pairwise orthogonal. Note that $M_\eta^* N = 0$ and since M_η is an isometric operator,

$$M_\eta^* \eta^k N = \eta^{k-1} N, k \geq 1.$$

Then $\tilde{M}$ is invariant for M_η^*, and also for M_η. Therefore, $\tilde{M}$ is a nontrivial reducing subspace of M_η.

To complete the proof, it suffices to show that $\{M_\eta\}' \subseteq \{M_{\varphi \circ \eta}\}'$. In fact, for each $\varphi \in H^\infty(\mathbb{D})$, there is a uniformly bounded sequence $\{p_n\}$ of polynomials such that p_n converges uniformly to φ on each compact subset of $\mathbb{D}$. For any $z \in \mathbb{D}$,

$$\lim_{n \to \infty} \| M_{p_n(\eta)}^* K_z - M_{\varphi(\eta)}^* K_z \| = 0.$$

Since $\{ \| M_{p_n(\eta)}^* \| \}$ is bounded and the span of all reproducing kernels is dense in $H^2(\mathbb{D})$, $M_{p_n(\eta)}^*$ converges strongly to $M_{\varphi(\eta)}^*$. For any $A \in \{M_\eta\}'$, we have $A \in \{M_{p_n(\eta)}\}'$, and hence

$$\langle A M_{p_n(\eta)} K_z, K_z \rangle = \langle M_{p_n(\eta)} A K_z, K_z \rangle, \quad z \in \mathbb{D}.$$

That is,

$$\langle K_z, M_{p_n(\eta)}^* A^* K_z \rangle = \langle A K_z, M_{p_n(\eta)}^* K_z \rangle, \quad z \in \mathbb{D}.$$

Letting n tends to infinity, we have

$$\langle K_z, M_{\varphi(\eta)}^* A^* K_z \rangle = \langle A K_z, M_{\varphi(\eta)}^* K_z \rangle, \quad z \in \mathbb{D}.$$

Equivalently,

$$\langle A M_{\varphi(\eta)} K_z, K_z \rangle = \langle A M_{\varphi(\eta)} K_z, K_z \rangle, \quad z \in \mathbb{D}.$$

Applying the property of Berezin transform (see Appendix A) gives

$$A M_{\varphi(\eta)} = M_{\varphi(\eta)} A,$$

completing the proof. □

It is worthwhile to mention that [Cow3] gave a somewhat complicated condition on ϕ for T_ϕ to have nontrivial reducing subspaces in the case of ϕ being some function close to a covering map.

Appendix C
List of Problems

In this context, some problems and conjectures are collected from this book, and they are also mentioned in [GH3].

If an $H^\infty(\mathbb{D})$-function ϕ enjoys the following property: for some λ in $\mathbb{D}$ the inner part of $\phi - \phi(\lambda)$ is a finite Blaschke product, then ϕ is called to belong to Cowen's class, see Chap. 3. By Theorems 3.1.1, if ϕ lies in Cowen's class, then for some finite Blaschke product B, $\{T_\phi\}' = \{T_B\}'$ holds on the Bergman space $L_a^2(\mathbb{D})$. Therefore, the corresponding von Neumann algebras are equal: $\mathcal{V}^*(\phi) = \mathcal{V}^*(B)$. Then by Douglas, Putinar and Wang's theorem (Theorem 4.2.1), both $\mathcal{V}^*(\phi)$ and $\mathcal{V}^*(B)$ are abelian. Their proof for Theorem 4.2.1 is very function-theoretic. But it seems that we are still a long way from a satisfactory understanding: an operator-theoretic approach is beyond our knowledge at present. We believe that any operator-theoretic consideration would help solve the following conjecture, also see Conjecture 8.1.4.

Conjecture C.1 *If ϕ is in the disk algebra $A(\mathbb{D})$, then the von Neumann algebra $\mathcal{V}^*(\phi)$ defined on $L_a^2(\mathbb{D})$ is abelian.*

It is known that any function that is holomorphic on a neighborhood of $\overline{\mathbb{D}}$ lies in Cowen's class. However, it is not known whether $A(\mathbb{D})$ is contained in Cowen's class.

Now we turn to functions in $H^\infty(\mathbb{D})$ without "good" geometric property. For example, any infinite Blaschke product does not lie in Cowen's class, see Chap. 3. Thus in this case it is difficult to use Thomson's method in [T1, T2] or Cowen's method in [Cow1]. In this book, two distinct classes of infinite Blaschke products B are discussed: thin Blaschke products and those Blaschke products which are also covering maps. However, there is dramatic difference between the structures of these von Neumann algebras $\mathcal{V}^*(B)$ where B run for these two classes. In more detail, for a thin Blaschke products B, $\mathcal{V}^*(B)$ is always abelian, and in most cases $\mathcal{V}^*(B)$ is trivial. But if a Blaschke product B is simultaneously a covering map, $\mathcal{V}^*(B)$ can be rather complicated. In fact, if E is a discrete subset of $\mathbb{D} - \{0\}$ and $\phi : \mathbb{D} \to \mathbb{D} - E$ is a holomorphic covering map, then ϕ is necessarily an interpolating

© Springer-Verlag Berlin Heidelberg 2015
K. Guo, H. Huang, *Multiplication Operators on the Bergman Space*,
Lecture Notes in Mathematics 2145, DOI 10.1007/978-3-662-46845-6

Blaschke product and $V^*(\phi)$ is $*$-isomorphic to the free group factor $\mathcal{L}(F_n)$, where n equals the cardinality $\sharp E$(if E is an infinite set, then $n = \infty$). As mentioned in the introduction, this makes significant contact with several areas, including algebraic topology, group theory, and operator theory. Most remarkably, it has a close connection with a long-standing open problem in von Neumann algebras: does it hold that

$$\mathcal{L}(F_n) \overset{*}{\cong} \mathcal{L}(F_m)$$

for $n \neq m$ and $n, m \geq 2$? As mentioned in Sect. 6.5 of Chap. 6, this problem is equivalent to the following:

Problem C.2 Let E and F be two subsets of $\mathbb{D} - \{0\}$ with $\sharp E = 2$ and $\sharp F = 3$, and let $B_1 : \mathbb{D} \to \mathbb{D} - E$ and $B_2 : \mathbb{D} \to \mathbb{D} - F$ be two holomorphic covering maps, which are necessarily Blaschke products. Then is $V^*(B_1)$ unitarily isomorphic (or $*$-isomorphic) to $V^*(B_2)$?

The study of von Neumann algebras $V^*(B)$ in Chaps. 4–6 naturally induces the following question.

Question C.3 For an infinite Blaschke product B, is there any connection between the commutativity of $V^*(B)$ on $L_a^2(\mathbb{D})$ and the density of the zero set $Z(B)$? If yes, how to describe it?

Also, we make the following conjecture.

Conjecture C.4 *For each Blaschke product B, the von Neumann algebra $V^*(B)$ on $L_a^2(\mathbb{D})$ is finite.*

Assuming the above conjecture is true, one can raise the following problem:
For each function $\phi \in H^\infty(\mathbb{D})$, is $V^*(\phi)$ on $L_a^2(\mathbb{D})$ a finite von Neumann algebra?
This problem can also be raised on other reproducing kernel Hilbert spaces of holomorphic functions and in multi-variable case. Recall that any von Neumann algebra $\mathcal{A}$ can be written as

$$\mathcal{A} = \mathcal{A}_1 \oplus \mathcal{A}_2 \oplus \mathcal{A}_3, \tag{C.1}$$

where $\mathcal{A}_1, \mathcal{A}_2, \mathcal{A}_3$ are of type I, II and III, respectively, and it is not necessary that all of them appear in the above decomposition. Now write $\mathcal{A} = V^*(\phi)$. If the answer of the above problem is affirmative, then $\mathcal{A}_3$ does not exist in (C.1) since a type III von Neumann algebra has no nonzero finite projection. That is, $V^*(\phi)$ has no direct summand of type III, and by Proposition 2.5.4, so does its commutant $W^*(\phi)$.

Furthermore, we provide the following. Note that $H^\infty(\mathbb{D})$ is a Banach space, and in any metric space the notion of first category makes sense. Recall that any countable union of nowhere dense sets is called *a set in the first category*.

Conjecture C.5 *There is a subset $\mathcal{E}$ of $H^\infty(\mathbb{D})$ in the first category, such that $V^*(\phi)$ is trivial provided that $\phi \in H^\infty(\mathbb{D}) - \mathcal{E}$. Similarly, in an appropriate sense, for "most" infinite Blaschke products B, $V^*(B)$ are trivial.*

There is a chance to give a geometric characterization for $V^*(\phi)$ being trivial. Such a characterization is possibly a key to the former part of Conjecture C.5. The following may be the condition for $V^*(\phi)$ being nontrivial: there is some local inverse ρ of ϕ admitting unrestricted continuation in an open subset U_0 of $\mathbb{D}$, such that $\mathbb{D} - U_0$ has zero capacity and $G[\rho]$ has only finite sheets. As done in Chap. 5, $G[\rho]$ denotes the graph of the equivalent class $[\rho]$.

It is worthwhile to note that if Conjecture C.5 holds, then the norm-closure of all irreducible multiplication operators M_h equals the algebra of all multiplication operators. In some sense, this gives a support for a conjecture in operator theory, which says that: on a separable infinite-dimensional Hilbert space H, the norm-closure of all irreducible operators equals the algebra $B(H)$.

Chapter 4 gives two conjectures associated with the commutativity of $V^*(\phi)$ defined on $L_a^2(\mathbb{D})$.

Conjecture C.6 (Conjecture 4.4.7) *Suppose that $\phi \in H^\infty(\mathbb{D})$ and $Z(\phi) \cap \mathbb{D} \neq \emptyset$. If M is a reducing subspace of M_ϕ satisfying $\dim M \ominus \phi M < \infty$, then P_M lies in the center of $V^*(\phi)$.*

As mentioned in Sect. 4.4 in Chap. 4, if Conjecture C.6 holds, then under the same condition of Conjecture C.6 the von Neumann algebra $P_M V^*(\phi)|_M$ is abelian. Special interest is focused on the case when M equals the whole space $L_a^2(\mathbb{D})$. If ϕ is a finite Blaschke product, then Conjecture C.6 holds, see Theorem 4.2.1. A bit more general version of Conjecture C.6 is listed as follows.

Conjecture C.7 (Conjecture 4.4.8) *Suppose $\dim V^*(\phi) < \infty$, then $V^*(\phi)$ is abelian. In general, if $V^*(\phi)$ is a type I von Neumann algebra, then is $V^*(\phi)$ abelian?*

In a similar way, $V^*(\Phi)$ can be defined in multi-variable case, see Sect. 2.6 in Chap. 2 and Sect. 7.3 in Chap. 7. Precisely, given a family Φ of bounded holomorphic functions on some bounded domain Ω, M_Φ denote the family $\{M_\phi : \phi \in \Phi\}$ of multiplication operators on $L_a^2(\Omega)$. Then put

$$V^*(\Phi) = W^*(\Phi)',$$

where $W^*(\Phi)$ denotes the von Neumann algebra generated by M_Φ. The following question is natural and basic.

Question C.8 Suppose $\Phi = (\phi_1, \cdots, \phi_n)$, and M_Φ acts on the Bergman space $L_a^2(\Omega)$, where Ω is a Reinhardt domain in $\mathbb{C}^d$. If $d > n$, is $\dim V^*(\Phi) = \infty$? If $d < n$, under what conditions do we have $V^*(\Phi) = \mathbb{C}I$? Also consider the structure of $V^*(\Phi)$ in the case of $d = n$.

Little is known about such questions for $d \geq 2$. The case $d = n$ may be of most interest.

Conjectures C.6 and C.7 fail in multi-variable case, see Examples 8.4.3 and 8.4.4. However, we can provide the following.

Conjecture C.9 *Let Φ be a family of polynomials. Given a Reinhardt domain Ω in $\mathbb{C}^d$, let $\mathcal{V}^*(\Phi)$ be the von Neumann algebra defined on the Bergman space $L_a^2(\Omega)$. If $\dim \mathcal{V}^*(\Phi) < 6$, then $\mathcal{V}^*(\Phi)$ is abelian.*

Conjecture C.10 *Suppose Φ is a family of holomorphic functions over the closure $\overline{\mathbb{B}_d}$ of the unit ball in $\mathbb{C}^d$. If $\dim \mathcal{V}^*(\Phi) < \infty$, then $\mathcal{V}^*(\Phi)$ is generated by U_ρ:*

$$U_\rho f = f \circ \rho J \rho, f \in L_a^2(\mathbb{B}_d),$$

where ρ are members in $\mathrm{Aut}(\mathbb{B}_d)$ satisfying $\phi \circ \rho = \phi$ for each ϕ in Φ, and $J\rho$ denote the determinants of the Jocobian matrices of ρ.

Recall that an open subset Ω in $\mathbb{C}^d$ is called *a Reinhardt domain* if $(z_1, \cdots, z_d) \in \Omega$ implies that $(e^{i\theta_1} z_1, \cdots, e^{i\theta_d} z_d) \in \Omega$ for all real numbers $\theta_1, \cdots, \theta_d$. For example, the polydisk $\mathbb{D}^d$ and the unit ball $\mathbb{B}_d$ are Reinhardt domains. Also, the Thullen domain

$$\{(z_1, z_2) \in \mathbb{C}^2 : |z_1|^2 + |z_2|^{\frac{2}{p}} < 1\} \ (p > 0 \text{ and } p \neq 1)$$

is a Reinhardt domain.

Bibliography

[A1]   M. Abrahamse, Analytic Toeplitz operators with automorphic symbol. Proc. Am. Math. Soc. **52**, 297–302 (1975)

[A2]   M. Abrahamse, Some examples of lifting the commutant of a subnormal operator. Ann. Polon. Math. **37**, 289–298 (1980)

[AB]   M. Abrahamse, J. Ball, Analytic Toeplitz operators with automorphic symbol II. Proc. Am. Math. Soc. **59**, 323–328 (1976)

[AC]   S. Axler, Z. Cuckovic, Commuting Toeplitz operators with harmonic symbols. Integr. Equ. Oper. Theory **14**, 1–12 (1991)

[ACR]   S. Axler, Z. Cuckovic, N. Rao, Commutants of analytic Toeplitz operators on the Bergman space. Proc. Am. Math. Soc. **128**, 1951–1953 (2000)

[AD]   M. Abrahamse, R. Douglas, A class of subnormal operators related to multiply-connected domains. Adv. Math. **19**, 106–148 (1976)

[Ap]   T. Apostal, Modular Functions and Dirichlet Series in Number Theory. Graduate Texts in Mathematics, vol. 41 (Springer, New York, 1976)

[Ar1]   W. Arveson, A Short Course on Spectral Theory. Graduate Texts in Mathematics, vol. 209 (Springer, New York, 2001)

[Ar2]   W. Arveson, *An Invitation to C*-Algebras*. Graduate Texts in Mathematics, vol. 39 (Springer, New York, 1998)

[Ar3]   W. Arveson, The probability of entanglement. Commun. Math. Phys. **286**, 283–312 (2009)

[Ar4]   W. Arveson, Quantum channels that preserve entanglement. Math. Ann. **343**, 757–771 (2009)

[Arm]   M. Armstrong, Basic Topology. Undergraduate Texts in Mathematics (Springer, New York, 1983) [Corrected reprint of the 1979 original]

[AS]   L. Ahlfors, L. Sario, Riemann Surfaces (Princeton Univercity Press, Princeton, 1960)

[Ba]   J. Ball, Hardy space expectation operators and reducing subspaces. Proc. Am. Math. Soc. **47**, 351–357 (1975)

[Be1]   H. Behncke, Structure of certain non-normal operators. J. Math. Mech. **18**, 103–107 (1968)

[Be2]   H. Behncke, A class of nonnormal operators. Preprint

[Bes]   A. Besicovitch, On sufficient condition for a function to be analytic and on behavior of analytic functions in the neighborhood of non-isolated singular points. Proc. Lond. Math. Soc. **32**, 1–9 (1931)

[BFP]   H. Bercovici, C. Foias, C. Pearcy, Dual algebras with applications to invariant subspaces and dilation theory, in *Conference Board of the Mathematical Sciences.*

© Springer-Verlag Berlin Heidelberg 2015

K. Guo, H. Huang, *Multiplication Operators on the Bergman Space*,
Lecture Notes in Mathematics 2145, DOI 10.1007/978-3-662-46845-6

Regional Conference Series in Mathematics, vol. 56 (American Mathematical Society, Rhode Island, 1985)

[BG]   C. Berenstein, R. Gay, *Complex Variables: An Introduction*. Graduate Texts in Mathematics, vol. 125 (Springer, New York, 1991)

[BDU]   I. Baker, J. Deddens, J. Ullman, A theorem on entire functions with applications to Toeplitz operators. Duke Math. J. **41**, 739–745 (1974)

[Bla]   B. Blackadar, *Operator Algebras*. Encyclopaedia of Mathematical Sciences, vol. 122 (Springer, Berlin, 2006) [Theory of C^*-algebras and von Neumann algebras, Operator Algebras and Non-commutative Geometry, III]

[Bli]   G. Bliss, *Algebraic Functions*, vol. 16 (American Mathematical Society Colloquium Publications, New York, 1933)

[BL]   M. Brodskii, M. Livsic, Spectral analysis of non self-adjoint operators and intermediate systems. Uspehi Mat. Nauk. **13**, 265–346 (1958)

[BMP]   M. Boileau, S. Maillot, J. Porti, Three-Dimensional Orbifolds and Their Geometric Structures. Panoramas et Syntheses, vol. 15 (Societe Mathematique De France, Paris, 2003)

[Br]   A. Brown, On a class of operators. Proc. Am. Math. Soc. **4**, 723–728 (1953)

[BSW]   D. Burns, S. Shnider, R. Wells, Deformation of strictly pseudoconvex domains. Invent. Math. **46**, 237–253 (1978)

[BTV]   J. Ball, T. Trent, V. Vinnikov, Interpolation and commutant lifting for multipliers on reproducing kernel Hilbert spaces. Oper. Theory Anal. **122**, 89–138 (2001)

[CaG]   L. Carleson, T. Gamelin, *Complex Dynamics* (Springer, Berlin, 1993)

[CDG]   L. Chen, R. Douglas, K. Guo, On the double commutant of Cowen-Douglas operators. J. Funct. Anal. **260**, 1925–1943 (2011)

[CFT]   I. Chalendar, E. Fricain, D. Timotin, Functional models and asymptotically orthonormal sequences. Ann. Inst. Fourier (Grenoble) **53**, 1527–1549 (2003)

[CG]   X. Chen, K. Guo, *Analytic Hilbert Modules*. π-Chapman & Hall/CRC Research Notes in Mathematics, vol. 433, (2003)

[CGW]   G. Cheng, K. Guo, K. Wang, Transitive algebras and reductive algebras on reproducing analytic Hilbert spaces. J. Funct. Anal. **258**, 4229–4250 (2010)

[Cl]   B. Cload, Toeplitz operators in the commutant of a composition operator. Studia Math. **133**, 187–196 (1999)

[CL]   E. Collingwood, A. Lohwater, *Theory of Cluster Sets* (Cambridge University Press, Cambridge, 1966)

[Cla1]   D. Clark, On Toeplitz operators with loops. J. Oper. Theory **4**, 37–54 (1980)

[Cla2]   D. Clark, On a similarity theory for rational Toeplitz operators. J. Riene Angew. Math. **320**, 6–31 (1980)

[Cla3]   D. Clark, On Toeplitz operators with loops II. J. Oper. Theory **7**, 109–123 (1982)

[Cla4]   D. Clark, Sz-Nagy-Foias theory and similarity for a class of Toeplitz operators, in *Proceedings of the 1977 Spectral theory Semester* (The Stefan Banach Mathematical Center, Warsaw, 1982)

[CLY]   Y. Chen, Y. Lee, T. Yu, Reducibility and unitary equivalence for a class of multiplication operators on the Dirichlet space. Compl. Anal. Oper. Theory **7**, 1897–1908 (2013)

[ClMo]   D. Clark, J. Morrel, On Toeplitz operators and similarity. Am. J. Math. **100**, 973–986 (1978)

[CMY]   R. Curto, P. Muhly, K. Yan, The C^*-algebra of an homogeneous ideal in two variables is type I, *Current Topics in Operator Algebras (Nara, 1990)* (World Scientific, River Edge, 1991), 130–136

[CoD]   M. Cowen, R. Douglas, Complex geometry and operator theory. Acta Math. **141**, 187–261 (1978)

[Con1]   J. Conway, *A Course in Operator Theory*. Graduate Studies in Mathematics, vol. 21 (American Mathematical Society, Rhode Island, 2000)

[Con2]   J. Conway, *Subnormal Operators*. Research Notes in Mathematics, vol. 51 (Pitman Advanced Publishing Program, Boston, 1981)

[Cow1] C. Cowen, The commutant of an analytic Toeplitz operator. Trans. Am. Math. Soc. **239**, 1–31 (1978)

[Cow2] C. Cowen, The commutant of an analytic Toeplitz operator, II. Indiana Univ. Math. J. **29**, 1–12 (1980)

[Cow3] C. Cowen, An analytic Toeplitz operator that commutes with a compact operator and a related class of Toeplitz operators. Funct. Anal. **36**, 169–184 (1980)

[Cow4] C. Cowen, On equivalence of Toeplitz operators. J. Oper. Theory **7**, 167–172 (1982)

[Cow5] C. Cowen, Finite Blaschke products as composition of other finite Blaschke products. arXiv: math.CV/ 1207.4010v1

[CW] C. Cowen, R. Wahl, Commutants of finite Blaschke product multiplication operators. Preprint

[CS] K. Chan, S. Seubert, Reducing subspaces of compressed analytic Toeplitz operators on the Hardy space. Integr. Equ. Oper. Theory **28**, 147–157 (1997)

[Cu] Z. Cuckovic, Commutants of Toeplitz operators on the Bergman space. Pac. J. Math. **162**, 277–285 (1994)

[Da] K. Davidson, *C*-Algebras by Example*. Fields Institute Monographs, vol. 6 (American Mathematical Society, Rhode Island, 1996)

[DF] D. Dummit, R. Foote, *Abstract Algebra*, 3rd edn (Wiley, Hoboken, 2004)

[DH] H. Dan, H. Huang, Multiplication operators defined by a class of polynomials on $L_a^2(\mathbb{D}^2)$. Integr. Equ. Oper. Theory **80**, 581–601 (2014) arXiv: math.OA/ 1404.5414v1

[DK] R. Douglas, Y. Kim, Reducing subspaces on the annulus. Integr. Equ. Oper. Theory **70**, 1–15 (2011)

[Di] J. Dixmier, *von Neumann Algebras* (North-Holland, Amsterdam, 1981)

[Dou] R. Douglas, Operator theory and complex geometry. Extracta Math. **24**, 135–165 (2009). arXiv: math.FA/ 0710.1880v2

[DP] R. Douglas, C. Pearcy, Spectral theory of generalized Toeplitz operators. Trans. Am. Math. Soc. **115**, 433–444 (1965)

[DPa] R. Douglas, V. Paulsen, *Hilbert Modules over Function Algebras*. Pitman Research Notes in Mathematics, vol. 217 (Longman Scientific & Technical, Harlow, 1989)

[DPW] R. Douglas, M. Putinar, K. Wang, Reducing subspaces for analytic multipliers of the Bergman space. J. Funct. Anal. **263**, 1744–1765 (2012) arXiv: math.FA/ 1110.4920v1

[DS] P. Duren, A. Schuster, *Bergman Spaces*. Mathematics Surveys and Monographs, vol. 100 (American Mathematical Society, Rhode Island, 2004)

[DSZ] R. Douglas, S. Sun, D. Zheng, Multiplication operators on the Bergman space via analytic continuation. Adv. Math. **226**, 541–583 (2011)

[Du] J. Dudziak, *Vitushkin' Conjecture for Removable Sets*. Universitext (Springer, New York, 2010)

[Dur] P. Duren, Extension of a result of Beurling on invariant subspaces. Trans. Am. Math. Soc. **99**, 320–324 (1961)

[DW] J. Deddens, T. Wong, The commutant of analytic Toeplitz operators. Trans. Am. Math. Soc. **184**, 261–273 (1973)

[Fr] O. Frostman, Potentiel d'equilibre et capacite des ensembles avec quelques applications à la theorie des fonctions. Medd. Lunds Mat. Sem. **3**, 1–118 (1935)

[Ga] J. Garnett, *Bounded Analytic Functions* (Academic, New York, 1981)

[GH1] K. Guo, H. Huang, On multiplication operators of the Bergman space: similarity, unitary equivalence and reducing subspaces. J. Oper. Theory **65**, 355–378 (2011)

[GH2] K. Guo, H. Huang, Multiplication operators defined by covering maps on the Bergman space: the connection between operator theory and von Neumann algebras. J. Funct. Anal. **260**, 1219–1255 (2011)

[GH3] K. Guo, H. Huang, Reducing subspaces of multiplication operators on function spaces: dedicated to the memory of Chen Kien-Kwong on the 120th anniversary of his birth. Appl. Math. J. Chinese Univ. **28**, 395–404 (2013)

[GH4] K. Guo, H. Huang, Geometric constructions of thin Blaschke products and reducing subspace problem. Proc. Lond. Math. Soc. **109**, 1050–1091 (2014)

[GH5]   K. Guo, H. Huang, Cowen's class and Thomson's class arXiv: math.CV/ 1312.7498v1

[GHa]   M. Greenberg, J. Harper, *Algebraic Topology: A First Course*. Mathematics Lecture
          Note Series, vol. 58 (Benjamin/Cummings, Reading, 1981)

[GHJ]   F. Goodman, P. Harpe, V. Jones, *Coxeter Graphs and Towers of Algebras* (Springer,
          New York, 1999)

[Gil1]   F. Gilfeather, The structure of non unitary opeartors and their asymptotic behavior.
          Ph.D. Thesis, University of California, Irvine, 1969

[Gil2]   F. Gilfeather, On the Suzuki structure theory for non self-adjoint operators on Hilbert
          space. Acta Sci. Math. (Szeged) **32**, 239–249 (1971)

[GK1]   R. Greene, S. Krantz, Stability properties of the Bergman kernel and curvature
          properties of bounded domains, in *Recent Progress in Several Complex Variables*
          (Princeton University Press, Princeton, 1982)

[GK2]   R. Greene, S. Krantz, Deformation of complex structures, estimates for the $\bar{\partial}$ equation,
          and stability of the Bergman kernel. Adv. Math. **43**, 1–86 (1982)

[GM1]   P. Gorkin, R. Mortini, Asymptotic interpolating sequences in uniform algebras. J. Lond.
          Math. Soc. **67**, 481–498 (2003)

[GM2]   P. Gorkin, R. Mortini, Value distribution of interpolating Blaschke product. J. Lond.
          Math. Soc. **72**, 151–168 (2005)

[GM3]   P. Gorkin, R. Mortini, Radial limits of interpolating Blaschke product. Math. Ann. **331**,
          417–444 (2005)

[Go]   C. Godbillon, *Elements de Topologie Algebrique* (Hermann, Paris, 1971)

[Gol]   G. Goluzin, *Geometric Theory of Functions of a Complex Variable*, vol. 26 (American
          Mathematical Society, 1969), p. 255

[Gr]   H. Griffiths, The fundamental group of a surface, and a theorem of Schreier. Acta Math.
          110, 1–17 (1963)

[Guo1]   K. Guo, Algebraic reduction for Hardy submodules over polydisk algebras. J. Oper.
          Theory **41**, 127–138 (1999)

[Guo2]   K. Guo, Characteristic spaces and rigidity of analytic Hilbert modules. J. Funct. Anal.
          163, 133–151 (1999)

[Guo3]   K. Guo, Equivalence of Hardy submodules generated by polynomials. J. Funct. Anal.
          178, 343–371 (2000)

[Guo4]   K. Guo, Defect operators, defect functions and defect indices for analytic submodules.
          J. Funct. Anal. **213**, 380–411 (2004)

[Guo5]   K. Guo, Operator theory and Von Neumann algebras (a manuscript)

[Guo6]   K. Guo, Defect operators for submodules of H_d^2. J. Reine Angew. Math. **573**, 181–209
          (2004)

[GuoW]   K. Guo, X. Wang, Reducing subspaces of tensor products of weighted shifts. Preprint

[GSZZ]   K. Guo, S. Sun, D. Zheng, C. Zhong, Multiplication operators on the Bergman space
          via the Hardy space of the bidisk. J. Reine Angew. Math. **629**, 129–168 (2009)

[GW1]   K. Guo, K. Wang, On operators which commute with analytic Toeplitz operators
          modulo the finite rank operators. Proc. Am. Math. Soc. **134**, 2571–2576 (2006)

[GW2]   K. Guo, K. Wang, Essentially normal Hilbert modules and K-homology II: quasi-
          homogeneous Hilbert modules over the two dimensional unit ball. J. Ramanujan Math.
          Soc. **22**, 259–281 (2007)

[GW3]   K. Guo, K. Wang, Essentially normal Hilbert modules and K-homology. Math. Ann.
          340, 907–934 (2008)

[GW4]   K. Guo, K. Wang, Beurling type quotient modules over the bidisk and boundary
          representations. J. Funct. Anal. **257**, 3218–3238 (2009)

[GZ]   K. Guo, D. Zheng, Rudin orthogonality problem on the Bergman space. J. Funct. Anal.
          261, 51–68 (2011)

[Hal1]   P. Halmos, *Measure Theory* (Van Nostrand, Princeton, 1950)

[Hal2]   P. Halmos, Shifts on Hilbert spaces. J. Reine Angew. Math. **208**, 102–112 (1961)

[Has]   W. Hastings, A Carleson measure theorem for Bergman spaces. Proc. Am. Math. Soc.
          52, 237–241 (1975)

[Hat]   A. Hatcher, *Algebraic Topology* (Cambridge University Press, Cambridge, 2002)
[HKZ]   H. Hedenmalm, B. Korenblum, K. Zhu, *Theory of Bergman Spaces* (Springer, New York, 2000)
[HNRR]   D. Hadwin, E. Nordgren, H. Radjavi, P. Rosenthal, An operator not satisfying Lomonosov's hypothesis. J. Funct. Anal. **38**, 410–415 (1980)
[Hof1]   K. Hoffman, *Banach Spaces of Analytic Functions* (Prentice-Hall, Englewood Cliffs, 1962)
[Hof2]   K. Hoffman, Bounded analytic functions and Gleason parts. Ann. Math. **86**, 74–111 (1967)
[Ho]   C. Horowitz, Facorization theorems for functions in the Bergman spaces. Duke Math. J. **44**, 201–213 (1977)
[Ho2]   C. Horowitz, Zeros of functions in the Bergman spaces. Duke Math. J. **41**, 693–710 (1974)
[Hor]   L. Hormander, *Introduction to Complex Analysis in Several Variables*, 3rd edn (North-Holland, Amsterdam, 1990)
[Huang1]   H. Huang, Maximal abelian von Neumann algebras and Toeplitz operators with separately radial symbols. Integr. Equ. Oper. Theory **64**, 381–398 (2009)
[Huang2]   H. Huang, von Neumann algebras generated by multiplication operators on the weighted Bergman space: a function-theory view into operator theory. Sci. China Ser. A **56**, 811–822 (2013)
[HSXY]   J. Hu, S. Sun, X. Xu, D. Yu, Reducing subspace of analytic Toeplitz operators on the Bergman space. Integr. Equ. Oper. Theory **49**, 387–395 (2004)
[Ja]   N. Jacobson, *Basic Algebra II* (San Francisco, Freeman, 1980)
[JL]   C. Jiang, Y. Li, The commutant and similarity invariant of analytic Toeplitz operators on Bergman space. Sci. China Ser. A **5**, 651–664 (2007)
[Jon]   V. Jones, *von Neumann Algebras*, UC Berkeley Mathematics (2009). http://math.berkeley.edu/~vfr/MATH20909/VonNeumann2009.pdf
[JZ]   C. Jiang, D. Zheng, Similarity of analytic Toeplitz operators on the Bergman spaces. J. Funct. Anal. **258**, 2961–2982 (2010)
[Kap]   M. Kapovich, *Hyperbolic Manifolds and Discrete Groups*. Progress in Mathematics, vol. 183 (Birkhauser, Boston, 2001)
[Ke]   J. Kelley, *General Topology*. Graduate Texts in Mathematics, vol. 27 (Springer, New York, 1955)
[Koo]   P. Koosis, *Introduction to H_p Spaces: With an Appendix on Wolff's Proof of the Corona Theorem*. London Mathematical Society, vol. 40 (Cambridge University Press, Cambridge, 1980)
[Kran]   S. Krantz, *Function Theory of Several Complex Variables*. Pure and Applied Mathematics (A Wiley-Interscience Publication, New York, 1982)
[La]   K. Lawson, Some lemmas on interpolating Blaschke products and a correction. Can. J. Math. **21**, 531–534 (1969)
[Lang1]   S. Lang, $SL_2(\mathbb{R})$. Graduate Texts in Mathematics, vol. 105 (Springer, New York, 1985)
[Lang2]   S. Lang, *Complex Analysis*. Graduate Texts in Mathematics, vol. 103 (Springer, New York, 1999)
[Le]   J. Lee, *Introduction to Topological Manifolds*. Graduate Texts in Mathematics, vol. 202 (Springer, New York, 2000)
[Lo]   V. Lomonosov, Invariant subspaces for the family of operators which commute with a completely continuous operator. Funct. Anal. Appl. **7**, 213–214 (1973)
[Lu]   D. Luecking, A technique for characterizing Carleson measures on Bergman spaces. Proc. Am. Math. Soc. **87**, 656–660 (1983)
[LZ]   Y. Lu, X. Zhou, Invariant subspaces and reducing subspaces of weighted Bergman space over bidisk. J. Math. Soc. Jpn. **62**, 745–765 (2010)
[Ma]   D. Marshall, Removable sets for bounded analytic functions. J. Math. Sci. **26**, 2232–2234 (1984)

[Mas]   J. Mashreghi, *Derivatives of Inner Functions*. The Fields Institute for Research in the Mathematical Sciences. Fields Institute Monographs, vol. 31 (Springer, New York, 2013)

[McS]   G. McDonald, C. Sundberg, Toeplitz operators on the disc. Indiana Univ. Math. J. **28**, 595–611 (1979)

[Mi]   J. Milnor, *Dynamics in One Complex Variable*. Annals of Mathematics Studies, vol. 160 (Princeton University Press, Princeton, 2006)

[Mil]   J. Milne, *Modular Functions and Modular Forms*. Lecture Notes (1990). http://jmilne.org/math/CourseNotes/MF.pdf

[MS]   G. Mostow, Y. Siu, A compact Kahler surface of negative curvature not coverd by the ball. Ann. Math. **112**, 321–360 (1980)

[MT]   R. Mazzeo, M. Taylor, Curvature and uniformization. Isr. J. Math. **130**, 323–346 (2002)

[Na]   I. Natanson, *Theory of Functions of a Real Variable* (Frederick Ungar Publishing Co., New York, 1955)

[Ne]   Z. Nehari, *Conformal Mapping* (McGraw-Hill, New York, 1952)

[Nev]   R. Nevanlinna, *Eindeutige analytische Funktionen*, 2te Aufl. (Springer, Berlin, 1953)

[Ni]   N. Nikolski, *Treatise on the Shift Operator*. Grundlehren der mathematischen Wissenschafte, vol. 273 (Springer, Berlin, 1986)

[Nor]   E. Nordgren, Reducing subspaces of analytic Toeplitz operators. Duke Math. J. **34**, 175–181 (1967)

[Og]   A. Ogg, *Modular Forms and Dirichlet Series* (Benjamin, New York, 1969)

[Oka]   K. Oka, Sur les fonctions de plusieurs variables II. Domaines d'holomorphie. J. Sci. Hiroshima Univ. **7**, 115–130 (1937)

[Pe]   G. Pederson, C^*-*Algebras and Their Automorphism Groups*. London Mathematical Society, vol. 14 (Academic, London, 1979)

[Pel]   V. Peller, *Hankel Operators and Their Applications*. Springer Monographs in Mathematics (Springer, New York, 2003)

[PS]   C. Pearcy, A. Shields, A survey of the Lomonosov technique in theory of invariant subspaces, *Topics in Operator Theory*. Mathematical Surveys, vol. 13 (American Mathematical Society, Providence, 1974), 219–229

[Rat]   J. Ratcliffe, *Foundations of Hyperbolic Manifolds*. Graduate Texts in Mathematics, vol. 149 (Springer, New York, 1994)

[Ri1]   J. Ritt, Prime and composite polynomials. Trans. Am. Math. Soc. **23**, 51–66 (1922)

[Ri2]   J. Ritt, Permutable rational functions. Trans. Am. Math. Soc. **25**, 399–448 (1923)

[Ro]   D. Robinson, *A Course in the Theory of Groups*. Graduate Texts in Mathematics, vol. 80 (Springer, New York, 1982)

[Ros]   P. Rosenthal, Completely reducible operators. Proc. Am. Math. Soc. **19**, 826–830 (1968)

[Ro]   B. Robati, On the commutant of multiplication operators with analytic polynomial symbols. Bull. Kor. Math. Soc. **44**, 683–389 (2007)

[Ru1]   W. Rudin, *Function Theory in Polydiscs* (Benjamin, New York, 1969)

[Ru2]   W. Rudin, *Function Theory in the UnIt ball of* $\mathbb{C}^n$. Grundlehren der Mathematischen, vol. 241 (Springer, New York, 1980)

[Ru3]   W. Rudin, *Real and Complex Analysis*, 3rd edn. (McGraw-Hill Book Co., New York, 1987)

[Ru4]   W. Rudin, A generalization of a theorem of Frostman. Math. Scand. **21**, 136–143 (1967)

[Sa]   D. Sarason, The H^p spaces of an annulus. Mem. Am. Math. Soc. **56**, 78 (1965)

[Sat]   I. Satake, The Gauss-Bonnet theorem for V-manifolds. J. Math. Soc. Jpn. **9**, 464–492 (1957)

[Sc]   P. Scott, The geometries of 3-manifols. Bull. Lond. Math. Soc. **15**, 401–487 (1983)

[Se1]   K. Seip, Beuling type density theorems in the unit disk. Invent. Math. **113**, 21–39 (1994)

[Se2]   K. Seip, On Korenblum's density condition for the zero sequences of $A^{-\alpha}$. J. Anal. Math. **67**, 307–322 (1995)

[Sh]   H. Shapiro, *Comparative Approximation in Two Topologies*. Approximation Theory, vol. 4 (Banach Center Publication, Warsaw, 1979), 225–232

[Shi]   A. Shields, *Weighted Shift Operators and Analytic Function Theory*. Mathematical Surveys, vol. 13 (American mathematical Society, Providence, 1974), 49–128

[SL]   Y. Shi and Y. Lu, Reducing subspaces for Toeplitz operators on the polydisk. Bull. Kor. Math. Soc. **50**, 687–696 (2013)

[St]   E. Stout, Bounded holomorphic functions on finite Riemann surfaces. Trans. Am. Math. Soc. **120**, 255–285 (1965)

[Str]   K. Stroethoff, The Berezin transform and operators on spaces of analytic functions. Linear Operators, **38**, 361–380 (1997)

[Su]   N. Suzuki, The algebraic structure of non self-adjoint operators. Acta Math. Sci. **27**, 173–184 (1966)

[Sun1]   S. Sun, On unitary equivalence of multiplication operators on Bergman space. Northeast. Math. J. **1**, 213–222 (1985)

[SW]   S.L. Sun, Y. Wang, Reducing subspaces of certain analytic Toeplitz operators on the Bergman space. Northeast. Math. J. **14**, 147–158 (1998)

[SWa]   A. Shields, L. Wallen, The commutants of certain Hilbert space operators. Indiana Univ. Math. J. **20**, 777–788 (1970/1971)

[SY]   S. Sun, D. Yu, On unitary equivalence of multiplication operators on Bergman space (II). Northeast. Math. J. **4**, 169–179 (1988)

[SZ1]   M. Stessin, K. Zhu, Reducing subspaces of weighted shift operators. Proc. Am. Math. Soc. **130**, 2631–2639 (2002)

[SZ2]   M. Stessin, K. Zhu, Generalized factorization in Hardy spaces and the commutant of Toeplitz operators. Can. J. Math. **55**, 379–400 (2003)

[SZZ1]   S. Sun, D. Zheng, C. Zhong, Classification of reducing subspaces of a class of multiplication operators via the Hardy space of the bidisk. Can. J. Math. **62**, 415–438 (2010)

[SZZ2]   S. Sun, D. Zheng, C. Zhong, Multiplication operators on the Bergman space and weighted shifts. J. Oper. Theory **59**, 435–452 (2008)

[T1]   J. Thomson, The commutant of a class of analytic Toeplitz operators. Am. J. Math. **99**, 522–529 (1977)

[T2]   J. Thomson, The commutant of a class of analytic Toeplitz operators II. Indiana Univ. Math. J. **25**, 793–800 (1976)

[T3]   J. Thomson, The commutant of certain analytic Toeplitz operators. Proc. Am. Math. Soc. **54**, 165–169 (1976)

[T4]   J. Thomson, Intersections of commutants of analytic Toeplitz operators. Proc. Am. Math. Soc. **52**, 305–310 (1975)

[V]   W. Veech, *A Second Course in Complex Analysis* (Benjamin, New York, 1967)

[Wa1]   J. Walsh, On the location of the roots of the jacobian of two binary forms, and of the derivative of a rational function. Trans. Am. Math. Soc. **19**, 291–298 (1918)

[Wa2]   J. Walsh, *The Location of Critical Points*, vol. 34 (American Mathematical Society Colloquium Publications, Rhode Island, 1950)

[WDH]   X. Wang, H. Dan, H. Huang, Reducing subspaces of multiplication operators with the symbol $\alpha z^k + \beta w^l$ on $L_a^2(\mathbb{D}^2)$. Sci. China Ser. A **58** (2015). doi:10.1007/s11425-015-4973-9

[Weil]   A. Weil, L'integrable de Cauchy et les fonctions de plusieurs variables. Math. Ann. **111**, 178–182 (1935)

[Wo]   B. Wong, Characterization of the unit ball in $\mathbb{C}^n$ by its automorphism group. Invent. Math. **41**, 253–257 (1977)

[XY]   A. Xu, C. Yan, Reducing subspace of analytic Toeplitz operators on weighted Bergman spaces. Chin. Ann. Math. Ser. A **30**, 639–646 (2009)

[Zhao]   L. Zhao, Reducing subspaces for a class of multiplication operators on the Dirichlet space. Proc. Am. Math. Soc. **139**, 3091–3097 (2009)

[Zhu1]   K. Zhu, Reducing subspaces for a class of multiplication operators. J. Lond. Math. Soc. **62**, 553–568 (2000)
[Zhu2]   K. Zhu, Irreducible multiplication operators on spaces of analytic functions. J. Oper. Theory **51**, 377–385 (2004)
[Zhu3]   K. Zhu, *Spaces of Holomorphic Functions in the Unit Ball*. Graduate Texts in Mathematics, vol. 226 (Springer, New York, 2005)

Index

C_{00}, 214
H^∞-removable, 27
$L^2_{a,\alpha}(\mathbb{D})$-interpolating sequence, 35
$\mathcal{L}(G)$, 218
$\mathcal{R}(G)$, 218
$\mathcal{W}^*(\phi)$, 2
ψ-glued domain with respect to w, 152
n-folds map, 8

abelian projection, 40
analytic continuation along a curve, 31
analytic Hilbert module, 297

Berezin transform, 301
Bergman space over Ω, 34
Bergman space over the upper plane, 250
Blaschke product, 9
Bochner's theorem, 155
Böttcher's theorem, 18
branched covering map, 21

capacity, 28
capacity zero, 27
central projection, 40
central support, 278
Cesaro mean, 25
cluster set, 11
commutant algebra, 2
complete local inverse, 129
completely reducible operator, 217
conformal, 81
conjugacy class, 215

convex hull, 24
coordinate operator, 1
countable, 58
covering map, 230
Cowen's class, 54
Cowen's condition, 54
critical point, 155
critical value set, 196
curve, 31

deck transformation group, 21
dihedral group, 293
direct continuations, 31
directly ψ-glued, 151
distinguished reducing subspace, 92

equivalent projections, 40
essential normal, 269
external direct product of two groups, 292

factor, 41
faithful, center-valued trace, 41
fiber, 21
finite von Neumann algebra, 40
free group on k generators, 195
free product of groups, 195
Frostman's theorem, 9
function element, 31
fundamental group, 194

gluable, 151
group-like von Neumann algebras, 247

© Springer-Verlag Berlin Heidelberg 2015
K. Guo, H. Huang, *Multiplication Operators on the Bergman Space*,
Lecture Notes in Mathematics 2145, DOI 10.1007/978-3-662-46845-6

holomorphic, 7
holomorphic covering map, 20

i.c.c. group, 215
identity component of S_B, 140
infinite dihedral group, 226
infinite von Neumann algebra, 40
inner function, 9
interpolating Blaschke product, 12
interpolating sequence, 34
interpolating sequence for $L_a^p(\mathbb{D})$, 35
invariant subspace, 1

Koebe Uniformization Theorem, 20
Kronecker delta, 135

Lebesgue's Dominated Convergence
        Theorem, 25
lift, 85
local inverse, 31
logarithmic potential, 27
loop, 31
Lusin's Theorem, 25

minimal normal extension, 172
minimal projection, 40
minimal reducing subspace, 47
modular function, 235
monodromy theorem, 33
multiplication operator, 1
multiplicity of $G[\rho]$, 140
multiplicity of the factor, 44

normal, 222

orbifold domain, 194
order of an inner function, 254

path, 31
perfect set, 27
positive capacity, 27
projection, 40
proper, 8
pseudohyperbolic disk, 116

pseudohyperbolic distance, 11
punctured disk, 7
punctured neighborhood, 72

ramified function, 194
rank of a projection, 40
reduced form, 226
reduced words, 195
reducing subspace, 1, 47
regular, 21
regular measure, 26
Reinhardt domain, 312
representing local inverse, 66
reproducing kernel, 1
reproducing kernel Hilbert space, 1
Riemann mapping theorem, 21

Seifert-Van Kampen Theorem, 194
Singular locus, 194
slit neighborhood, 72
standard form of type II factors, 224
subspace, 1
symbol, 1

thin Blaschke product, 12
Thomson's condition, 54
trivial component of S_B, 140
type I_n factor, 44

uniformly discrete, 36
uniformly separated, 34
unit ball, 7
unit polydisk, 7
unitarily isomorphic, 44
univalent, 8
universal covering of an orbifold domain, 194
universal divisor, 36

von Neumann Bicommutant Theorem, 45
von Neumann algebra, 40

weighted Bergman space, 34

Zhukovski function, 229

LECTURE NOTES IN MATHEMATICS  Springer

Edited by J.-M. Morel, B. Teissier; P.K. Maini

Editorial Policy (for the publication of monographs)

1. Lecture Notes aim to report new developments in all areas of mathematics and their applications - quickly, informally and at a high level. Mathematical texts analysing new developments in modelling and numerical simulation are welcome.

   Monograph manuscripts should be reasonably self-contained and rounded off. Thus they may, and often will, present not only results of the author but also related work by other people. They may be based on specialised lecture courses. Furthermore, the manuscripts should provide sufficient motivation, examples and applications. This clearly distinguishes Lecture Notes from journal articles or technical reports which normally are very concise. Articles intended for a journal but too long to be accepted by most journals, usually do not have this "lecture notes" character. For similar reasons it is unusual for doctoral theses to be accepted for the Lecture Notes series, though habilitation theses may be appropriate.

2. Manuscripts should be submitted either online at www.editorialmanager.com/lnm to Springer's mathematics editorial in Heidelberg, or to one of the series editors. In general, manuscripts will be sent out to 2 external referees for evaluation. If a decision cannot yet be reached on the basis of the first 2 reports, further referees may be contacted: The author will be informed of this. A final decision to publish can be made only on the basis of the complete manuscript, however a refereeing process leading to a preliminary decision can be based on a pre-final or incomplete manuscript. The strict minimum amount of material that will be considered should include a detailed outline describing the planned contents of each chapter, a bibliography and several sample chapters.

   Authors should be aware that incomplete or insufficiently close to final manuscripts almost always result in longer refereeing times and nevertheless unclear referees' recommendations, making further refereeing of a final draft necessary.

   Authors should also be aware that parallel submission of their manuscript to another publisher while under consideration for LNM will in general lead to immediate rejection.

3. Manuscripts should in general be submitted in English. Final manuscripts should contain at least 100 pages of mathematical text and should always include

   - a table of contents;
   - an informative introduction, with adequate motivation and perhaps some historical remarks: it should be accessible to a reader not intimately familiar with the topic treated;
   - a subject index: as a rule this is genuinely helpful for the reader.

   For evaluation purposes, manuscripts may be submitted in print or electronic form (print form is still preferred by most referees), in the latter case preferably as pdf- or zipped ps-files. Lecture Notes volumes are, as a rule, printed digitally from the authors' files. To ensure best results, authors are asked to use the LaTeX2e style files available from Springer's web-server at:

   ftp://ftp.springer.de/pub/tex/latex/svmonot1/ (for monographs) and
   ftp://ftp.springer.de/pub/tex/latex/svmultt1/ (for summer schools/tutorials).

Additional technical instructions, if necessary, are available on request from lnm@springer.com.

4. Careful preparation of the manuscripts will help keep production time short besides ensuring satisfactory appearance of the finished book in print and online. After acceptance of the manuscript authors will be asked to prepare the final LaTeX source files and also the corresponding dvi-, pdf- or zipped ps-file. The LaTeX source files are essential for producing the full-text online version of the book (see http://www.springerlink.com/openurl.asp?genre=journal&issn=0075-8434 for the existing online volumes of LNM). The actual production of a Lecture Notes volume takes approximately 12 weeks.

5. Authors receive a total of 50 free copies of their volume, but no royalties. They are entitled to a discount of 33.3 % on the price of Springer books purchased for their personal use, if ordering directly from Springer.

6. Commitment to publish is made by letter of intent rather than by signing a formal contract. Springer-Verlag secures the copyright for each volume. Authors are free to reuse material contained in their LNM volumes in later publications: a brief written (or e-mail) request for formal permission is sufficient.

Addresses:

Professor J.-M. Morel, CMLA,
École Normale Supérieure de Cachan,
61 Avenue du Président Wilson, 94235 Cachan Cedex, France
E-mail: morel@cmla.ens-cachan.fr

Professor B. Teissier, Institut Mathématique de Jussieu,
UMR 7586 du CNRS, Équipe "Géométrie et Dynamique",
175 rue du Chevaleret
75013 Paris, France
E-mail: teissier@math.jussieu.fr

For the "Mathematical Biosciences Subseries" of LNM:

Professor P. K. Maini, Center for Mathematical Biology,
Mathematical Institute, 24-29 St Giles,
Oxford OX1 3LP, UK
E-mail: maini@maths.ox.ac.uk

Springer, Mathematics Editorial, Tiergartenstr. 17,
69121 Heidelberg, Germany,
Tel.: +49 (6221) 4876-8259

Fax: +49 (6221) 4876-8259
E-mail: lnm@springer.com

Printed in the United States
by Bookmasters

Printed in the United States
By Bookmasters